Processes in Microbial Ecology

Processes in Microbial Ecology

David L. Kirchman

School of Marine Science and Policy
University of Delaware, USA

OXFORD
UNIVERSITY PRESS

OXFORD

UNIVERSITY PRESS

Great Clarendon Street, Oxford, OX2 6DP,
United Kingdom

Oxford University Press is a department of the University of Oxford.
It furthers the University's objective of excellence in research, scholarship,
and education by publishing worldwide. Oxford is a registered trade mark of
Oxford University Press in the UK and in certain other countries

First published 2012

Corrected 2012

Published in the United States of America by Oxford University Press
198 Madison Avenue, New York, NY 10016, United States of America

British Library Cataloguing in Publication Data

Data available

Library of Congress Cataloging in Publication

Data available

ISBN 978–0–19–958692–9

Preface

Microbial ecology is now examining some of the most important questions in science and is helping to solve some of the most serious environmental problems facing our society today. To those wishing to learn more about the field, this textbook is for you.

The book targets students, young and old, interested in learning some basic principles of microbial ecology and the processes carried out by microbes in nature. One basis for this book is a course I teach in marine microbial ecology. The students enrolled in that course are usually a diverse lot of biologists, rarely microbiologists, often inorganic marine chemists, even the occasional geologist. The biologists may have a strong background in biochemistry, but know little about geochemistry. The chemists are comfortable talking about chemical reactions in the environment, but they know little biology. Senior colleagues working with other organisms or in geochemistry or geology ask me questions about microbes affecting the organisms and processes they are examining. These students and colleagues are the people I had in mind as I put this book together.

The book attempts to cover both land and the sea, all types of microbes, and all kinds of microbial metabolisms important in nature. It tries to do it all. While at times writing the book seemed a foolhardy endeavor reeking with hubris, most of the time it was liberating to focus on the most important processes and microbes, to attempt to identify what a student must know about a particular geochemical reaction and the microbes mediating it. Of course, there is huge diversity and complexity in the microbial world. What is surprising is to learn about the similarities among environments and microbes,

and the insights gained by comparing disparate habitats and organisms. For these reasons, I hope experienced soil microbial ecologists will learn something about microbes in lakes and the oceans, and likewise for aquatic people and terrestrial systems. These people already know in a general sense about the importance of microbes other than their favorites, and about how microbes interact with larger organisms, but this book should help them gain a better appreciation of all microbes in all environments.

One name may appear on the cover of this book, but many people made the book possible. I want to thank especially Erland Bååth, Gordon Wolfe, and Mrina Nikrad, who commented on several chapters, Tom Hanson, who answered my endless bacterial physiology questions, in addition to reviewing a couple chapters, and Mary Thaler, who read nearly the entire book. Chapters were also reviewed by the following people, including several students: Ruth Anderson, J.-C. Auguet, Albert Barberan, Ron Benner, Mya Breitbart, Alison Buchan, Claire Campbell, Andy Canion, Doug Capone, E. Casamayor, Colleen Cavanaugh, Matt Church, D. C. Coleman, Nathan Cude, L. de Brabandere, Angela Douglas, Dan Durall, Bryndan Durham, Ashley Frank, Jed Fuhrman, Rich Geider, Rodger Harvey, Kelly Hondula, Dave Hutchins, Puja Jasrotia, Bethany Jenkins, Kurt Konhauser, K. Konstantinidis, Joel Kostka, Raphael Lami, Jay Lennon, Ramon Massana, George McManus, Jim Mitchell, Mary Ann Moran, M. Muscarella, Diana Nemergut, N. Nikita, Brady Olson, Mike Pace, Rachael Porestky, Jim Prosser, B. Rodriguez-Mueller, Ned Ruby, Ashley Shaw, Claire Smith, Roman Stocker, Suzanne Strom, Z. Sylvain, Brad Tebo, Bo Thamdrup, Bill Ullman, D. Wall, Flex Weber, Markus Weinbauer, Steve

Wilhem, and Eric Wommack. Thanks (and sorry) to any one I have missed. Several people, acknowledged in the book, provided information, data, or pictures, and Clara Chan tracked down a couple photomicrographs at the last minute. I greatly appreciate all their help. Any errors in the book are because I didn't listen to these people for inexplicable reasons. If you see a mistake or have a comment or suggestion, please email me (kirchman@udel.edu).

I also would like to thank Ana Dittel for keeping me organized, Peggy Conlon for formatting help and copyediting, and Helen Eaton and Ian Sherman at Oxford University Press for bearing with my questions and missed deadlines. Finally, I would like to acknowledge financial assistance from the National Science Foundation, the Department of Energy, and the University of Delaware. A large part of this book was written while I was on sabbatical at the Institut de Ciències del Mar, Barcelona (thanks to Pep Gasol) with support from AGAUR - Generalitat de Catalunya and C.C. Grácia. This book was made possible by these people and institutions.

Table of Contents

Chapter 14   Symbiosis and microbes

CHAPTER 1

Introduction

Microbes make up an unseen world, unseen at least by the naked eye. In the pages that follow, we will explore this world and the creatures that inhabit it. We will discover that processes carried out by microbes in the unseen world affect our visible world. These processes include virtually every chemical reaction occurring in nature, making up the great elemental cycles of carbon, nitrogen, and the rest of the elements in the biosphere. The processes also involve interactions between organisms, both among microbes and between microbes and large organisms.

This chapter will introduce the types of microbes found in nature and some basic terms used throughout the book. It will also discuss why we should care about microbes in nature. The answers will give some flavor of what microbial ecology is all about. Chapters 2 and 3 continue the introduction to microbes and their environment. But first, we need to look at a few definitions.

What is a microbe?

The microbial world is populated by a diverse collection of organisms, many of which having nothing in common except their small size. Microbes include by definition all organisms that can be observed only with a microscope and are smaller than about 100 μm. Microbes and its synonym "microorganisms" include bacteria, archaea, fungi, and other types of eukaryotes (Fig. 1.1). The microbial world also houses viruses, though arguably they are not alive and are not microbes. Ignoring viruses for now, bacteria and archaea are the simplest and usually the smallest microbes in nature.

Bacteria and archaea often look quite similar under the microscope, and in fact archaea were once thought to be a type of bacteria. An old name for archaea is "archaebacteria" while bacteria were once referred to as "eubacteria", the "true" bacteria. Now we know that bacteria and archaea occupy separate kingdoms, accounting for two of the three kingdoms of life found on earth. The third kingdom, Eukarya (sometimes spelled Eucarya), includes microbes, such as fungi, protozoa, and algae (but not blue-green algae, more appropriately called cyanobacteria), as well as higher plants and animals. Prominent members of the third kingdom are protists, which are single-cell eukaryotes.

The diversity in the types of microbes found in nature is matched by the diversity of processes they carry out. Microbes do some things that are similar to the functions of plants and animals in the visible world. Some microbes are primary producers and carry out photosynthesis similar to plants, some are herbivores that graze on microbial primary producers, and still others are carnivores that prey on herbivores. But microbes do many more things that have no counterparts among large organisms. These things, these processes, are essential for life on this planet.

Why study microbial ecology?

The main reason has already been implied: microbes mediate many processes essential to the operation of the biosphere. But there are other reasons for studying microbial ecology. The following list of seven reasons

1

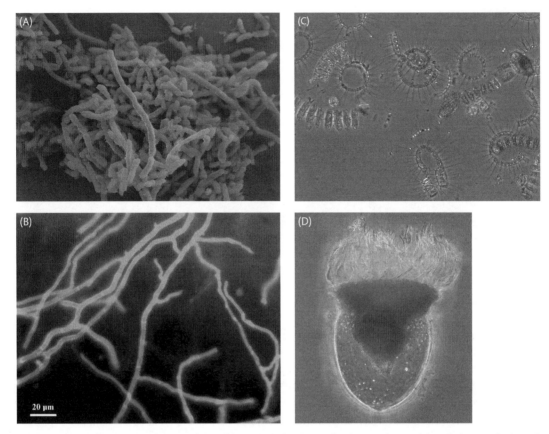

Figure 1.1 Examples of some microbes. Panel A: Soil bacteria belonging to the Gemmatimonadetes phylum, each about 1 μm wide. Image courtesy of Mark Radosevich. Panel B: Fungal hyphae. Image courtesy of David Ellis. Panel C: Various eukaryotic algae from the summer, Narragansett Bay, 50–100 μm. Image courtesy of Susanne Menden-Deuer. Panel D: A marine ciliate, *Cyttarocylis encercryphalus*, about 100 μm. Image courtesy of John Dolan.

starts with those that a layperson might give, if asked why we should learn about microbes.

Microbes cause diseases of macroscopic organisms, including humans

Most people probably think "germs" when asked how microbes affect their life. Of course, some microbes do cause diseases of humans and other macroscopic organisms. The role of infectious diseases in controlling population levels of macroscopic plants and animals in nature is recognized to be important (Ostfeld et al., 2008), but its impact probably is still underestimated. Sick animals in nature are most likely to be killed off by predators, or simply disappear before

being counted as being ill. We know less about the impact of diseases on smaller organisms, such as the zooplankton in aquatic systems (Fig. 1.2) or invertebrates in soils. These small organisms are crucial for maintaining the health of natural ecosystems, which is now being threatened on many fronts by climate change. There is some evidence that diseases in the ocean are becoming more common (Lafferty et al., 2004), and amphibians on land are now declining worldwide due to infections by chytrid fungi, perhaps linked to global warming (Rohr and Raffel, 2010).

But pathogens are exceptions rather than the rule. The microbiologist John Ingraham pointed out that there are more murders among humans than pathogens among microbes (Ingraham, 2010). The vast majority of microbes

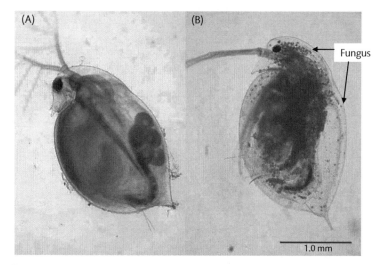

Figure 1.2 A common freshwater zooplankton *Daphnia pulicaria* (the common name is "water flea"), uninfected (panel A), and infected (panel B) by a fungus. The fungi are the small numerous dots visible in the transparent body cavity of the *Daphnia*. Taken from Johnson et al. (2006). Used with permission from the Ecological Society of America.

in nature are not pathogenic, including those living on and in us. The human body is host to abundant and diverse microbial communities. In fact, most of the cells in the human body are not human, but rather are bacteria. An average adult has about 1×10^{14} microbes, tenfold more than the number of human cells. Microbes inhabiting our skin and mucous membranes help to prevent invasion by pathogens, and the bacteria in the gastrointestinal tract do the same as well as aiding digestion. Disruption of the microbial community in the colon, for example, allows the pathogen *Clostridium difficile* to flourish, which often leads to severe diarrhea. One cure is "bacteriotheorapy", also called fecal transplantation, in which a normal microbial community is "transplanted" into the colon of a diarrhea-suffering patient (Khoruts et al., 2010). A huge project is now examining the genomes of human-associated microbes, the "human microbiome" (http://www.hmpdacc.org/), using metagenomic approaches (Chapter 10) first designed for soils and oceans.

Much of our food depends on microbes
Microbes produce several things that we eat and drink every day, including yogurt, wine, and cheese. Some of

the first microbiologists, who could be called microbial ecologists, worked on topics in what we now call food microbiology. Louis Pasteur (1822–1895) was hired by the wine industry to figure out why some wines turned sour and became undrinkable. The problem, as Pasteur found out, was a classic one of competition between two types of microbes, one that produced alcohol (good wine) and the other organic acids (sour, undrinkable wine). To this day, food microbiologists try to understand the complex microbial interactions and processes that affect our favorite things to eat and drink, an important job in applied microbial ecology. Microbes are also involved in meat and dairy products. Ruminants, such as cows, goats, and sheep, depend on complex microbial consortia to digest the polysaccharides in the grasses they eat (Chapter 14). A branch of microbial ecology can be traced to microbiologists such as Robert Hungate (1906–2004), who studied microbe-ruminant interactions and the mainly anaerobic processes carried out by these microbes (Hungate, 1966).

Microbes are also important in supporting life in lakes and the oceans, and eventually make possible the fish we may eat. Microbes take over the role of macroscopic plants in aquatic environments and are the main pri-

---

Box 1.1 Two founders of microbiology

One of the founders of microbiology, Pasteur, made many contributions to chemistry and biology during the early days of these fields. His more important contributions include work showing that life does not arise from spontaneous generation, a theory held in the mid-nineteenth century to explain organic matter degradation. Decomposition of organic material is still examined today in microbial ecology (Chapter 5). Pasteur also explored the role of bacteria in causing diseases, but it was a contemporary of Pasteur, Robert Koch (1843–1910), who defined the criteria, now known as Koch's postulates, for showing that a particular microbe causes a disease. Koch, another founder of microbiology, developed the agar plate method for isolating bacteria, a method still used today in microbial ecology.

---

mary producers, meaning they use light energy to convert carbon dioxide to organic material (Chapter 4). These microbes, the "phytoplankton", include cyanobacteria and eukaryotic algae. Phytoplankton are not directly eaten by fish, not even by the small, young stages of fish, the fish larvae. Rather, mostly microscopic animals (zooplankton) and protists are the main herbivores in lakes and oceans. Zooplankton and protists in turn are eaten by still larger zooplankton and fish larvae, as part of a food chain leading eventually to adult fish (Fig. 1.3). There can be more direct connections between microbes and fish. In some aquaculture farms shrimp feed on "bioflocs" which form from bacteria growing on added wheat flour and ammonium from the shrimp. The simple, linear food chain shown in Figure 1.3 is accurate in only some aquatic systems. But even in waters with more complex food webs, microbes are the base upon which the fisheries of the world depend. Consequently, there is a general relationship between microbial production and fishery yields (Conti and Scardi, 2010).

Another important connection with our food is the role of microbes in producing the inorganic nutrients that are essential for growth and biomass production by higher plants in terrestrial environments and by phytoplankton in aquatic environments. Essential inorganic nutrients, such as ammonium and phosphate, come from microbes as they degrade organic material (Chapter 5). Other microbes change ("fix") nitrogen gas, which cannot be used by plants as a nitrogen source, into ammonium, which can (Chapter 12). The fertility of soils depends on microbes in other ways. Organic material from higher plants, partially degraded by microbes, and other organic compounds directly from microbes make up soil organic material. This organic component of soil contains essential plant nutrients and affects water flow, fluxes of oxygen and other gases, pH, and many other physical-chemical properties of soils that directly contribute to the growth of crop plants. So, our food indirectly and directly depends on microbes and what they do.

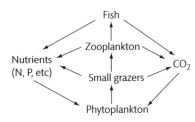

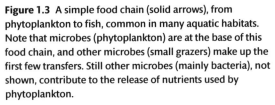

Figure 1.3 A simple food chain (solid arrows), from phytoplankton to fish, common in many aquatic habitats. Note that microbes (phytoplankton) are at the base of this food chain, and other microbes (small grazers) make up the first few transfers. Still other microbes (mainly bacteria), not shown, contribute to the release of nutrients used by phytoplankton.

Microbes degrade and detoxify pollutants
The modern environmental movement is often said to have started with the publication of *Silent Spring* in 1962 by Rachael Carson (1907-1964). The book chronicled the damage to wildlife and ecosystems caused by the pesticide dichloro-diphenyl-trichloroethane, better

known by its initials, DDT. Fortunately, the concentrations of DDT have been decreasing over time, in part due to regulations banning it in most developed countries, following publication of *Silent Spring*. In addition, microbes, mostly bacteria, degrade DDT and other organic pollutants to innocuous compounds and eventually to CO_2 (Alexander, 1999) in spite of many organic pollutants being recalcitrant and difficult to degrade because of complex chemical structures. With very few exceptions, bacteria and fungi are quite adept at degrading organic compounds, even those quite toxic to macroscopic organisms.

Inorganic pollutants, such as heavy metals, cannot be removed by microbial activity, but microbes can change the electrostatic charge of these pollutants which affects their mobility through the environment. An example of this process is the action of the bacterium *Geobacter* on the spread of uranium in groundwater and subsurface environments near waste dumps for radioactive material (Lovley, 2003). In this case, the most oxidized form of uranium, U(VI), moves easily through subsurface environments. When U(VI) is reduced by *Geobacter* and probably other bacteria, the resulting U(IV) is less mobile. So, while microbial activity does not remove the contamination in this case, it can reduce its spread.

Microbes can be useful model systems for exploring general principles in ecology and evolution

Microbes have served as models for exploring many questions in biochemistry, physiology, and molecular biology. They are good models because they grow rapidly and can be manipulated easily in laboratory experiments. For similar reasons, microbes also are used as models to explore general questions in ecology, population genetics, and evolution. Virus-bacteria interactions, for example, have been used to examine questions and models of predator-prey interactions (Chapter 8). Experiments with protozoa and bacteria were crucial for establishing Gause's competitive exclusion principle (Fig. 1.4), which states that only one species can occupy a niche at a time. Experiments with both bacteria and fungi have demonstrated basic principles about natural selection and adaptation in varying environments (Beaumont et al., 2009, Schoustra et al., 2009). Richard Lenski and colleagues have explored the evolution of *Escherichia coli* over 50 000 generations by following this bacterium in cultures that have been transferred into fresh media every day, including weekends and holidays, since 1988 (Lenski, 2011, Woods et al., 2011). Genome sequencing (Chapter 10) has revealed exactly how these organisms have changed over time, providing insights into evolution not possible with large organisms.

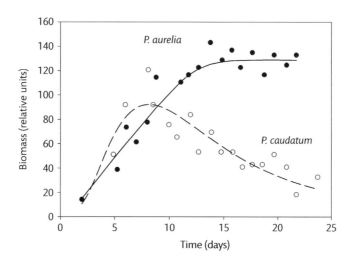

Figure 1.4 Experimental evidence for the competitive exclusion principle, which states that no two species can occupy the same niche at the same time. Here, two species of the protozoan *Paramecium* are forced to compete for the same food source, the bacterial prey (*Bacillus pyocyaneus*). Only one *Paramecium* species wins. Data from Gause (1964).

Just as we can learn about large organisms from microbes, the flow of ideas can go the other way. General theories developed for exploring the ecology of plants and animals often are useful for exploring questions in microbial ecology. For example, microbial ecologists have used island biogeography theory, which was first conceived for large animals (MacArthur and Wilson, 1967), to examine the dispersal of microbes and relationships between microbial diversity and habitat size (Chapter 9). Likewise, models about stability and diversity developed for animal communities are now being applied to microbial communities and processes. Microbial ecologists look at microbial diversity for patterns that have been seen for plant and animal diversity, such as how diversity varies with latitude. Not all large organism-based theories are applicable to thinking about microbes, but many are.

Some microbes are examples of early life on earth and perhaps of life on other planets

Microbial ecologists examine microbial processes now occurring in various environments in order to understand how today's biosphere operates and to predict how it may be altered in the future due to climate change. But what we learn about microbes living today can also help us understand life in the distant past. The first life form on earth undoubtedly was a microbe-like creature, and its microbial descendants went on to rule the planet without large organisms for the first three billion years of earth's history (Fig. 1.5). Multicellular animals and plants did not appear until about a billion years ago, two to three billion years after microbes had invented via evolution most of the various strategies now known for existing on earth. We can gain insights into the early evolution of life by looking at microbes in today's environments that may mimic those on early earth (Chapter 13).

In addition to looking at life millions of years ago, today's microbes may provide insights into life on planets millions of kilometers from earth. Studying microbes in extraterrestrial-like environments on earth is the main focus of the field of "astrobiology". These environments are extreme ones where only microbes, and often only bacteria and archaea, the "extremophiles", survive (Chapter 3). Microbes live in hot springs and deserts, polar ice, permafrost of the tundra, and within rocks—unworldly habitats where it is hard to imagine life existing. Perhaps these earthly extremophiles are similar to life on other planets and perhaps insights gained from astrobiological studies on this planet will help in the search for life on other planets. But the work would be worthwhile even if there are no extraterrestrial microbes. Extreme environments and extremophiles are often bizarre and always fascinating.

Microbes mediate many biogeochemical processes that affect global climate

This reason for studying microbial ecology is arguably the most important one. It shapes many topics appearing in this book. The role of microbes in degrading pollut-

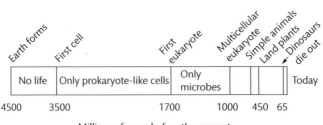

Figure 1.5 A few key dates during the history of life. For much of this history only microbes were present, as multicellular eukaryotes appeared only one billion years ago, after 3.5 billion years had passed. "Prokaryote" refers to bacteria and archaea. Based on Czaja (2010), Humphreys et al. (2010), Payne et al. (2009), and Rasmussen et al. (2008).

ants was already mentioned, but microbes are involved in an even more serious "pollution" problem.

Humans have been polluting the earth's atmosphere with various gases that affect our climate. These gases are called "greenhouse gases" because they trap long wave radiation, better known as heat, from the sun. Most of these gases also have natural sources, and the earth always has had greenhouse gases, fortunately. Because of greenhouse gases, the average global temperature is 16 °C (Schlesinger, 1997), much warmer than the chilly −21 °C earth would be without them. Mars does not have

any greenhouse gases and is even colder (−55 °C). High concentrations of greenhouse gases, mainly carbon dioxide, in Venus's atmosphere, along with its proximity to the sun, explain why that planet has an average surface temperature of 460 °C. Greenhouse gases have been increasing in earth's atmosphere since the start of the industrial revolution in the early 1800s (Fig. 1.6). Water vapor is the dominant greenhouse gas, but human society has a bigger, more direct impact on other gases, most notably carbon dioxide. Other important greenhouse gases include methane (CH_4) and nitrous oxide (N_2O)

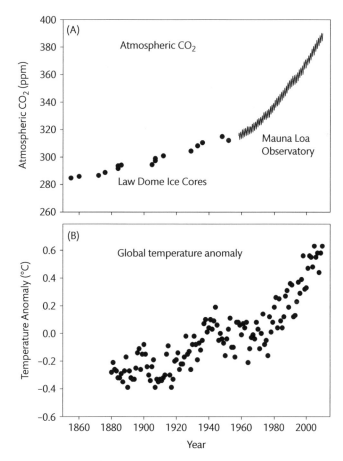

Figure 1.6 Atmospheric CO_2 concentrations (A) and global temperature anomaly (B) since the nineteenth century. Concentrations were estimated from ice cores (Etheridge et al. 1996) or measured directly at the Mauna Loa Observatory, Hawaii, provided by Pieter Tans at the NOAA Earth System Research Laboratory and used with permission (http://www.esrl. noaa.gov/gmd/ccgg/trends/#mlo_data). The global temperature is expressed as the difference between the average temperature for a year minus the average over 1951–1980. Data from Hansen et al. (2006).

Table 1.1 Some greenhouse gases and how they are affected by microbes. Concentrations are for 2005 and are expressed as parts per million (ppm), per billion (ppb), or per trillion (ppt). Data from Forster et al. (2007).

Gas	Concentrations	Greenhouse effect*	Microbes or Process
Carbon dioxide (CO_2)	379 ppm	1	Algae and heterotrophic microbes
Methane (CH_4)	1774 ppb	21	Methanogens and methanotrophs
Nitrous oxide (N_6O)	319 ppb	270	Denitrification and nitrification
Halocarbons **	3–538 ppt	5– >10 000	Degradation by heterotrophs?

* Relative to CO_2
** Examples include CFC-11 and CF_4

(Table 1.1). Although the concentrations of these gases in the atmosphere are much lower, they trap more heat per molecule than does CO_2. Because of higher greenhouse gases, average temperatures for the planet are about 1 °C warmer now than in the nineteenth century (Fig. 1.6).

Microbial ecology has an essential role in understanding the impact of greenhouse gases on our climate and the response of ecosystems to climate change, one reason being that nearly all of these gases are either used or produced or both by microbes (Table 1.1). Carbon dioxide, for example, is used by higher plants on land and by phytoplankton in aquatic ecosystems. This gas is released by heterotrophic microbes in both terrestrial and aquatic ecosystems. The impact of this biological activity can be seen in the yearly oscillations of carbon dioxide in Figure 1.4; it goes down in the summer when plant growth is high and it increases in winter when carbon dioxide produced by respiration exceeds carbon dioxide use by plant growth. Fluxes of methane, another gas that has been increasing in the atmosphere, are nearly entirely controlled by microbes (Chapter 11). Methane and nitrous oxide are both produced in anoxic environments which have increased over the years, mainly due to the growth in agriculture.

What complicates our understanding of these greenhouse gases is that nearly all are produced and consumed by natural processes mediated by microbes, in addition to the anthropogenetic inputs. For nearly all of these gases, the natural processes are much larger than the human-driven ones, although that is changing. Production of the important plant nutrient ammonium, for example, directly by humans (fertilizer synthesis) or aided by humans (microbial production in agriculture) rivals the natural production of ammonium by microbes (Chapter 12). To complicate things further, greenhouse gases vary

with the seasons, as already seen for carbon dioxide, and have varied greatly over geological time, independent of human intervention. So, the challenge is to separate the natural changes from those affected by humans and to understand the consequences of these changes.

Microbial ecologists cannot solve the greenhouse problem. But many of the topics discussed in this book can help us understand the problem. One job of microbial ecologists and other scientists studying the earth system is to figure out the impact of increasing greenhouse gases and other global changes on the biosphere. How will an increase in global temperature affect the balance between photosynthesis and respiration? How will aquatic ecosystems respond to increases in dissolved CO_2 and resulting decreases in pH? How much CO_2 and CH_4 will be released if the permafrost of the tundra in Alaska and Siberia melt? Answering these and other questions depends on the work of microbial ecologists.

Microbes are everywhere, doing nearly everything
The reasons discussed so far for studying microbial ecology have focused on practical problems facing human society. But microbial ecology would be an exciting field even if all of these problems were solved tomorrow. One overall goal of this book is to show the importance of microbial ecology in explaining basic processes in the biosphere even if they may appear to be far from any practical problem facing us today. We should want to know about the most numerous and diverse organisms on the planet, the microbes.

As a general rule, the smaller the organism, the more numerous it is (Fig. 1.7). Viruses are the smallest and also the most abundant biological entity in both aquatic habitats and soils, whereas large organisms, such as zooplankton and earthworms, are rare, being 10^{10} less abundant than viruses. A typical milliliter of water from the surface of a lake or the oceans contains about 10^7 viruses, 10^6 bacteria, 10^4 protists, and 10^3 or fewer phytoplankton cells, depending on the environment. A typical gram of soil or sediment likewise contains about 10^{10} viruses, 10^9 bacteria, and so on for larger organisms. Even deep environments, kilometers below the earth's surface, have thousands of microbes. The deep ocean and probably deep subsurface environments also have relatively large numbers of archaea. Even seemingly impenetrable rocks can harbor dense microbial communities.

We already heard about the many microbes living on and in macroscopic organisms, including humans. Overall, the biomass of bacteria and archaea rivals that of all plants in the biosphere (Table 1.2), and certainly is greater than animal biomass.

Microbes are found where macroscopic organisms are not, in environments with extremes in temperatures, pH, or pressure: "extreme" for humans, but quite normal for many microbes (Chapter 3). Some hyperthermophilic bacteria and archaea thrive in near boiling water (>80 °C), which kills all other organisms, including eukaryotic microbes. The hot springs of Yellowstone are famous for

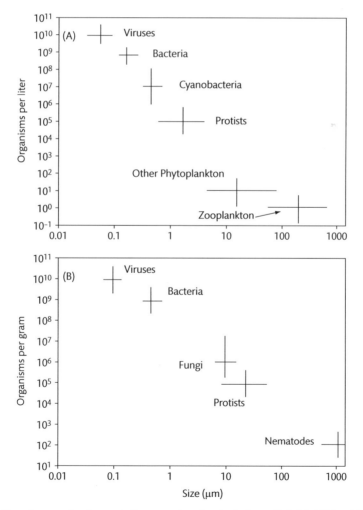

Figure 1.7 Size distribution of some microbes and other organisms in a typical aquatic habitat (A) and in soils (B). The size given for fungi is the diameter of the hyphae. Some fungi can be several meters long.

Table 1.2 Biomass of bacteria and archaea versus plants in the biosphere. Taken from Whitman et al. (1998). "Pg" is petagrams or 10^{15} grams.

Organism	Habitat	No. of cells ($\times 10^{28}$)	Pg of carbon
Bacteria and archaea	Aquatic	12	2.2
	Oceanic subsurface	355	303
	Soil	26	26
	Terrestrial subsurface	25–250	22–215
	Total	415–640	355–546
Plants	Terrestrial	–	560
	Marine	–	51

harboring dense and exotic microbial communities that not only thrive at high temperatures but also low pH; these microbes live in boiling acid baths. At the other extreme, both eukaryotic and prokaryotic microbes live in the brine channels of sea ice where water is still liquid but very salty (20% versus 3.5% for seawater) and cold (–20 °C). The deep ocean may be extreme to us with its high hydrostatic pressure, one hundredfold higher at 1000 m than at sea level, and perennially cold temperatures (about 3 °C), but this is one of the largest ecosystems on the planet; 71% of the globe is covered by the oceans of which 75% (by volume) is deeper than 1000 m. Many microbes thrive and grow albeit slowly in these deep waters.

In addition to being numerous, there are many different types of microbes, some with strange and weird (from our biased perspective) metabolisms on which the biosphere depends. In addition to plant- or animal-like metabolism, some microbes can live without oxygen and "breathe" with nitrate (NO_3^-) or sulfate (SO_4^{2-}) (Chapter 11). Compounds like hydrogen sulfide (H_2S) are deadly to macroscopic organisms, but they are essential comestibles for some microbes. Several metabolic reactions, such as methane production and the synthesis of ammonium from nitrogen gas, are carried out only by microbes. Other microbes are capable of producing chemicals, like acetone and butane, that seem incompatible with life. Microbes are truly capable of doing nearly everything.

How do we study microbes in nature?

The facts about microbes in nature discussed above came from many studies using many approaches and

methods. It is a great intellectual puzzle to figure out the actions and creatures in the unseen world and how they affect our visible world. This book will introduce some of the methods used in microbial ecology so that readers can gain deeper insights and appreciation of the boundaries between the known and unknown. By learning a bit about the methodology of microbial ecology, readers will also understand better why some seemingly simple questions are difficult to answer.

Here we start with one of the most basic questions: how many bacteria are in an environment? One of the first answers came from the plate count method, which consists of growing or cultivating organisms on solid agar media (Fig. 1.8). (The terms "to culture" and "to cul-

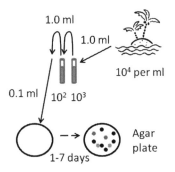

Figure 1.8 The plate count method. A sample from the environment is usually diluted first by adding 1.0 ml or 0.1 g of soil to 9.0 ml of an appropriate buffer, then diluted again by adding 1.0 ml of the first dilution to a new tube containing 9.0 ml of the buffer. Then 0.1 ml from the second dilution tube is spread on an agar plate. After a few days, if 10 colonies grew up after two dilutions (more may be needed for some environments), we could deduce that there were 10^4 culturable bacteria per ml or gram in the original habitat.

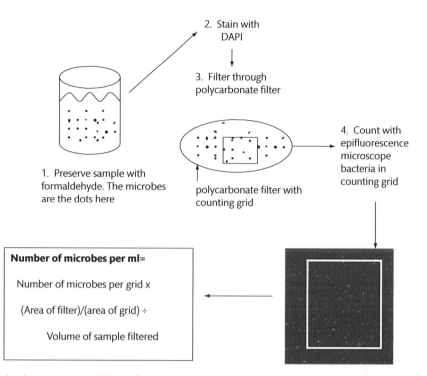

Figure 1.9 How microbes are examined by epifluorescence microscopy, the direct count method. "DAPI" is 4′,6-diamidino-2-phenylindole, a stain specific for double-stranded nucleic acids. While under the microscope, the sample is exposed (the DAPI is "excited") to UV light (in the case of DAPI staining) and cells stained by DAPI fluoresce—they give off light, resulting in bright spots of light on a black background. The "epi" part of epifluorescence comes from the fact that the excitation light is above rather than below the sample.

Box 1.2 Able assistance with agar plates

Agar plates are made by pouring molten agar amended with various compounds into Petri dishes. Once cool, the agar solidifies and becomes a porous support on which microbes grow to form macroscopic colonies. The added compounds may provide necessary organic material for microbial growth or they may inhibit growth of some microbes, allowing only the targeted microbes to grow up. Although the approach is usually attributed to Robert Koch, two assistants of Koch came up with the key parts. Petri dishes were thought up by Julius Richard Petri while Koch's wife, Fannie Hesse, suggested agar, which was used at the time to make jam.

tivate" mean the same). The assumption behind the "plate count method" is that each microbe in the original sample will grow on the solid media and form a macroscopic clump of cells or colony, which can be counted by eye or with a low power microscope. The bacterium, now isolated on the agar plate, can be identified by examining its response to a battery of biochemical tests. These tests provide some of the first clues about the bacterium's physiology and thus its ecological and biogeochemical roles in nature. The physiology and genetics of isolated bacteria in "pure culture" (cultures with only a single microbe) can then be examined in great detail.

The problem is that most microbes are very difficult to isolate and to grow on agar plates. This problem became apparent when the abundance of bacteria determined by the plate count method was found to be orders of magnitude lower than the abundance determined by direct microscopic observations. In seawater, for example, the plate count method indicates that there is about 10^3 bacteria per milliliter, a thousand-fold less than the number determined by direct microscopic counts (Jannasch and Jones, 1959). This difference has been called the "Great Plate Anomaly". One explanation for the anomaly is that the uncultured microbes are dead, since the microbe must be viable and capable of growing enough to form a macroscopic colony if it is to be counted by the plate count method. For this reason, the plate count method sometimes is called misleadingly the "viable count method". In contrast, a particle must only have DNA to be included in the direct count method (Fig. 1.9). Dead bacteria could still have DNA and be counted. So, the discrepancy between direct and plate counts was first thought to be due to large numbers of dead or at least inactive bacteria that were included in direct count methods but not by the plate count method. There are similar problems with other microbes, although the methods differ (Chapter 9).

In fact, to explain the discrepancy, nearly all bacteria would have to be dead. If most bacteria were dead or dormant, it would have huge consequences for understanding the role of microbes in nature.

We now know that the discrepancy between plate and direct counts is not due mainly to dead or dormant bacteria (Chapter 5). Microbial ecologists still argue about the numbers of viable, dormant, and dead bacteria, but problems with the plate count method explain most of the discrepancy. The basic problem is that an agar plate is a very foreign habitat for most bacteria and other microbes. Even plates with "minimal media", to cite one problem, have organic compounds in concentrations much higher than encountered by microbes in nature. Also, many microbes are not adapted to grow in aggregates and to form macroscopic colonies, necessary for a microbe to be counted by the traditional plate count method. There are some problems also with the direct count method, such as confusing inert particles with real microbes due to non-specific staining. But overall, many particles observed by epifluorescence microscopy are active bacteria, other microbes, or viruses.

Regardless of what explains the Great Plate Anomaly, the difficulties in isolating microbes from nature and growing them in the lab have many consequences for microbial ecology. For starters, it means that most microbes cannot be identified by traditional methods. Even if they can be identified by other methods (Chapter 9), the physiology of these microbes cannot be studied by traditional laboratory approaches. This lack of information about physiology hinders understanding the ecological and biogeochemical roles of specific microbes in nature. Fortunately, much can be learned about microbes as a whole in nature by using approaches that examine processes and bulk properties of microbes. For example, methods are available to examine the contribution of bacteria and fungi versus larger organisms in degrading organic material (Chapter 5). The methods include those that yield estimates of bacterial and fungal biomass and activity, although the identity of the bacteria and fungi remains unknown. This approach is sometimes called "black box microbial ecology" because bacteria and fungi, in this case, are being treated as black boxes whose contents (the types of microbes) we cannot see. Opening up the black box and connecting specific microbes with specific processes or functions is an important topic in microbial ecology today.

The three kingdoms of life: Bacteria, Archaea, and Eukarya

One solution to the problem of identifying microbes without cultivation is to use sequences of genes. These genes are often called "phylogenetic markers", because the gene sequences are also used to deduce evolutionary relationships among microbes in addition to determining their taxonomy. For reasons discussed in Chapter 9, the favorite phylogenetic marker of microbial ecologists and microbiologists is the gene coding for a type of ribosomal RNA (rRNA) found in the small subunit of ribosomes (SSU rRNA). More specifically, the 16S rRNA gene is used for bacteria and archaea while the 18S rRNA gene and others are used for eukaryotes. Before the development of rRNA gene-based methods to identify uncultivated microbes (Chapter 9), rRNA genes from cultivated microbes were sequenced.

In the 1970s, Carl Woese first championed the use of rRNA molecules for categorizing microbes (Woese and

Fox, 1977). Although he started by examining 5S rRNA molecules, he soon switched to 16S rRNA because its larger size made it more informative than the smaller 5S molecule (1500 versus 120 nucleotides). Using these rRNA gene sequences, Woese divided all life into three kingdoms: Bacteria, Archaea, and Eukarya (Fig. 1.10). Bacteria and Archaea make up the prokaryotes (see Box 1.3), which are those organisms without a nucleus. All other organisms are in the Eukarya kingdom. The term "archaea" came from early ideas about when these microbes first appeared on the planet. Even before Woese's work on rRNA sequences, microbiologists knew that what we now call archaea had strange metabolisms that seemed to be advantageous for life on early earth. For this reason, Woese called these microbes "archae-bacteria", derived from the geological term Archaean,

which is an early stage in the Precambrian, some 2000–4000 million years ago. We now know that Archaea are not any more ancient than Bacteria, but the name stuck anyway.

The difference between prokaryotes and eukaryotes is easily seen microscopically. A prokaryotic cell appears to be empty when viewed by light microscopy, and in some sense it is because it lacks a nucleus and all other organelles. The genome of prokaryotes is usually in a single circular piece of DNA in the cytoplasm (Chapter 10). The genome of a eukaryote, in contrast, is contained within a nucleus ("karyote" in Greek) and is organized into chromosomes. In addition to nuclei, eukaryotes have compartmentalized some metabolic functions into organelles, such as mitochondria and chloroplasts, which are absent in prokaryotes. These organelles are visible

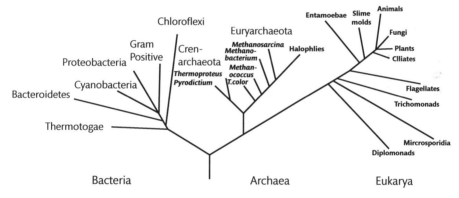

Figure 1.10 A phylogenetic tree showing the three domains of life: Archaea, Bacteria, and Eukarya (eukaryotes). All bacteria and archaea are microbes, as are many organisms in the Eukarya domain. Tree based on Olsen and Woese (1993) with updated names for bacterial phyla. "Gram positive" now includes two phyla: Actinobacteria and Firmicutes. Many phyla are not shown to keep this tree simple.

Box 1.3 Is prokaryotes a bad word?

A prominent microbial ecologist, Norman Pace, has argued strongly that the term "prokaryotes" should not be used because any similarities shared by bacteria and archaea are superficial (Pace, 2006). There are counter-arguments justifying use of "prokaryotes", however (Whitman, 2009). Here, "prokaryotes" will be used because it is useful shorthand for "bacteria and archaea", especially for describing cells in nature, about which nothing is known except that they are clearly not eukaryotes. The terms "microbes" and "microorganisms" are useful even if the organisms covered by these terms are even more diverse and phylogenetically disparate than the prokaryotes.

Table 1.3 Characteristics defining Bacteria, Archaea, and Eukarya (eukaryotes).

Characteristic	Bacteria	Archaea	Eukarya
1. Membrane-bound nucleus	Absent	Absent	Present
2. Cell wall	Muramic acid	Muramic acid absent	Muramic acid absent
3. Membrane lipids	Ester linked	Ether linked	Ester linked
4. Ribosomes	70S	70S	80S (in cytoplasm)
5. Initiator tRNA	Formylmethionine	Methionine	Methionine
6. Introns in tRNA genes	Rare	Yes	Yes
7. RNA polymerases	One (4 subunits)	Several (8–12 subunits each)	Three (12–14 subunits each)
Sensitivity to:			
8. Diphtheria toxin	No	Yes	Yes
9. Chloramphceniocol, streptomycin, and kanamycin	Yes	No	No (cytoplasm)

under standard light microscopy and fill up the eukaryotic cell. Of special interest, the nucleus of a eukaryote is easily seen with epifluorescence microscopy after staining for DNA. In contrast, in the same epifluorescence photomicrograph, prokaryotes appear as solid dots with no internal structure. When aggregated, the DNA of prokaryotic cells forms a dense body, the nucleoid, which is visible by light microscopy.

Among the other important characteristics distinguishing prokaryotes and eukaryotes is size (Table 1.3). Because of the space needed for the nucleus and other organelles, eukaryotic cells, even microbial eukaryotes, are generally bigger than prokaryotes. There are some exceptionally large prokaryotic cells (Schulz and Jorgensen, 2001), but these microbes have a vacuole that pushes the cytoplasm to the outer perimeter of the cell, making the effective volume of these giants more like a typical bacterium. Size is not a useful taxonomic trait for distinguishing among organisms, as there is much overlap in size among prokaryotes and eukaryotes. However, cell size has a huge impact in ecological interactions, such as in predator-prey interactions (Chapter 7) and in the success of bacteria in competing with eukaryotes for dissolved nutrients.

Most bacteria and archaea are small, on the order of a micron, whereas most microbial eukaryotes are >3 µm, although the marine alga *Ostreococcus* is <1 µm (Lopez-Garcia et al., 2001). Bacteria in the laboratory are often bigger, depending on the growth stage and media; the common lab bacterium *E. coli*, for example, is about 1 × 3 µm and has a rod or bacillus shape. Other bacteria are spheres or coccus-shaped (cocci is the plural), and the

vibrios have a comma-like appearance. In contrast, bacteria and archaea in most natural environments are much smaller, about 0.5 µm, and usually appear as simple cocci. There are even reports of even smaller bacteria, called nanobacteria, with cells on the order of 0.1 µm. The Martian meteorite ALH84001 was initially thought to have fossilized nanobacteria (McKay et al., 1996), but this was later disproven (Jull et al., 1998). It is hard to fit all cellular components necessary for a free-living organism into a 0.1 µm cell. Just one important component, a ribosome, is typically about 25 nm in diameter. The size of microbes is illustrated in Figure 1.11.

The final general characteristic distinguishing prokaryotes and eukaryotes is metabolic diversity. Eukaryotes have two basic types of metabolism, one found in plants

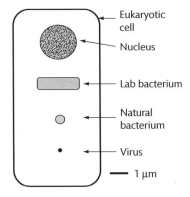

Figure 1.11 Approximate size of eukaryotes, bacteria, and viruses. All of these organisms and viruses vary greatly in size and shape. "Lab bacteria" grown in nutrient-rich media in the laboratory are much bigger usually than those bacteria found in natural environments. Inspired by a similar figure in Madigan et al. (2003).

(autotrophy) and the other in animals (aerobic heterotrophy). In addition to these two metabolisms, prokaryotes have many variations of autotrophy and heterotrophy and many unusual pathways with no analogues in eukaryotes. These pathways include the reduction of nitrogen gas to ammonium (nitrogen fixation is discussed in Chapter 12) and the synthesis of methane (see Chapter 11 for more on methanogenesis). The metabolic diversity of prokaryotes is vast and important in driving a great variety of biogeochemical processes in the biosphere.

Functional groups of microbes

An entirely different way to divide up the microbial world is to sort microbes into various groups based on their metabolic capacity and physiology (Fig. 1.12). The metabolism of a particular group will then help define its role in the ecosystem—its function. Groups defined by physiology and ecosystem function may include microbes only distantly related, if at all, by phylogenetic criteria. For example, both cyanobacteria and eukaryotic algae synthesize organic material that is degraded by both bacteria and fungi. These organisms are related by function (primary production and organic material degradation, respectively), but

they could not be more different phylogenetically, being in separate kingdoms. A major problem in microbial ecology is to determine the relationship between function and "structure". Structure is the taxonomic and phylogenetic make-up of microbial communities.

Before discussing the broad categories that describe microbial metabolisms, it is useful to step back and remember what organisms need to survive and to reproduce. In the most basic terms, organisms need the raw materials that make up a cell, the most abundant being carbon. A microbe also needs a source of energy from which ATP can be synthesized. ATP, the universal currency of energy, is used to drive biosynthetic reactions that turn raw starting compounds into proper cellular components and ultimately more cells. Finally, microbes also need select chemicals for various oxidation-reduction ("redox") reactions which transfer electrons from one compound to another. Biosynthesis sometimes requires elements in the starting material to be reduced. In this case, electrons from an electron donor are transferred to the starting material in a redox reaction. The most important example is the reduction of CO_2 to organic carbon, a reaction that all autotrophs carry out by definition. Other microbes need compounds to

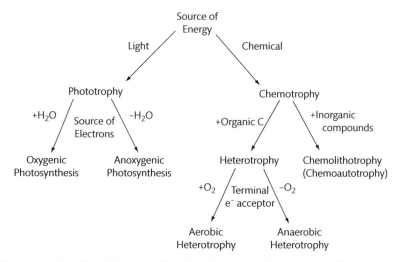

Figure 1.12 Microbial metabolisms that define many of the functional groups found in natural environments. The two types of phototrophy differ in the source of electrons used in photosynthesis. Oxygenic photosynthesis uses water and evolves oxygen while anoxygenic photosynthesis does not use water and does not evolve oxygen. Many chemolithotrophs use carbon dioxide as a carbon source, making them chemoautotrophs. Adapted from Fenchel and Blackburn (1979).

accept electrons, an example being oxygen which receives the electrons produced by the oxidation of organic material. How microbes satisfy these three needs—carbon, energy, and redox compounds—defines their functional group and role in the ecosystem.

Autotroph versus heterotroph

These terms refer to the source of carbon and are equivalent to "plant" and "animal" in the macroscopic world. However, "plant" and "animal" are not useful for describing microbes. Rather than plant and animal, microbial ecologists use autotroph and heterotroph. The etymology of these words helps in remembering their definitions. "Auto" and "hetero" are from the Greek meaning "self" and "different", respectively, while "troph" refers to food. So, an autotroph uses CO_2 and makes its own organic carbon whereas a heterotroph uses organic carbon made by other organisms. All heterotrophs depend directly (herbivores) or indirectly (carnivores and higher trophic levels) on autotrophs.

Phototroph versus chemotroph

The other major characteristic distinguishing microbes is the source of energy used for biomass synthesis. Phototrophs have devised means to capture light energy and convert it to chemical energy. Many phototrophs are also autotrophs, whereas some microbes are mainly heterotrophic and only supplement their energy supply with light energy; these microbes are called photoheterotrophs. Microbes using light energy to fix CO_2 are photoautotrophs. Photoautotrophs can be further subdivided depending on the source of the electrons used for reducing CO_2 to organic material. Higher plants and cyanobacteria use water and produce oxygen, making them oxygenic photoautotrophs. Other bacteria do not use water and do not evolve oxygen; these are called anoxygenic photoautotrophs. An example is the photoautotrophic use of hydrogen sulfide (Chapter 11).

The other major source of energy is from the oxidation of reduced compounds. Organisms living off these

oxidation reactions are chemotrophs. Microbes using organic compounds are chemoorganotrophs, although heterotroph is the term used more frequently. *Homo sapiens* and other animals are chemoorganotrophs, specifically aerobic heterotrophs. Much less common are those organisms capable of harvesting ATP from the oxidation of reduced inorganic compounds. This type of metabolism, chemolithotrophy, is restricted to the prokaryotes, and no eukaryote is known to use it for ATP synthesis. Since the middle syllable, "litho", is derived from the Greek for "rock", chemolithotrophs could be called "rock-eaters", although the compounds used by these microbes are not rocks. Rather, chemolithotrophs use compounds like ammonium and hydrogen sulfide (Chapter 11).

Sources of background information

The diversity of microbes and processes outlined above gives some indication of the breadth of microbial ecology and of this book. Because of this breadth, microbial ecology touches on many other fields, such as microbiology, biogeochemistry, and aquatic and soil chemistry. For students coming from other fields, this book attempts to provide enough background information, so that the microbes and the processes being discussed can be understood without recourse to other textbooks, websites, or original research papers. That is the ideal, at least. In reality, the beginning student may need to brush up on some topics or even learn totally new ones in order to understand some of the items explored here. A selection of textbooks that may be useful and that are certainly relevant to the study of microbial ecology are listed in Table 1.4.

Fortunately, the other textbooks are not needed to understand the most important points about microbes and the unseen world. Those points, already touched on here, are that microbes are the most numerous and diverse organisms on the planet, that they carry out many essential processes which keep ecosystems running and the biosphere operating. Exploring these microbe-driven processes is the heart of this book.

Table 1.4 Textbooks with background information relevant to microbial ecology.

Title	Year	Authors or Editors	Comments
Aquatic Geomicrobiology	2005	D.E. Canfield, B. Thamdrup, E. Kristensen	Many biogeochemical processes occurring in all environments are covered in this book
Biology of the Prokaryotes	1999	J.W. Lengeler, G. Drews, H.G. Schlegel	For questions about bacterial and archaeal physiology, this book provides many answers
The Biology of Soil: A Community and Ecosystem Approach	2005	R.D. Bardgett	This book is a concise yet thorough overview of the soil environment
Biological Oceanography	2004	C.B. Miller	Much of modern biological oceanography is about microbes
Brock Biology of Microorganisms	2011	M.T. Madigan, J.M. Martinko, D. Stahl, D.P. Clark	This general microbiology textbook is now in its 13th edition. T.D. Brock was the sole author of the first edition in 1970
Environmental Microbiology: From Genomes to Biogeochemistry	2008	E. Madsen	Environmental microbiology is virtually the same as microbial ecology
Introduction to Geomicrobiology	2007	K.O. Konhauser	Geomicrobiology combines microbial ecology and geology, as discussed in Chapter 13
Limnology: Inland Water Ecosystems	2002	Jacob Kalff	Microbes are also important in lakes
Soil Microbiology and Biochemistry	2007	E.A. Paul	Now in its third edition, this book is an excellent starting point for learning more about soil microbial ecology

Summary

1. There are many reasons for studying microbial ecology, ranging from human health to the degradation of organic pollutants. The work of microbial ecologists is also important in examining the impact of greenhouse gases and other climate change issues.

2. Studying microbes and microbial ecology is also important because microbes, especially bacteria, are the most numerous organisms on the planet and mediate many essential biogeochemical processes.

3. Traditional microbiological approaches are often not successful for examining microbes in nature because many microbes cannot be isolated and maintained as pure cultures in the laboratory.

4. Sequences of SSU rRNA genes (16S rRNA in prokaryotes, 18S rRNA in eukaryotes) are used to define phylogenetic groups of Bacteria, Archaea, and Eukarya. These organisms also differ in several key aspects, notably the composition of their cell walls and membranes.

5. Microbes can also be divided into functional groups, depending on the mechanisms for acquiring carbon and energy and on the compounds used in redox reactions. Photoautotrophs use light energy and CO_2 whereas chemoorganotrophs, or more simply, heterotrophs, obtain both energy and carbon from organic compounds.

CHAPTER 2

Elements, biochemicals, and structures of microbes

The first chapter defined microbes by size, sketched out how microbial ecologists identify these organisms, and gave an overview of microbial physiology. This chapter will continue to introduce microbes by discussing the composition of bacteria, fungi, and protists. Here "composition" includes virtually everything found in microbes, ranging from elements to complex macromolecular structures. This information will be used in later chapters to understand the ecology of microbes and to explain the role of particular microbes in biogeochemical cycles. Microbiologists usually discuss the composition of microbes in terms of macromolecules, such as protein, RNA, and DNA. Microbial ecologists and biogeochemists often think of composition in terms of elements (carbon (C), nitrogen (N), and phosphorus (P), for example) and ratios of elements, most commonly the C:N ratio. Both approaches are used here for a complete picture of what makes up a microbial cell. Some of the information presented here is from basic microbiology and experiments with laboratory-grown microbes. Although we can learn much from these experiments, we will see that the composition of microbes grown in the lab differs from that of microbes in natural environments. These differences give clues about how microbes make a living in nature.

Composition also helps to explain the imprint of microbes on the environment. The contents of microbial cells are released by various processes and left behind in soils and aquatic habitats. These remains can contribute to large geological formations, such as the White Cliffs of Dover (Chapter 4). More common is the impact of microbial cellular components on the organic compounds found in aquatic habitats and soils. Although not as obvious as a cliff, this contribution by microbes has a large impact on the environment and on biogeochemical cycles. Understanding these effects starts with knowing what makes up a microbe.

Elemental composition of microbes

Figure 2.1 illustrates how the relative abundance of elements found in cells differs from the abundance of these elements in the earth's crust, the ultimate source of the building blocks for life. The earth's crust has high amounts of silicon (Si), but only some algae (diatoms—see Chapter 4) and a few other protists use Si. Although sodium (Na) and magnesium (Mg) are abundant in the earth's crust and as major cations in natural environments, they are not large components of biochemical structures in microbes.

Microbes do require these cations for growth in order to maintain osmotic balance and for some enzyme and membrane functions, but they make up a very small fraction of the average microbial cell. In total, inorganic ions make up only about 1% of the dry weight of a microbial cell. Use of these cations by microbes is also insignificant compared to the large concentrations usually found in natural environments. Another abundant cation, calcium (Ca^{2+}), is used only by select algae (coccolithophorids—see Chapter 4), but again not by bacteria, archaea, and fungi, except as cationic bridges among polymers. The major biogenic elements are C, N, P, and sulfur (S) (Table 2.1).

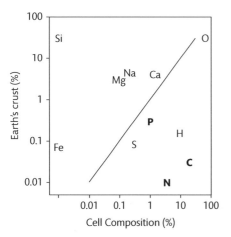

Figure 2.1 Elemental composition of the earth's crust and of a typical cell. The line indicates equal percentages in both cells and the crust and the three elements in bold (C, N, and P) are those that commonly limit microbial growth in nature. Note that some elements are highly enriched in cells compared to the inert world whereas others are present in only low amounts. Silicon is used in cell walls of some algae (diatoms) but not by bacteria. Inspired by a similar graph in Brock and Madigan (1991).

Several other elements occur only in trace amounts in the earth's crust and in cells. Although only vanishingly small amounts are needed, these trace elements or micronutrients are essential for microbial growth. Metals like zinc (Zn) and cobalt (Co) are important co-factors in some enzymes, such as the enzyme mediating urea degradation (urease needs Zn) and the vitamin B_{12}-requiring enzymes, which contain Co. A few microbes use more rare and strange metals such as tungsten (W) and nickel (Ni). The most important micronutrient is iron (Fe).

Iron is abundant in the earth's crust and is present in all cells. But microbes need only relatively low amounts (the C:Fe ratio is on the order of 10 000), mostly for electron transfer reactions, such as in the respiratory pathway. Consequently, iron concentrations are sufficient for microbial growth in most environments, with the important exception of the open oceans, where uptake by microbes reduces Fe concentrations to very low levels (10^{-12}M or pM). The low concentration is also due to the insolubility of Fe oxides (FeIII) in oxygenated water at near-neutral pH (Chapter 3). In some oceans, most notably the high nutrient-low chlorophyll (HNLC)

Table 2.1 Major and trace biogenic elements used by microbes in nature. The order roughly reflects the abundance in microbes. Data from Kirchman (2002b).

Element	Chemical form in nature[a]	Location or use in cell
Major Biogenic Elements		
C	HCO_3^-	All organic compounds
N	N_2, NO_3^-	Protein, nucleic acids
P	PO_4^{3-}	Nucleic acids, phospholipids
S	SO_4^{2-}	Protein
Si	$Si(OH)_4$	Diatom frustules
Trace Biogenic Elements		
Fe	Fe^{3+} organic	Electron transfer system
Mn	Mn^{2+}, MnO_2, $MnOOH$	Superoxide dismutase
Mg	Mg^{2+}	Chlorophyll
Ni	Ni^{2+} organic	Urease; hydrogenase
Zn	Zn^{2+} organic	Carbonic anhydrase, protease, alkaline phosphatase
Cu	Cu^{2+} organic	Electron transfer system, superoxide dismutase
Co	Co^{2+} organic	Vitamin B_{12}
Se	SeO_4^{2-}	Formate dehydrogenase
Mo	MoO_4^{2-}	Nitrogenases
Cd	Cd^{2+} organic	Carbonic anhydrase
I	IO_3^-	Electron acceptor
W	WO_4^{2-}	Hyperthermophilic enzymes
V	$H_2VO_4^-$	Nitrogenases

[a] Those metals with "organic" occur mainly in organic complexes.

oceans and in some upwelling regions, iron limits primary production and the growth of many microbes. In contrast, iron is much more abundant in soils, varying from 0.05% in coarse-textured soils to >10% in highly weathered soils (oxisols) in the tropics.

Of the six most abundant elements in bacteria, two (oxygen and hydrogen) are readily obtained from water, and a third (sulfur) from a major anion in natural environments (sulfate; SO_4^{2-}). Except in some anaerobic environments, microbes easily obtain sufficient S from assimilatory sulfate reduction; "assimilatory" implies that the end product is assimilated and used for biosynthesis, in contrast to dissimilatory sulfate reduction (Chapter 11). The reduced sulfur from assimilatory sulfate reduction is used mostly in the synthesis of two sulfur amino acids, methionine and cysteine. The remaining three elements, C, N, and P, are those thought to limit microbial growth most frequently in natural environments.

Elemental ratios in biogeochemical studies

Ecologists and biogeochemists often use elemental ratios to explore various questions in food web dynamics and in elemental cycles in the biosphere (Sterner and Elser, 2002). The use of elemental ratios to examine biogeochemical processes started with Alfred Redfield (1890–1983) who was one of the first to compare the composition of free-floating organisms (the plankton) in seawater and of major nutrients (Redfield, 1958). Redfield found that the ratio of C:N:P was 106:16:1 (in atoms) in the plankton and that the N:P ratio in the plankton was very similar to the ratio of nitrate to phosphate concentrations in the deep ocean. This observation was crucial in establishing the role of microbes in influencing the chemistry of the oceans. It is a great example of microbes molding their environment. Since then, the "Redfield ratios" of C:N = 6.6:1 and C:P = 106:1 have been used extensively in oceanography, limnology, and other aquatic sciences.

There is some work suggesting that elemental ratios of microbes in soils are remarkably close to Redfield (Cleveland and Liptzin, 2007). The C:N:P ratios for soil and soil microbes are 186:13:1 and 60:7:1, respectively. They vary depending on the soil type, vegetation, and climate regime (latitude) of the soils. These ratios are statistically different from the Redfield ratio and the elemental ratios found in aquatic systems, reflecting the vast differences in growth conditions between the two ecosystems. What is more remarkable, however, are the similarities. All things considered, the elemental ratios of soils and soil microbes are not that different from the Redfield ratio of freshwaters and the oceans. The implications of this observation have not been fully explored.

Elemental ratios can be used to examine a variety of biogeochemical and microbial processes in both aquatic and terrestrial ecosystems. For example, high C:P ratios could imply microbial growth is limited by P availability, and similarly for high C:N ratios and N limitation. Microbes do have some capacity, termed "homeostasis", however, to maintain elemental ratios even as availability changes (Fig. 2.2), and this has profound implications for understanding controls of microbial metabolism in nature. Another example discussed in Chapter 12 is the use of C:N ratios to deduce whether heterotrophic microbes are net producers or consumers of ammonium. Finally, ratios of nitrate to phosphate concentrations have been used to identify regions of the oceans where denitrification (loss of nitrate to N gases) and nitrogen fixation are common (Chapter 12).

Perhaps most elegantly, the Redfield ratio can be used to explore how primary production and respiration affect the concentrations of the major nutrients in

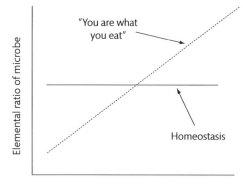

Figure 2.2 Potential relationships between the elemental ratio (e.g. C:N) of the resource used by a microbial consumer and the ratio of the consumer itself. This figure illustrates one case of homeostasis (the horizontal line) in which a microbe is able to maintain a constant elemental ratio even though the resource varies. The other extreme is the lack of any control, such that the elemental ratio of the microbe varies proportionally with the ratio of the resource. In this case, "you are what you eat". Modified from Sterner and Elser (2002).

aerobic ecosystems. The following equation is built on the Redfield ratio:

Primary production

$$106\ CO_2 + 16HNO_3 + H_3PO_4 + 122\ H_2O \rightarrow$$
$$(CH_2O)_{106}(NH_3)_{16}(H_3PO_4) + 138\ O_2 \qquad (2.1).$$
$$\leftarrow$$

Respiration

Equation 2.1 is not entirely accurate, and it hides some critical reactions. For example, Equation 2.1 implies that nitrate (NO_3^-) is released during respiration and degradation of organic material, but in fact ammonium (NH_4^+) is the main nitrogenous compound produced during mineralization of organic material, which is the conversion of organic material back to inorganic compounds. A couple additional steps, which are part of nitrification (Chapter 12), are needed to oxidize NH_4^+ to NO_3^-. Another potential inaccuracy in Equation 2.1 is the stoichiometry implied by it, such as the amount of oxygen needed to oxidize one mole of organic carbon (Kortzinger et al., 2001), a crucial ratio for interpreting respiration rates. Nonetheless, Equation 2.1 remains a very powerful and succinct description of the interactions between microbes and the geochemistry of natural environments.

C:N and C:P ratios for various microbes

The ratios C:N and C:P are used most frequently in microbial ecology and biogeochemistry. It is worthwhile emphasizing that these ratios vary substantially among species of a particular microbial group and also because of nutrient status. As mentioned above, nitrogen starvation can lead to high C:N ratios, and phosphorus starvation has the same effect on C:P ratios. Here we consider some grand averages in order to see if there are any fundamental differences among types of microbes.

Microbes share the common trait of being more nitrogen- and phosphate-rich than macroscopic organisms, but there are important differences among microbes (Table 2.2). Heterotrophic bacteria are generally thought to be nitrogen-rich compared to algae. The C:N of bacteria is often assumed to be <5, lower (more N) than that of algae and of the Redfield C:N ratio, and it is true that bacteria in the lab can have very low C:N ratios. Natural bacterial assemblages, however, have C:N ratios of 5.3–9.1 in coastal and oceanic waters (Gundersen et al.,

Table 2.2 Elemental ratios of some microbes, expressed in moles. *Synechococcus* is a cyanobacterium common in the oceans and some lakes. Data from Bertilsson et al. (2003); Caron et al. (1990); Cleveland and Liptzin (2007); Cross et al. (2005); Geider and La Roche (2002); Goldman et al. (1987); Hunt et al. (1987); Van Nieuwerburgh et al. (2004).

Microbe	C:N	C:P
Aquatic heterotrophic bacteria	3.8 – 6.3	26 – 50
Soil microbes (all)	8.6	59.5
Fungi	5 – 17	300 – 1190
Synechococcus	5.4 – 7	130 – 165
Protozoa	6.7 ± 0.9	102 ± 58
Eukaryotic algae	7.7 ± 2.6	75 ± 31
Zooplankton	5 – 11	80 – 242
Nematodes	8 – 12	?

2002), which do not differ significantly from the ratio for algae. Likewise, coccoid cyanobacteria appear to have C:N ratios similar to the Redfield ratio. In pure cultures, the C:N ratios for strains of two cyanobacterial genera, *Synechococcus* and *Prochlorococcus*, range from 5.4 to nearly 10 (Table 2.2).

Heterotrophic bacteria are also phosphorus-rich compared to algae and fungi. For example, the C:P ratio of algal biomass is about 75, much higher (much less P) than that for heterotrophic bacteria grown in the lab (26–50). Few investigations have examined the P content of both bacteria and algae simultaneously in natural communities, but freshwater studies have confirmed the higher P content in natural bacterial assemblages (Vadstein, 1998). Heterotrophic bacteria in the Sargasso Sea, on the other hand, have C:P ratios of 59–143 that are as high or higher than the Redfield ratio, reflecting potential P limitation in these waters. Fungi also have very high C:P ratios (300–>1000) and are much less phosphorous-rich than bacteria.

The phosphorus content of cyanobacteria differs from that of heterotrophic bacteria. The C:P ratios for two common coccoid cyanobacterial genera, *Synechococcus* and *Prochlorococcus*, for example, are much higher (less P) than the ratios for either heterotrophic bacteria or for eukaryotic algae. The C:P ratio is 121–215 for cyanobacteria compared with 75 and 26–50 for eukaryotic algae and heterotrophic bacteria, respectively. Total P amounts are low in cyanobacteria because their membranes have less phosphorus than the membranes of other microbes.

Rather than the standard phospholipids typically found in membranes, some cyanobacteria have sulfolipids (Van Mooy et al., 2006). The substitution of sulfur for phosphorus is one reason why *Synechococcus* and *Prochlorococcus* are so abundant in oligotrophic oceans where phosphate concentrations are very low (<10 nM).

Biochemical composition of bacteria

The elemental composition of organisms is mainly determined by the composition of the organic compounds making up the main macromolecules of cells. The main macromolecules of a cell—protein, nucleic acids, polysaccharides and lipids—contain >97% of the elements found in microbes (Table 2.3). Concentrations of small compounds, like monosaccharides (e.g. glucose), amino acids and other monomers, and salts, are low (<3%), although fluxes through the pools of these compounds can be quite high. Can the biochemical composition of microbes explain the observed Redfield ratios of the biota and the C:N and C:P ratios of bacteria and algae? The answer is, yes, more or less. We first discuss the biochemical composition of heterotrophic bacteria.

Protein is the largest component of a microbial cell, making about 55% of the dry weight of bacteria (Table 2.3). Protein, the main metabolic machinery of all cells, occurs mostly as enzymes that catalyze reactions both within a cell and in the micron-scale environment surrounding the cell. In addition to enzymes, other proteins mediate transport of compounds across cell membranes (active transport proteins), while others make up flagella, the propeller used by motile microbes for moving through aqueous environments. About 55 different proteins are in a ribosome, which is the site of protein synthesis in all cells. A typical cell may have over 1000 different protein molecules, each ranging in abundance from only a few "copies" to several thousands per cell. The abundance of different proteins can vary greatly with growth rate, exposure to different substrates, and other environmental conditions, but the relative amount of total protein per bacterial cell is roughly constant at 55%, regardless of growth rate.

Unlike protein, growth rate has a large impact on the relative amount of the other macromolecules of bacteria, especially on RNA and DNA content (Fig. 2.3). Slow-growing bacteria have relatively more DNA than rapidly

Table 2.3 Biochemical composition of a "typical" bacterial cell (*E.coli*) growing rapidly (40 minutes doubling time) in the lab. Data from Neidhardt et al. (1990).

Component	% of dry wt	Weight per cell 10^{-15} g	Number of molecules per cell Total	Number of molecules per cell Unique molecules
Protein	55.0	155	2 360 000	1050
RNA	20.5	59		
23S rRNA		31	18 700	1
16S rRNA		16	18 700	1
5S rRNA		1	18 700	1
tRNA		8.6	205 000	60
mRNA		2.4	1380	400
DNA	3.1	9	2	1
Lipid	9.1	26	22 000 000	4
Lipopolysaccharides	3.4	10	1 200 000	1
Peptidoglycan	2.5	7	1	1
Glycogen	2.5	7	4360	1
Macromolecules	96.1	273		
Soluble pool	2.9	8		
Inorganic ions	1	3		
Total dry weight	100	284		
Water (70%)		670		
Total weight		954		

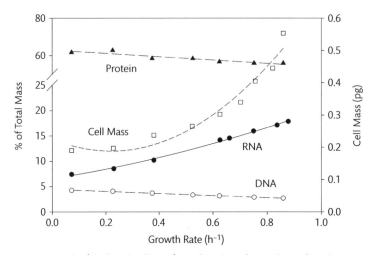

Figure 2.3 Variation in macromolecules (% of total cell mass) as a function of growth rate for a heterotrophic bacterium growing in pure culture in the lab. Data from Mandelstam et al. (1982).

growing bacteria: 10% of dry weight for slow-growing bacteria versus 3% for fast-growing bacteria (Table 2.4). The absolute amount of DNA per cell of slow-growing bacteria is not any bigger; in fact, bacteria which cannot grow quickly and are adapted to oligotrophic environments often have low absolute amounts of DNA per cell; that is, small genomes (Chapter 10). Still, these intrinsically slow-growing bacteria must put all of the genes necessary for independent existence into a small cell, leading to a high fraction of total biomass present as DNA. In contrast, even though fast-growing cells may have two or more copies of their genome per cell, their cell size is disproportionately larger, so that the ratio of DNA to cellular biomass is lower for fast-growing bacteria.

The second major effect of growth rate is on RNA content. Bacteria growing at high rates in the lab typically have about 20% of dry weight as RNA, compared to the estimate of 14% for bacteria growing slowly in nature (Table 2.4). This difference arises because RNA amount per cell increases with growth rate. Fast growth requires high rates of protein synthesis which in turn necessitates large numbers of ribosomes and ribosomal RNA (rRNA); rRNA is about 80% of total RNA. Fast-growing *E.coli*, for example, devote over 70% of transcription to the production of rRNA (Vieira-Silva and Rocha, 2010). The amount of RNA per cell may be used to estimate growth rates of not only microbes (Kemp et al., 1993) but macroscopic organisms as well.

Table 2.4 Biochemical composition of a bacterium growing fast (generation times of 1 hour or less) or slow (day timescale) and for a eukaryotic (yeast) cell growing with a generation time of about 7 hours. Data from Kirchman (2000b) and Forster et al. (2003).

	% of Dry Weight		
	Bacteria		
Biochemical component	Fast	Slow	Eukaryote
Protein	55.0	55.0	45.0
RNA	20.0	13.7	6.3
Lipids	9.0	12.0	2.9
Lipopolysaccharides (LPS)	3.4	3.3	0
Cell wall (peptidoglycan or chitin)	2.5	4.1	<1
C Storage (glycogen)	2.5	0.0	8.4
Other polysaccharides	<1	<1	31.5
DNA	3.0	10.0	0.4
Monomers (e.g. sugars, inorganic ions)	4.0	2.1	6
Total	99.4	100.2	100.5

Biochemical composition of eukaryotic microbes

It is more difficult to make generalizations about the biochemical composition of eukaryotic microbes because of the huge range of cell sizes for these organisms. Table 2.4 gives one example of a eukaryotic microbe, the yeast *Saccharomyces cerevisiae*. Though it is not abundant in natural environments, this microbe was chosen because much is known about it. What is noteworthy is the low nucleic acid content of this microbe—low relative to the rest of its cellular mass. RNA and DNA make up <10% and <1%, respectively, of total dry weight for yeast, much lower than the values for bacteria (Table 2.4). Eukaryotic algae have a similarly low fraction of their dry weight devoted to RNA and DNA (Geider and La Roche, 2002). The dinoflagellate *Amphidinium carterae*, for example, consists of 2.5% RNA and 0.7% DNA.

Eukaryotic microbes have low amounts of DNA relative to the rest of the cell, but the absolute amounts of DNA, the genome sizes of these microbes, are larger than those of prokaryotes (Chapter 10). For example, the genome of a simple eukaryotic microbe, a yeast, is >twofold larger than the genome of the archetypical bacterium, *E.coli*, with its 4.6×10^6 base pairs (4.6 Mb). Another example is a eukaryotic alga, a diatom, which has a 34.5 Mb genome. Although bigger than prokaryotes, genome sizes of eukaryotic microbes are substantially smaller than those of more complicated, macroscopic eukaryotes, such as *Homo sapiens* (3000 Mb).

Explaining elemental ratios

Now knowing the composition of microbes in terms of elements like C, N, and P, and of biochemicals, like protein and DNA, it should be possible to put the two approaches together. The few relevant studies found some agreement between the two approaches, although not completely. One study found that the biochemical composition implied by Equation 2.1 was 52% protein, 35% carbohydrate, and 12% lipid (Kortzinger et al., 2001). Another study used NMR to measure elemental ratios directly and deduced the biochemical composition to be 65% protein, 19% lipid, and 16% carbohydrate (Hedges et al., 2002). The deduced protein amount (52–65%) brackets the values actually observed in microbes (Table 2.4), but the other two biochemical components are higher than the observed values. One problem is that previous studies assumed that nucleic acids were negligible, even though it is known that nucleic acids are rich in both N and P and can be abundant in some microbes. The

C:N:P of ATP, for example, is 3:1.7:1. The exclusion of nucleic acids may explain why the deduced lipid and carbohydrate amounts are higher than those actually observed for microbes.

The biochemical composition of microbes sheds some light on why C:P ratios of bacteria differ from algae and fungi. The C:P ratio of bacteria is lower (more P) than that of most eukaryotic algae and fungi because of differences in the relative contribution of nucleic acids. Bacteria have almost three-fold more nucleic acids, either as DNA or RNA, depending on growth rates, than large algae and fungi, when expressed as a percentage of total cell mass (Table 2.4), whereas these microbial groups have nearly equal relative amounts of protein. Since nucleic acids are P-rich whereas protein has no P, the difference results in much lower C:P ratios for bacteria. Another contributing factor is the differences in cell size and the ratio of surface area to volume (Fig. 2.4). Since most of microbial C is in protein of the cytoplasm, C will vary with cell size because of cell volume. In contrast, P in membrane lipids increases with surface area. From basic equations for surface area and volume, we can see that C increases with the cube of cell radius (r^3) whereas P increases only as the square of the radius (r^2). Consequently, the ratio of surface area to volume decreases with increasing r. The end result is that the C:P of small cells like bacteria is lower than that of larger cells like many algae and fungi.

The biochemical composition data also help explain why C:P ratios vary more than C:N ratios. Because a large fraction (80%) of cellular N is in protein, and because protein as a fraction of cell mass changes little with growth rate, C:N ratios of bacteria do not vary much with environmental conditions. There are some important exceptions. Bacteria excreting carbon-rich extracellular polymers (see "slime", below) would have much higher C:N ratios than normal, but this excretion is probably only possible in carbon-rich environments. More common is high C:N ratios in some algae due again to extracellular or intracellular polysaccharides. Algae are not generally C-limited, as they can always obtain C from abundant inorganic C pools.

In contrast to C:N ratios, C:P ratios appear to vary greatly in bacteria, algae, and fungi. Unlike protein, the macromolecules rich in P, nucleic acids (RNA) and lipids, vary with growth rate. Cell size can also vary with growth

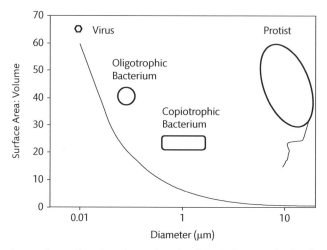

Figure 2.4 Surface area to volume ratio as a function of organism size. A bacterium growing in oligotrophic environments with very low nutrients may be half or less than the size of a copiotrophic bacterium living with high nutrient concentrations. Protists can be as small as about a micron and have much higher surface area to volume ratios than bigger protists. Viruses being the smallest biological entity in the biosphere have the highest surface area to volume ratios.

rate, and consequently, changes in the surface area to volume ratio result in changes in lipids per total cellular carbon. Polyphosphate is another P-containing biochemical that may contribute to variability in C:P ratios; polyphosphate bodies store P when concentrations and supply are high in P-rich environments such as some sediments and soils. Changes in RNA content may be offset partially by changes in relative DNA and lipid content (Table 2.4), but clearly variability in all P-containing biochemicals, especially RNA, leads to variability in C:P ratios.

Architecture of a microbial cell

The internal structure of microbes is fairly simple, at least at the level that needs to be considered here. Prokaryotes are especially simple in appearance. Bacteria and archaea are sometimes called "bags of enzymes", due to their lack of organelles and simple cellular design. In contrast, eukaryotic microbes have various organelles, such as chloroplasts (the site of photosynthesis) and mitochondria (respiration and energy production), as mentioned in Chapter 1. They also have a nucleus (sometimes more than one) where the genetic material (DNA) is kept. Heterotrophic protists form food vacuoles, which are membrane-lined

pouches where ingested food particles are digested and degraded during phagocytosis (Chapter 7).

Membranes of microbes and active transport
All cells, whether microbial or those in macroscopic organisms, must have a barrier, a membrane, that separates cellular components in the cytoplasm from the outside environment. This cell membrane, which is on the order of 8 nm thick, keeps cytoplasmic components from leaking out while preventing the entry of unwanted chemicals from the environment. The overall structure of membranes is remarkably similar for bacteria and eukaryotes, as mentioned in Chapter 1. This type of membrane consists of a phospholipid bilayer, with each layer having a hydrophilic part (glycerol and phosphate) pointing to the aqueous environment or cytoplasm and a hydrophobic interior (fatty acids). In both bacteria and eukaryotes, the glycerol and hydrocarbon chain are linked by an ester bond (Fig. 2.5), whereas in archaea, the two are linked by an ether bond. In eukaryotes and bacteria, the hydrocarbons are straight chains of an even number of fatty acids, with the occasional double bond or minor branch. In Archaea, however, these hydrocarbon chains are often highly branched with saturated isoprenes and complete rings. Ether lipids are more stable

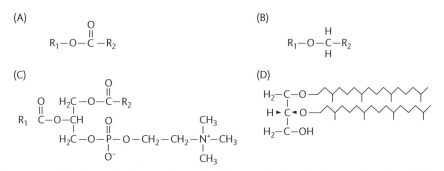

Figure 2.5 Types of lipids found in microbes. A) General structure for ester-linked lipids, which are found in bacteria and eukaryotes; B) General structure for ether-linked lipids, which are found in archaea. In both panels A and B, number of carbon and bond types represented by R_1 and R_2 vary among microbes. C) An ester-linked lipid (phosphatidylcholine); and D) an ether-linked lipid.

at high temperatures than ester lipids and may give archaea a selective advantage in hot environments (Valentine, 2007).

Some small hydrophobic molecules and gases may pass the lipid bilayers, but hydrophilic or charged compounds cannot. Ammonia (NH_3), for example, readily diffuses across membranes, but ammonium (NH_4^+) does not. Very few compounds needed by microbes for growth are both small and without a charge. To facilitate the transport of molecules across membranes, all cells have membrane proteins that span the phospholipid bilayer. Some of these membrane proteins are non-specific porins or "holes" that allow into the cell all compounds below a particular size. Several transport proteins are designed to transport specific compounds across the membrane. If membranes consisted only of phospholipids bilayers, cells would soon starve and die.

Microbes can rely on diffusion to bring only a few compounds into the cell. For most compounds, concentrations are higher inside the cell than outside and diffusion will not work. For these compounds, cells must use energy-requiring, active transport systems to transport the compound against the concentration gradient, from low concentrations outside the cell to high concentrations inside. The form of the energy driving the system defines the type of active transport. Simple transport relies on the proton motive force whereas transport by an ABC (*ATP binding cassettes*) system is fueled by ATP hydrolysis. Compounds transported by group translocation are modified during transport, a classic example being glucose transport by *E. coli*. In nearly all cases,

several membrane and often cytoplasmic proteins are involved in the transport process. These transport systems are specific for a particular compound (glucose, for example) or a class of related compounds (branched-chain amino acids, for example). Many microbes have more than one system for a particular compound that differ in affinity and energetic costs. Although the transport systems seem redundant, microbes can switch among systems depending on concentrations to maximize transport while minimizing costs.

Transport by ABC systems may be particularly relevant in thinking about bacteria in nature, because this transport mechanism is able to bring in compounds or substrates found in very low concentrations in the external environment. For Gram-negative bacteria, a key is the presence of proteins in the periplasm (periplasmic binding proteins—periplasm is defined below) that bind to the substrate with high affinity. The periplasmic binding protein then hands over the substrate to a designated membrane transporter protein. Final transport into the cytoplasm is via an ATP-hydrolyzing protein. Gram-positive bacteria, which do not have a periplasm, have different types of ABC transporters.

Cell walls in prokaryotes and eukaryotes
In addition to membranes, many cells have a cell wall, which confers a more rigid structure and shape to the cell than is possible with only a membrane. As the word implies, cell walls offer some protection for the cell while also helping to prevent the cell from breaking apart due

Table 2.5 Cell wall constituents of some microbes and selected macroscopic organisms.

Microbe	Cell wall	Main constituents
Bacteria	Peptidoglycan	N-acetyl-muramic acid, N-acetyl-glucosamine, amino acids
Archaea	Protein, pseudomurein	Amino acids, N-acetylglucosamine, N-acetylalosaminuronic acid
Algae (chlorophytes and dinoflagellates)	Cellulose	Glucose
Algae (diatoms)	Frustules	Silicate
Other algae	Various	Glucose, other sugars
Fungi	Chitin	N-acetyl-glucosamine
Insects	Chitin	N-acetyl-glucosamine
Crustaceans	Chitin	N-acetyl-glucosamine

to self-generated turgor pressure. In bacteria, for example, the concentration of chemicals dissolved in the cytoplasm creates a turgor pressure of about 2 atmospheres, equivalent to that of an automobile tire. Higher plants, fungi, and most prokaryotes have cell walls whereas animals and some protists do not.

Cell walls vary greatly in eukaryotic microbes (Table 2.5). Yeasts and fungi have chitin, a β 1, 4- linked polymer of N-acetylglucosamine, in their cell walls. Similar to higher plants, the cell wall of some algae is composed of cellulose, a β 1,4-linked polymer of glucose, whereas the polysaccharides of other algal cell walls consist of sugars other than glucose. The cell wall of dinoflagellates, a complex group of heterotrophic and autotrophic protists (Chapter 7), is made of cellulose. Diatoms, another algal group important in freshwaters, coastal marine waters, and soils, are encased in a glass house or frustule, consisting mostly of silica. The composition of the cell wall of many protists is not known.

Peptidoglycan, or more specifically murein, is the main constituent of the cell wall in bacteria. The backbone of peptidoglycan is a polysaccharide (glycan), consisting of another β 1,4- linked polymer, this time of N-acetyl-glucosamine alternating with N-acetyl-muramic acid. (Note that this linkage (β 1,4) is found in the three major polysaccharides of cell walls and exoskeletons: cellulose, chitin, and peptidoglycan). The glycan strands of peptidoglycan are cross-linked by peptide chains of a few amino acids, usually L-alanine, D-alanine, D-glutamic acid, and either lysine or diaminopimelic acid. D-amino acids are unusual because they are not used in proteins. Variation in these few amino acids making up the peptide chain lead to subtle changes in structure and is the main reason why there are some 100 types of peptidoglycan. Although amino acids other than those listed above can be in the peptide chain, the following amino acids are never found: branched-chain amino acids, aromatic amino acids, sulfur-containing amino acids, and histidine, arginine, and proline.

There are two basic designs for the organization of the wall and membranes found in bacteria: Gram-positive and Gram-negative (Fig. 2.6). The names refer to how bacteria react to a stain (the Gram stain) devised by the Danish physician H. C. J. Gram (1853–1938). Later work found that the reaction of a bacterium to the Gram stain depended on its cell wall and membrane architecture. Gram-positive bacteria have a thick cell wall consisting of peptidoglycan and no other external membrane, although they may have capsules and other less well-defined extracellular coverings. Gram-negative bacteria also have these extracellular coverings and a cell wall, although it is thinner than that of Gram-positive bacteria. In addition, Gram-negative bacteria have an outer membrane containing the macromolecule lipopolysaccharide (LPS), which is found only in Gram-negative bacteria. The space between this outer membrane and the cytoplasmic membrane is the periplasm or periplasmic space.

Cyanobacteria, which are often abundant and important primary producers (Chapter 4), have a Gram-negative-type cell wall and membrane, with some important differences (Hoiczyk and Hansel, 2000). The peptidoglycan layer of cyanobacteria is much thicker than is typical of Gram-negative bacteria, and the cross-linking between peptidoglycan chains is higher. While cross-linking between peptidoglycan chains is 20–33% in Gram-negative bacteria, cross-linking in cyanobacteria approaches that of Gram-positive bacteria (56–63%). Nevertheless, the amino acids involved in the cross link are more similar to that of Gram-negative bacteria than of Gram-positive bacteria, and like Gram-negative bacteria, cyanobacteria have LPS and the O-antigen. The latter may contribute to the toxicity of some cyanobacterial strains, as it does in Gram-negative pathogens. In addition to LPS and the O-antigen, the outer membrane

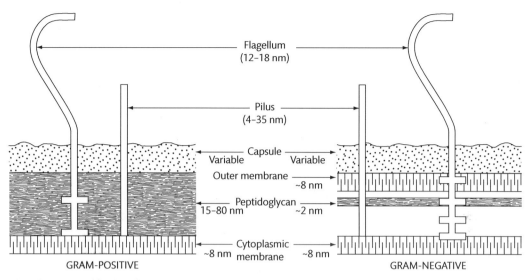

Figure 2.6 The two organizations for cell wall and membranes of bacteria. Taken from the first edition of Neidhardt et al. (1990) and used with permission from F. C. Neidhardt.

of cyanobacteria has other constituents not found in typical Gram-negative bacteria, including carotenoids and unusual fatty acids, such as β-hydroxypalmitic acid.

Neither LPS nor muramic acid is present in the cell walls and membranes of archaea. Instead, these prokaryotes have a variety of cell wall types (Mayer, 1999). Some have pseudomurein or pseudopeptidoglycan, an acidic heteropolysaccharide sharing some characteristics with peptidoglycan in bacteria. Pseudomurein has N-acetylglucosamine, but it has N-acetyltalosaminuronic acid instead of N-acetyl-muramic acid. Also, the glycosidic bonds are β 1,3 rather than β 1,4. Other archaea, such as many halophiles and methanogens, have a protein or glycoprotein coat (S-layer) and no pseudomurein. Still other archaea do not have any cell wall and survive with only a membrane separating the cytoplasm from the external environment. This wall-less trait is also found in a genus of bacteria, *Mycoplasma*, which includes some human pathogens, such as *M. pneumoniae*, the causative agent of pneumonia and other respiratory disorders.

Components of microbial cells as biomarkers

We have seen that the elemental and biochemical composition of microbes can be used to gain some insights into the growth state of microbes in nature. There are other uses for data on microbial composition. Compounds known to be specific for microbes can be used to estimate microbial biomass and to understand the sources of organic material found in natural environments (Bianchi and Canuel, 2011). These compounds are often called biomarkers (Table 2.6).

The size of a population, both in terms of cell number and biomass, is a fundamental parameter for understanding the ecology of microbes, as it is for the ecology of all organisms. The most common approach for assessing microbial numbers and biomass in environments uses microscopy or flow cytometry (Chapters 1 and 4), but two components of bacterial cell walls and membranes, LPS and muramic acid, have been used to estimate bacterial biomass in natural environments. As mentioned earlier, both compounds are unique to bacteria. Muramic acid is found in nearly all bacteria whereas only Gram-negative bacteria have LPS. Data on muramic acid were important in the early studies of bacteria in sediments (Moriarty, 1977, King and White, 1977), and likewise LPS concentrations were used to confirm that the cells counted by direct microscopy in the oceans were in fact bacteria (Watson et al., 1977). The LPS and direct count estimates are similar because Gram-positive bacteria, which do not have LPS, are not common in marine waters (Chapter 9).

Organic geochemists examine many other biomarkers to gain some insights into sources of organic material encountered in natural environments (Table 2.6). Nearly all of these microbial biomarkers are associated with microbial membranes and walls. Some notable biomarkers include various phospholipid-linked fatty acids (PLFA). Some PLFA are associated with specific groups of bacteria, such as sulfate reducers (i17:1 and 10Me16:0) and *Actinomycetes* (10Me17:0 and 10Me18:0), while others are found in eukaryotic algae (20:5 o3). Archaea can be traced by their unique ether lipids. The number of cyclopentane rings in archaeal membrane lipids ("TEX$_{86}$") is used to estimate temperature over geological timescales (Wuchter et al., 2004).

The biomarker-based approaches can be combined with ^{13}C stable isotope analysis of both the entire (bulk) organic carbon pool and of the specific biomarker. The natural abundance of ^{13}C can be used to deduce sources of organic carbon used by an organism because of the variation in relative amounts of ^{13}C, usually expressed as $\delta^{13}C$, among primary producers (Fig. 2.7). In brief, the $\delta^{13}C$ of C4 plants found on land is about −14‰ whereas it is −21‰ for algae and −27‰ for C3 land plants. (The symbol "‰" is interpreted as parts per thousand, analogous to "%" for parts per hundred). The food chain relying on these forms of primary production will have similar $\delta^{13}C$ values, perhaps separated by only 1‰. In short, "you are what you eat". For example, if an

Table 2.6 Some biomarkers for microbes. Abbreviations: PLFA = phospholipid-linked fatty acid, and LPS = lipopolysaccharides. "G-" is Gram-negative. Anammox is anaerobic ammonium oxidizers (Chapter 12).

Biomarker	Cellular Component	Organism	Comments
Muramic acid	Cell wall	Bacteria	Biomass indicator
D-amino acids	Cell wall	Bacteria	Eukaryotic sources?
LPS	Membrane	G- bacteria	Biomass indicator
i14:0	Membrane PLFA	Bacteria	See text for other PLFAs
20:5 ω 3	Membrane PLFA	Algae (diatoms)	
Dinosterol	Membrane PLFA	Dinoflagellates	
Sterols	Membrane	Eukaryotes	
Ergosterol	Membrane	Fungi	Type of sterol
Bacteriohopane polyols	Membrane	Bacteria	
Glycolipids	Heterocyst membrane	Cyanobacteria	See Bauersachs et al. (2010)
Ether lipids	Membrane	Archaea	
Ladderane	Membrane	Planctoymycetes	Anammox

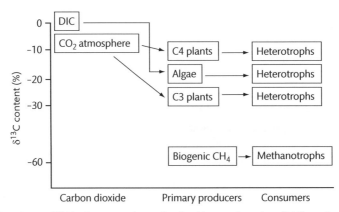

Figure 2.7 The relative abundance of ^{13}C in the atmosphere, dissolved inorganic carbon (DIC), methane, plants, and microbes. Biogenic methane comes from methanogenic archaea. In contrast, thermogenic methane is produced by geothermal heating of organic material and has a much higher (less negative) $\delta^{13}C$ value than biogenic methane. Adapted from Boschker and Middelburg (2002).

organism is found to have a $\delta\,^{13}C$ value of about –15‰ we would conclude that it is a herbivore feeding on C4 plants. In contrast, the difference in $\delta\,^{15}N$ values or fractionation between trophic levels is about 3‰.

The same general rules can apply to biomarkers and other individual organic compounds, but the initial starting point may differ from the bulk values because of fractionation during biosynthesis of the compound. For example, lipids are depleted in ^{13}C by 2–6‰ relative to bulk ^{13}C values; if the bulk $\delta\,^{13}C$ is –26‰, then the $\delta\,^{13}C$ of lipids may be –28 to –32‰. In addition to using the natural variation in ^{13}C values to deduce sources of organic material and trophic interactions, compounds greatly enriched with ^{13}C can be added to microbial assemblages, incubated over time to allow the added ^{13}C-labeled compound to be taken up, and the rate of ^{13}C uptake into bulk carbon or into specific biomarkers can be traced.

Extracellular structures

In addition to cell membranes and walls, many microbes have other structures and macromolecules that are attached to the cell but extend beyond the outer cell membrane or wall. These extracellular structures and macromolecules serve a great variety of functions for microbes, from propelling microbes around to keeping them stuck in one place.

Extracellular polymers of microbes

Depending on the environment and growth state, bacteria and other microbes can secrete a complex suite of extracellular polymers often dominated by polysaccharides. For bacteria, these polymers can be organized to form a defined layer, the capsule, around the cell (Fig. 2.8A) while other bacteria attached to surfaces produce less coherent and more extensive networks of polymers (Fig. 2.8B). Some terms used to describe these extracellular polymers include glycocalyx, extracellular polysaccharides, and extracellular polymeric substances (EPS). Sometimes the simple word, "slime" is most appropriate. Because of carbon limitation, free-living bacteria in natural environments probably do not form thick capsules, but attached bacteria are often associated with extracellular polymers.

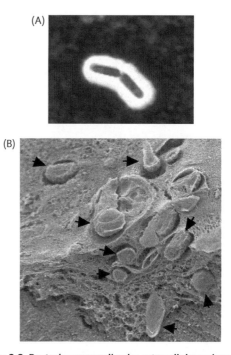

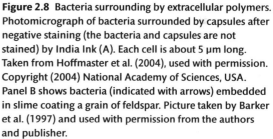

Figure 2.8 Bacteria surrounding by extracellular polymers. Photomicrograph of bacteria surrounded by capsules after negative staining (the bacteria and capsules are not stained) by India Ink (A). Each cell is about 5 µm long. Taken from Hoffmaster et al. (2004), used with permission. Copyright (2004) National Academy of Sciences, USA. Panel B shows bacteria (indicated with arrows) embedded in slime coating a grain of feldspar. Picture taken by Barker et al. (1997) and used with permission from the authors and publisher.

Regardless of the name, these polymers have several potential functions for microbes. They may serve as a carbon source when environmental conditions change from being carbon-replete to carbon-limited. The polymers help glue attached microbes to surfaces, while also providing some protection against ingestion by protist grazers. Complex polymers are important components of the symbiosis between bacteria and root nodules of legumes (Chapter 14). Pathogenic bacteria encased in extracellular polymers are protected from antibiotics, limiting their effectiveness in controlling infections. The polymers themselves can contribute to diseases caused by microbes, such as in the case of the bacterium *Pseudomonas aeruginosa* and cystic fibrosis. Eukaryotic microbes can also secrete extracellular polymers, and

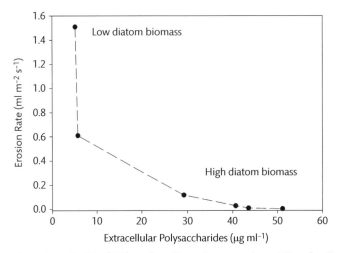

Figure 2.9 Effect of extracellular polysaccharides (EPS) produced by a diatom on the stability of sediments. Other microbes are likely to contribute to EPS production and sediment stability. Data from Sutherland et al. (1998).

their function is probably quite similar to that for prokaryotic microbes. As with prokaryotic microbes, these extracellular polymers are often dominated by carbohydrates. The carbohydrate composition of the polymers produced by algae varies greatly among different species, although glucose is the most common monosaccharide (Biersmith and Benner, 1998). Some diatoms secrete chitin strands extending several cell lengths away from the diatom. These strands may help protect the diatom from predation or keep it afloat in the water column.

In addition to being important to microbes, extracellular polymers are important to other organisms and the environment. While polymers can protect microbes from prediction by some organisms, metazoans that eat detritus and associated microbes ("detritivores") ingest and often digest the extracellular polymers along with the microbes. Polymers released by microbes contribute to aggregates in aquatic habitats and soils. This organic material is so important to soil quality that artificial polymers are added to soils to retard erosion and promote water and nutrient retention. In rivers and other aquatic habitats, hydrated extracellular polymers contribute to the physical stability of sediments (Gerbersdorf et al., 2008). In marine systems, for example, growth of benthic diatoms correlates with extracellular polysaccharides and decreases in erosion rates of material from sediments (Fig. 2.9).

Flagella, cilia, fimbriae, and pili
Many microbes are motile and can swim quite rapidly through aqueous environments, including the aqueous microhabitats of soils. Many motile prokaryotes and eukaryotic microbes are propelled by hair-like structures, the flagella, sticking out from the cell. Some microbes have one or more flagella attached to one pole of a microbe (polar flagella), while others have flagella sticking out from all sides (peritrichous flagella).

Although the same term is used for both prokaryotes and eukaryotes, and flagella have analogous functions for both types of microbes, the biochemical structures of flagella in prokaryotes and eukaryotes differ greatly. In bacteria, the flagellum consists of a single strand of a protein, flagellin, which is highly conserved among bacteria. The archaeal flagellum is analogous to that of bacteria, but the protein and attachment to the cell are not the same as for bacteria. In eukaryotes, the flagellum is a complex bundle of nine proteins surrounding an inner core of two proteins, all of which is covered by an extension of the cytoplasmic membrane. When a microbe has several short flagella, they are called "cilia", which have the same structure as flagella, but are just shorter. Some microbes can move without any flagella by gliding along solid surfaces. These microbes include pennate diatoms (shaped like a cigar), which glide by excreting polymers through a hole (the raphe) in the frustule touching the surface. Some filamentous cyanobacteria also glide by rotating and flexing of cells in the filament. Other bacteria, such as myxobacteria and some members of the Bacteroidetes phylum, are well known for gliding, but the mechanisms remain unclear. Gliding by one member of the Bacteroidetes, *Flavobacterium johnsoniae*, appears to be due to 5 nm-wide tufts of filaments attached to the outer membrane of this bacterium (Liu et al., 2007).

Summary

1. The elemental and biochemical composition of microbes has large impacts on the chemistry of natural environments. Information about microbial composition can be used to examine several biogeochemical processes.

2. The elemental and biochemical composition of microbes varies with growth conditions and the type of microbe. This variation can be used to deduce microbial growth conditions.

3. The cell walls of prokaryotes, fungi, and protists differ greatly in composition, but they often contain β 1,3-linked polymers. Cell walls of bacteria have unusual components, including muramic acid and D-amino acids, which make up peptidoglycan.

4. The membranes of bacteria and eukaryotes are similar and consist of ester-linked lipids. In contrast, archaea have ether-linked lipids. All membranes have transport proteins for bringing desired compounds into the cell.

5. Microbes can release extracellular polymers often dominated by polysaccharides. These polymers help microbes when attached to surfaces and impact the physical environment surrounding microbes.

CHAPTER 3

Physical-chemical environment of microbes

As with all organisms, microbes are affected by many environmental properties that help to determine their diversity and abundance, their chemical composition, growth rates, and metabolic functions. Some environmental properties, such as temperature and pH, are also important in thinking about the biology of macroscopic organisms, which in turn affects microbial communities. For example, extremes in temperature and pH affect microbes both directly, and indirectly due to the exclusion of many large organisms. For many of these environmental properties, the same mechanisms and equations describe how both microbes and macroscopic organisms are affected. Other properties, however, are unique to life at the micron scale. This chapter will discuss several physical and chemical properties of the microbial environment and how they affect microbes.

To help with visualizing life at the scale of a microbe, Figure 3.1 compares the size of organisms in the microbial world with organisms and other things in the macroscopic world. It shows why a 100 μm ciliate or alga is huge in the microbial world. Some interactions among microbes and between microbes and the environment can be "scaled up" or simply enlarged to our more familiar macroscopic world. For example, several predator-prey relationships in the microbial world are similar to those in macroscopic world. However, other aspects of life at the microbial scale are radically different from what is encountered in a macroscopic world. A challenge of microbial ecology is to understand how things happening at the scale of microns and molecules have

such huge impacts on the biosphere and the entire planet.

Water

The search for life on other planets often focuses on water because we know that on earth, where there is water, there is life. All cells, whether microbial or metazoan, are about 70% water by weight (Chapter 2), and all organisms require water for growth. Some microbes can survive without water by forming a resting stage, called spores or cysts, depending on the microbial type, but none can grow while completely desiccated. Even in soils, microbes need an aqueous environment at the micron scale to metabolize and grow. Water has a huge impact on the types, abundance, and growth of microbes in soils.

A pioneer in microbial ecology, T.D. Brock (1926–), pointed out that the unusual properties of water explain its "admirable utility as a menstruum for the evolution of living creatures" (Brock, 1966). Among its 63 anomalous properties (Kivelson and Tarjus, 2001), water is polar, has a high dielectric constant, and is a small molecule, making it an excellent solvent for many biologically important compounds. It is also a viscous liquid, an important feature for understanding life at the scale inhabited by microbes. Water is much more viscous than air. This seemingly trivial difference between water and air has many fundamental implications for understanding how life differs at the microbial scale from how we and other macroscopic organisms experience

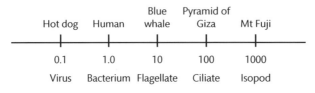

Figure 3.1 Scaling of sizes in the macroscopic and microscopic worlds. The organisms in both worlds vary greatly in size (length) and even more so in volume and mass. Even hot dogs vary from dainty to foot-long. The items serve as examples to illustrate the order-of-magnitude differences in size in both worlds.

it. It affects the movement of gases and organisms, and explains the bones of animals and the wood of trees. While air supplies the oxygen and carbon dioxide needed by macroscopic organisms, water is the medium of microbes even if present only as a thin film in soils.

Temperature

Microbes can live or at least survive temperatures from well below freezing to well above 100 °C. At the lower end, microbes are found in Antarctica where temperatures dip down routinely to –60 °C with the record being about –90 °C. While some metabolism may continue in solid ice, liquid water is needed for any substantial microbial activity, which is possible even at very low temperatures. One example is the brine channels of Arctic sea ice where temperatures can be –20 °C. Water can remain liquid at this temperature only because salinity is high, reaching 20% or nearly tenfold higher than that of seawater, as mentioned in Chapter 1. At the other end of the thermometer, microbes proliferate in hot springs on land and in hydrothermal vents found at the bottom of some oceans. Microbes have been found in >150 °C waters coming out of hydrothermal vents (Chapter 14), but it is not clear whether they are metabolically active; they may come from cooler waters mixed into the hot hydrothermal water. Currently, the temperature record is held by an iron-reducing bacterium isolated from the Juan de Fuca Ridge in the Northeast Pacific Ocean. This microbe can grow at 121 °C (Kashefi and Lovley, 2003). Water can be liquid at temperatures >100 °C only because of the high pressure in the deep ocean. It has been hypothesized that liquid water, rather than temperature per se,

sets the limits at which life can exist. Figure 3.2 summarizes some of the terms used to describe the microbes growing within various temperature ranges.

Eukaryotic microbes are found in very cold environments, but not in extremely hot ones. Diatoms, other eukaryotic algae, and heterotrophic protists live in the cold brine channels of sea ice, and fungi are commonly isolated from Antarctic soil. Light, probably more so than temperature, limits the growth of phototrophic microbes in sea ice, and the grazing of phagotrophic protists (Chapter 7) on bacteria and other microbes may be physically inhibited by the small confines of brine channels. In contrast, the upper limit of eukaryotes is about 65 °C, well below the >100 °C record of some prokaryotes. The maximum temperature for growth by phototrophic eukaryotes is even lower than that for heterotrophic ones. The hottest water in which a eukaryotic alga (*Cyanidium caldarium*) can grow is 55 °C, whereas cyanobacteria can thrive in waters >70 °C.

Table 3.1 lists the terms for describing the temperature preference of microbes and analogous terms for other environmental properties. Many of these terms end with "phile", which comes from the Greek meaning "loving". A psychrophile, for example, grows at low temperatures (about 10 °C) whereas a thermophile is best suited for about 40 °C. Neither of these organisms grows well at "normal" temperatures, at least normal by human standards. Some thermophilic bacteria appear to have diverged earlier than other bacteria in evolution, suggesting that life arose in hot environments such as hydrothermal vents. Hyperthermophiles grow at temperatures above about 60 °C and include many archaea.

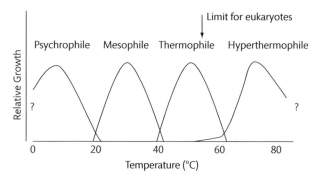

Figure 3.2 Definition of terms used to describe organisms with different temperature ranges.

There can be disparities between the optimal temperature for a microbe in the lab versus the microbe's original environment. Often the microbe grows best in the lab at temperatures higher than it ever sees in its natural environment. For example, the Arctic heterotrophic bacterium *Colwellia psychrerythraea* grows optimally at 8 °C in the lab, and it can grow in waters as hot as 19 °C (Methe et al., 2005), but mean temperatures in the Arctic Ocean are below 5 °C. One explanation is that it may be advantageous for the microbe to have the capacity to take advantage of rare high temperatures in its environment as long as it grows well enough at low temperatures. Another explanation is that the mismatch between temperature optimum and the environment is an artifice. The cultivated microbes may not be representative of the uncultivated microbes abundant in cold environments. Whatever the explanation, microbes are probably adapted to the temperature range found in their environment.

The effect of temperature on reaction rates
Of all environmental parameters, temperature has one of the most profound effects on microbial activity because of its immediate impact on enzymatic reactions being carried out by microbes and on abiotic reactions in the microbial environment. The rate of all chemical reactions increases with temperature following a well-defined rule encapsulated in the Arrhenius equation. It describes how a reaction rate (k, with units of per time) varies as a function of temperature (T, expressed in Kelvin):

$$k = Ae^{-E/RT} \tag{3.1}$$

Table 3.1 Terms used to describe organisms growing under various environmental conditions.

Environmental property	Organism	Optimal growth conditions
Temperature	Psychrophile	<15 °C
	Mesophile	15–40 °C
	Thermophile	45–80 °C
	Hyperthermophile	>80 °C
pH	Acidophile	pH <5
	Neutrophile	pH 6–8
	Alkaliphile	pH >8
Salt	Mild halophile	1–6% NaCl
	Moderate halophile	6–15% NaCl
	Extreme halophile	>15% NaCl
Pressure	Piezotolerant	Survival but no growth above atmospheric pressure
	Piezophile	Growth under moderate pressure (10–80 MPa)
	Hyperpiezophile	Growth under high pressure (>80 MPa)

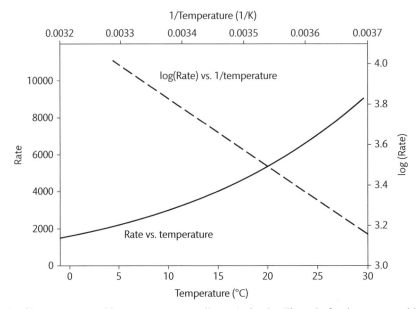

Figure 3.3 Example of how rates vary with temperature according to Arrhenius. The units for the rate are arbitrary. The activation energy for this reaction was set at 40 kJ mol^{-1}, which is roughly equivalent to a Q$_{10}$ of 2 near 20 °C. The top axis is the inverse of temperature expressed as Kelvin, the units required in the Arrhenius equation.

where R is the gas constant (8.29 kJ^{-1} mol^{-1} K^{-1}), A is an arbitrary constant, and E is the activation energy, an important defining characteristic of the reaction. The Arrhenius equation predicts that reaction rates increase exponentially with temperature (Fig. 3.3).

Although accurate for many reactions, microbes sometimes do not follow the Arrhenius equation, perhaps due to variation in the activation energy (E) with temperature, and microbial ecologists often use other ways to express how rates vary as a function of temperature. In soils, experiments have showed that microbial rates vary as the square root of temperature. This transformation is simpler than the Arrhenius equation but it lacks a mechanistic foundation. The Arrhenius equation can be derived from first principles governing how molecules interact as a function of temperature. Another commonly used expression is Q$_{10}$, which is the factor at which a rate increases with a 10 °C increase in temperature. Many reactions in biology have a Q$_{10}$ of 2, an important number to remember. For example, a microbe growing at 0.5 d^{-1} at 15 °C would be expected to grow at 1.0 d^{-1} at 25 °C if Q$_{10}$ = 2. Experimentally, Q$_{10}$ is often measured at

> **Box 3.1 Arrhenius and the greenhouse effect**
>
> Svante Arrhenius (1859–1927) was awarded the Nobel Prize in 1903 for his work in electrochemistry, his main research field. However, one of Arrhenius's first scientific contributions was to understanding the effect of greenhouse gases on our climate. He published a paper in 1896 titled, *On the Influence of Carbonic Acid in the Air upon the Temperature of the Ground*, in which he argued that global temperatures would increase by 5°C if CO$_2$ concentrations increased by two to threefold. Arrhenius was trying to explain the comings and goings of ice ages, but his calculations are quite relevant to understanding climate change issues. Arrhenius's estimate for the sensitivity of the climate system to CO$_2$ is remarkably similar to current estimates derived by much more complicated modeling studies and field observations.

temperature intervals other than 10°C. In this case, the Q_{10} can be calculated using the following equation:

$$Q_{10} = (r2/r1)^{(10/T2-T1)} \tag{3.2}$$

where r1 and r2 are the rates measured at two temperatures, T1 and T2, respectively. The Q_{10} is related to the activation energy (E) from the Arrhenius equation by

$$Q_{10} = \exp(10E/R \cdot T_1 \cdot T_2) \tag{3.3}$$

where $T_2 = T_1 + 10$ with temperature expressed in Kelvin.

Q_{10} is a convenient, easy way to express temperature effects, but it also lacks a mechanistic foundation. It may not be constant over a temperature range, even if the effect of temperature on a process is the same for 5–10 °C as for 25 –30 °C, for example. If the temperature effect for a process is accurately described by the Arrhenius equation, Q_{10} changes when over a large range in temperature is examined.

It is sometimes said that heterotrophic microbes are affected more by temperature than phototrophic ones, and in fact models examining climate change often assume that primary production will not be directly affected by increases in temperature but heterotrophy and respiration in soils will (Bardgett et al., 2008). One reason given is that unlike heterotrophic reactions, the light reaction of phototrophy is independent of temperature. However, phototrophic organisms carry out many chemical reactions that are not light-dependent and thus would be sensitive to temperature. In fact, there is probably no difference in how temperature affects metabolic rates of heterotrophic and phototrophic microbes (Li and Dickie, 1987). The relationship between the maximum growth rate of phytoplankton and temperature is often described by the "Eppley curve" (Eppley, 1972), found by the biological oceanographer, R.W. Eppley (1931–). This curve suggests that the maximum growth rate (G_{max}) increases exponentially with temperature (T) according to the following equation:

$$G_{max} = 0.59e^{0.0633T} \tag{3.4}$$

This curve was compiled from studies of about 130 phytoplankton strains. While the average growth rates of all phytoplankton followed an exponential relationship, the response of individual phytoplankton species to temperature varies substantially.

Any differences in how temperature affects heterotrophic organisms versus primary producers or aquatic versus terrestrial organisms would have huge implications for understanding the impact of climate change on the carbon cycle and the rest of the biosphere. The question is complicated by the many indirect effects of temperature, such as those on the hydrological cycle, which is the movement of water from the atmosphere to and from land, lakes, and the oceans. Even though it has been examined since the days of Arrhenius in the nineteenth century, the temperature effect on microbes remains an active research topic today (Kirschbaum, 2006).

pH

The pH has nearly as great an effect on microbes and their environment as does temperature. Microbes able to grow in various pH ranges are described by terms analogous to those used for temperature. Acidophilic microbes grow in waters and soils with pH of 1–3 while at the other extreme, alkaliphiles prefer a pH of 9–11. The pH of marine waters is about 8 while many lakes have a neutral pH or are slightly acidic. Non-marine geothermal waters, however, are extremely acidic (pH<5), as are drainage waters from coal and metal mines, with devastating effects on neighboring environments. Lakes can become acidic because of "acid rain" caused by pollution from upwind power plants. Many soils are also naturally acidic, with pH<4. Although large organisms cannot survive extremely low pHs (pH<4), some eukaryotic microbes can. For example, eukaryotic algae dominate the Rio Tinto in south-western Spain, which is acidic (pH = 2.3) because its watershed has large iron and copper sulfide deposits, which have been mined for millennia. Some bacteria, such as those that rely on iron oxidation for energy (see Chapter 13), can only flourish in low pH environments.

Some examples of alkaline aquatic environments (pH>10) include Mono Lake (California), the Great Salt Lake (Utah), and some lakes in the Rift Valley of Africa. These alkaline lakes have very high salt concentrations, ranging from 30 g liter^{-1}, about the level of seawater, to >300 g liter^{-1}. Like acidic environments, the diversity of biological communities in alkaline environments is low and consists of only a few metazoans but potentially several types of microbes. Mono Lake, for example, has

Box 3.2 Acid rain versus acidic rain

"Acid rain" is used in the popular press to describe rain contaminated by sulfur and nitrogen oxides from industrial activity. However, all rain is acidic, even if not affected by human activity. Because it is not well-buffered, even pristine water in equilibrium with the atmosphere has a pH of 5.2 due to carbon dioxide and carbonic acid. Any SO_2, which has natural sources, would reduce pH further.

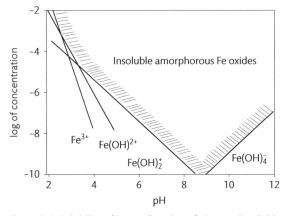

Figure 3.4 Solubility of iron as function of pH. Iron is soluble (present as Fe^{+3} and other charged oxide species) for very low pH (pH<3). At pH>3, iron is present mostly as solid (insoluble) amorphorous iron oxides, $Fe(OH)_3$. Data from Stumm and Morgan (1996).

no fish, but is famous for a species of a small brine shrimp (*Artemia monica*) that is food for several migrating birds. The brine shrimp feed off a productive algal community consisting of a few species of eukaryotic photoautotrophs and cyanobacteria (Roesler et al., 2002).

Alkaline soils have large amounts of limestone ($CaCO_3$) and clay. As in aquatic habitats, soils with high pH often but not always have high concentrations of the major cations, notably calcium (Ca^{2+}), along with magnesium (Mg^{2+}), potassium (K^+), and sodium (Na^+). Soil pH is also affected by water content, with low rainfall leading to high pH. For this reason, arid regions of the world, such as many areas of the western USA, have alkaline soils. Many plants have difficulty growing in soils with high pH because it reduces the availability of critical nutrients. Overall, pH has a huge impact on microbial diversity in soils (Chapter 9).

The pH also affects the chemical state of several important compounds and elements. For example, the most abundant form of iron in aerobic environments, Fe(III), is soluble in acidic environments (pH<3) but forms insoluble iron oxides at the pH of most environments (Fig. 3.4). The adsorption of essential nutrients such as phosphate and nitrate to soil and sediment particles is governed by pH and is determined by the charge of cations in these particles. The charge of other compounds varies also as a function of pH. One example is ammonium, a key source of nitrogen for many microbes. The exchange between ammonia (NH_3) and ammonium (NH_4^+) is described as follows:

$$NH_3 + H^+ \Leftrightarrow NH_4^+ \qquad (3.5).$$

Since the pKa of this reaction is 9.3, ammonium predominates in nearly all environments, except those with a high pH. The seemingly simple addition of a single proton changes an uncharged molecule (ammonia), which can easily pass through cell membranes, to a charged molecule (ammonium) that requires specialized transport mechanisms in order for it to be used by microbes. The effect of pH on ammonia versus ammonium has implications for understanding how a key reaction in the nitrogen cycle, nitrification (Chapter 12), may be affected by ocean acidification (Chapter 4).

Salt and osmotic balance

The salt concentration of microbial environments varies from distilled water levels to brines and ponds near saturation (35% for NaCl). Microbes vary in their capacity to survive and grow in these environments. Halophilic microbes require at least some NaCl if not other salts for growth, whereas other microbes cannot survive with any appreciable salt. Salt curing preserves meat by inhibiting microbial growth, although some extreme halophiles can grow even under these conditions. Extreme halophiles, which include some interesting types of archaea, can dominate alkaline lakes and evaporation ponds, coloring the water a brilliant red with their pigments.

The problem facing microbes and all organisms is not salt per se but the relative amount of water—more precisely, water activity—in the cell relative to the environment. With the exception of some extreme halophiles, water activity is lower and solute concentrations are higher in a cell than in the external environment, resulting in the net flow of needed water into the cell. This gradient is relatively easy for cells to maintain in low salinity environments. However, as salinity increases and thus water activity decreases, cells face the problem of retaining water. To do so, they need to raise their internal solute concentrations by either pumping in inorganic ions (such as K^+) or by synthesizing organic solutes. Whether inorganic or organic, these solutes, called the compatible solutes, must not disrupt normal cellular biochemical reactions. The compatible organic solutes include glycine betaine, proline, glutamate, glycerol, and dimethysulfoniopropionate (DMSP). DMSP is used as a sulfur source by marine bacteria and can be broken down to dimethylsulfide (DMS), which contributes to negative feedbacks between oceanic biology and climate change. Figure 3.5 provides examples of organic compatible solutes.

There are advantages and disadvantages for cells using organic versus inorganic compatible solutes (Oren, 1999). Cells using inorganic solutes have to have enzymes and other proteins specially adapted to high salt concentrations. In contrast, cells using organic solutes do not need

especially designed enzymes and proteins because organic compatible solutes are either uncharged or zwitterionic at the physiological pH. Consequently, only a few microbes, such as some extreme halophiles, use inorganic compatible solutes. However, synthesizing organic solutes is energetically expensive. Energetics may explain why bacteria and archaea relying on low energy-yielding metabolisms, such as methanogenesis (Chapter 11) and ammonia oxidation (Chapter 12), have not been isolated from high-saline environments. One organic solute, glycerol, can be synthesized cheaply but is used only by some eukaryotes, perhaps because membranes have to be modified to retain this small, uncharged molecule.

Oxygen and redox potential

All metazoans and nearly all eukaryotic microbes (except yeasts and a few protists) require oxygen for survival and growth. Many bacteria and archaea are also obligate or strict aerobes, meaning they require oxygen. Many other prokaryotes, however, can grow in the absence of oxygen and either are facultative or strict anaerobes, depending on whether they can or cannot tolerate oxygen. We return in Chapter 11 to discuss how oxygen controls microbial community structure and growth in many natural environments. Oxygen and other oxidants contribute to the redox potential of an environment.

The redox state of water and soils can be measured with a platinum electrode relative to the half-potential of hydrogen (H^+/H_2). The redox potential for an individual redox reaction is defined by the Nernst equation

$$E_h = E° - (0.0591/n)\log[\text{reductants}]/[\text{oxidants}] + (0.059m/n)pH \tag{3.6}$$

where $E°$ is the standard half-cell potential, n the number of electrons transferred, m the protons exchanged, and [reductants] and [oxidants] are the concentrations of reduced and oxidized compounds, respectively. Table 3.2 gives the oxidized and reduced forms of some compounds important to microbes and the biosphere. By definition, oxidized compounds can take on electrons and become more reduced whereas the opposite is the case for reduced compounds. Oxidized compounds, most notably oxygen, are abundant in oxidizing environments (positive E_h), while

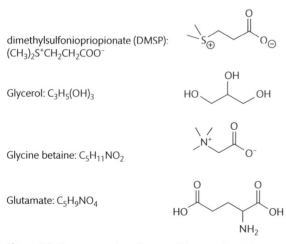

dimethylsulfoniopriopionate (DMSP): $(CH_3)_2S^+CH_2CH_2COO^-$

Glycerol: $C_3H_5(OH)_3$

Glycine betaine: $C_5H_{11}NO_2$

Glutamate: $C_5H_9NO_4$

Figure 3.5 Some examples of compatible organic solutes found in microbes.

Table 3.2 Some examples of oxidized and reduced forms of key elements in microbial environments. The E_h is for the half-reaction, relative to hydrogen.

Element	Oxidized	Reduced	E_h (mV)	Comments
Hydrogen	H^+	H_2	0	$E_h = 0$ by definition
Oxygen	O_2	H_2O	+600 to +400	See oxygenic photosynthesis
Nitrogen	NO_3^-	N_2, NH_4^+	+250	Many other reduced forms, including in organic compounds
Manganese	Mn^{+4}	Mn^{+2}	+225	Mn^{+3} in some environments
Iron	Fe^{3+}	Fe^{2+}	+100 to –100	Speciation also depends on pH
Sulfur	SO_4^{2-}	S^{2-}	–100 to –200	Sulfide (S^{2-}) usually occurs as HS^-, depending on pH
Carbon	CO_2	CH_4	< –200	Many other reduced forms, including organic compounds

reduced compounds are more common in a reducing environment (negative E_h).

From many possible examples, let us consider again iron and the form it takes as a function of redox potential of an environment. It is convenient to express E_h in a form analogous to pH:

$$pE_h = -\log(E_h) \qquad (3.7).$$

Figure 3.6 illustrates how Fe^{+2} dominates in reducing environments with low E_h (high pE_h) whereas Fe^{+3} is the main form of iron in oxidizing environments with high E_h (low pE). Exchanges between Fe^{+2} and Fe^{+3} and indeed all redox reactions are governed by thermodynamics. Some of these reactions are mediated by some microbes in

order to synthesize ATP, whereas others occur abiotically without direct involvement of microbes. The relative importance of microbial versus abiotic processes is still unknown for some redox reactions.

Light

Light provides energy for phototrophic microbes and for the synthesis of organic carbon via fixation of carbon dioxide by photoautotrophic microbes (Chapter 4). Light also affects heterotrophic microbes directly by damaging macromolecules and indirectly by affecting organic and inorganic compounds used by microbes. The effect of light varies with wavelength (color) and

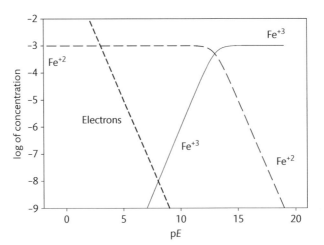

Figure 3.6 Relationship between redox potential (pE) and the concentrations of different Fe species. Data taken from Stumm and Morgan (1996).

Box 3.3 Ozone hole

The most damaging and shortest wavelengths of light, including UV-C (200–280 nm), are absorbed by ozone in the atmosphere. Unfortunately, atmospheric ozone concentrations can be depleted, forming ozone holes in the high latitudes, especially over Antarctica. Ozone has been declining in the stratosphere over the last few decades. Ozone is destroyed by chlorine and bromine. These halogens come from chlorofluorocarbons (CFCs) that were once heavily used in refrigeration and other industrial applications, but are now banned by the Montreal Protocol established in 1989. Microbial ecologists have noticed substantial differences in microbial activity under ozone holes in Antarctica (Pakulski et al., 2008, Nisbet and Sleep, 2001).

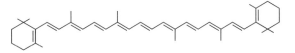

Figure 3.7 Example of a double-bond-rich compound (beta-carotene) which is common in microbes. This and similar compounds absorb light and give color to even heterotrophic microbes without pigments involved in photosynthesis.

enable light absorption and give them their characteristic color. Colonies of these pigmented microbes growing on solid media often are intensely colored. Of several repair mechanisms, RecA repairs breaks in the DNA strand. Many microbes have other enzymes, such as peroxidases and super-oxide dismutase, to disarm reactive oxygen species before they can damage cellular components. These enzymes are common in aerobic microbes but not in anaerobes which do not see any light (or oxygen) (Chapter 11).

Pressure

The largest biomes on the planet include the deep ocean and the deep subsurface environment—the geological formations deep below the earth's surface. These biomes are usually thought of as being extreme environments where exceptional microbes live. However, non-extreme environments—as defined by where humans live—may be the exceptional ones, given that the volume of our environment is much smaller than that of the deep ocean and deep subsurface. More of life is subjected to high pressure than to atmospheric pressure.

Pressure inhibits the activity of microbes that are normally found at atmospheric pressure, but even some of these microbes may tolerate high pressure and are able to resume growth when the pressure returns to normal (Fang et al., 2010). Piezophiles (also called barophiles) grow only at high pressures and hyperpiezophiles are adapted to grow at >60 megapascals (MPa). These microbes are found at the bottom of the oceans, the deepest spot (11 km) being the Pacific Ocean's Mariana Trench where pressures are about 110 MPa or over a thousand-fold atmospheric pressure. Unlike extremes of temperature and pH, some fish and other metazoans can survive the high pressure of the deep ocean, although

thus its energy. The most energetic colors of light are in the ultraviolet (UV) range with very short wavelengths. UV-C (200–280 nm) is absorbed in the atmosphere (see Box 3.3), but UV-B (280–315 nm) and UV-A (315–400 nm) does reach the earth's surface and has several impacts on natural environments. Even visible light with short wavelengths can also affect heterotrophic microbes and chemicals.

Light can directly damage DNA and other key macromolecules in microbes. One light effect is the cross-linking between adjacent pyrimidine bases of DNA (cytosine and thymine), forming pyrimidine dimers. UV light can also cause the formation of reactive oxygen species, such as perioxides (H_2O_2) and super-oxide radicals (O_2^-), which oxidize DNA, proteins, and other macromolecules in cells. Light energy can be transformed into heat which also can damage cellular components. Damage to DNA is especially important. If left unrepaired, DNA damage causes mutations and changes in the genetic make-up of the affected microbe.

Nearly all microbes have various mechanisms for preventing or correcting the damage caused by light. Some microbes have sunscreens, pigments such as carotenoids that absorb light and minimize its damage. These pigments have alternating double-bonds (Fig. 3.7) that none are found in deep subsurface environments (Bartlett, 2002).

An accident provided one of the first glimpses into the effect of pressure on microbial activity. On 16 October 1968, the research submersible Alvin sank off the coast of Massachusetts and came to rest on the sea floor 1540 m below the surface (Jannasch et al., 1971). The crew of three escaped safely, but their lunch of bologna sandwiches and apples was left behind. When Alvin was retrieved from the bottom eight months later, the lunch seemed still edible. In contrast, bologna sandwiches kept at the deep sea temperature (3 °C) spoil within weeks. The implication of this "experiment" is that decomposition of the starch and protein making up the sandwich bread and meat is inhibited by pressure, not just the cold temperatures found at the sea floor.

Piezophiles are thought to have evolved from low-pressure psychrophiles found in high-latitude environments (Lauro et al., 2007). Both types of microbes have similar adaptations to dealing with their respective extreme environments. The lipids of both piezophiles and psychrophiles are highly unsaturated, and high pressure and low temperatures lead to similar alterations in protein and DNA structures.

The consequences of being small

The physical factors discussed so far are all familiar, more or less, to those in the macroscopic world. But smallness itself imposes constraints on what microbes can and cannot do. It would be trivial to say that microbes are small, if not for all the consequences of being small. Cell size affects transport of limiting nutrients, predator-prey interactions, macromolecular composition, and many other aspects of microbial biology and ecology.

The world of small cells is fundamentally different than that of macro-organisms. One measure of that difference is the Reynolds number (Re), a dimension-less parameter that is the ratio of inertial forces to viscous forces and is defined by:

$$Re = D \cdot v \cdot \rho / \mu \tag{3.8}$$

where D is the characteristic length scale, v the velocity, ρ the density of the fluid and μ the dynamic or absolute viscosity. The Reynolds number for the world inhabited by humans is huge (10^4). If values from the bacterial world are plugged into the equation (D = 1 micron, v = 30 micron/sec, ρ = 1 g cm^{-3}; μ = 10^{-2} cm^2 sec^{-1}), the resulting Reynolds number is low (<1). For this reason, microbes are said to live in a "low Reynolds number environment". Unlike our world, viscous forces dominate over the inertial forces in the low Reynolds number environment of microbes. As the physicist E.M. Purcell once pointed out (Purcell, 1977), for humans to live in a low Reynolds number environment would feel like swimming in molasses. A person pushed while in molasses would glide much less than 10 nm. Figure 3.8 gives actual motility rates as a function of organism size in various Reynolds number worlds.

One consequence of being in a low Reynolds number environment is that mixing of molecules is governed by diffusion (a gradient-driven process) whereas in the macroscopic world, mixing is dominated by turbulence (an inertia-driven process). In a diffusion-dominated world, mixing is the result of countless, random collusions between molecules. A measure of how readily this diffusion-driven movement and mixing occurs is diffusivity (D). Movement or flux (J) of a compound as function of distance (z) follows Fick's first law:

$$J = -D_c \cdot dC / dz \tag{3.9}$$

where D_c is the diffusion constant for a particular compound with a concentration of C. In words, the flux due to diffusion is a product of the gradient in the concentration (dC/dz) and the diffusion constant (D_c). The flux is always from high concentrations to low, hence the negative sign in Equation 3.9.

The diffusion constant varies with the phase (water, air, or solid), temperature and the chemical itself. All things being equal, small compounds diffuse more quickly than large compounds, as do uncharged

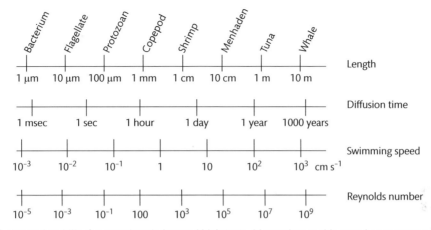

Figure 3.8 Diffusion and mobility for organisms in low and high Reynolds number worlds. Based on Jørgensen (2000).

molecules compared to charged ones. Proteins have higher diffusivity than polysaccharides such as dextrans because they tend to be more hydrophobic and more compact than dextrans with similar molecular weights (Table 3.3). For gases, diffusion increases with temperature and decreases with pressure according to the following equation:

$$D_n = D \cdot P / P_n \cdot \left(T_n / T \right)^{3/2} \qquad (3.10)$$

where D_n is the new diffusion constant at the new or changed temperature (T_n) and pressure (P_n) compared with the same values at the original conditions (D, T, and P) (Logan, 1999). For aqueous solutions, diffusion also increases with temperature, but this is partially offset by a decrease in the viscosity of water as temperature increases. For example, the diffusion constant for oxygen in water is 0.157×10^{-4} and 0.210×10^{-4} cm^2 sec^{-1} at 10

and 20 °C, respectively, only a 57% increase for a doubling of temperature.

It is possible to use the Stokes-Einstein relationship to get some feel for the time (T) and length scales (L) of diffusion for compounds with different diffusion constants (D):

$$T = L^2 / 2D \qquad (3.11).$$

As one example, consider oxygen and glucose in water at 10 °C (Table 3.4). These two compounds move the length of a bacterium (about 1 μm) within fractions of a second, but take several seconds to move 100 μm, and years before oxygen and glucose spread over a meter.

Diffusion sets an upper limit for how fast a compound can be taken up by a microbe. If every molecule arriving at a cell surface is taken up, then the flux (J) to a spherical cell with a radius r is:

Table 3.3 Some diffusion coefficients of chemicals and a virus. Data from Logan (1999).

Compound	Molecular weight (Daltons)	Diffusivity (cm^2 s^{-1} X 10^8)
Ammonia (NH$_3$)	17	2200
Glucose	220	673
Dextran	60 200	35
Serum albumin	70 000	61
Tobacco mosaic virus	31 400 000	5.3

Table 3.4 Time and distance scales for oxygen and glucose moving by diffusion. Data from Jørgensen (2000).

Distance	Oxygen	Glucose
1 μm	0.34 ms	1.1 ms
10 μm	34 ms	110 ms
100 μm	3.4 s	10 s
1 mm	5.7 min	19 min
1 cm	9.5 h	1.3 d
10 cm	40 d	130 d
1 m	10.8 y	35 y

$$J = 4\pi D \cdot r \cdot C \qquad (3.12)$$

where C is the concentration in the bulk solution infinitely far from the cell. This flux has units of mass per unit surface area per unit time. In the next chapter, we build on Equation 3.12 to discuss why small cells are better at taking up dissolved compounds than large cells, which in turn explains why microbes in oligotrophic environments, like the open oceans, are small.

Microbial life in natural aquatic habitats

The rules of a low Reynolds number environment apply to all microbial habitats, including soils and sediments and even pure cultures in the laboratory. These habitats perhaps share more similarities than differences, given the importance of size in structuring them. However, there are important differences between the water column of aquatic environments and soils and sediments. The water column does not have the solids found in soils or sediments, and consequently the microbial environment may seem simple and sparse, as illustrated by the calculations to follow. However, we will soon see that even aquatic environments may be more complex than appearances suggest.

A typical milliliter of water from natural aquatic environments contains about 1 million bacteria (Chapter 1). Although this density may seem high, it is much less than can be achieved with rich laboratory media (about 10^8 cells ml^{-1}). At a density of 10^6 cells ml^{-1}, each bacte-

rium would be surrounded by an empty sphere of 10^6 μm^3 assuming an even distribution of microbes (Fig. 3.9). If so, then the distance between bacteria is on the order of 60 μm, much more than the 10 μm for bacteria growing in rich media in the lab. The distance to the nearest bacterium of the same species is even greater, depending on the definition of species and its relative abundance (Chapter 9). The great distances between bacteria in freshwaters and seawater explains why these aquatic microbes appear to lack mechanisms for sensing the presence of one another ("quorum sensing"—see Chapter 14) and are "uncommunicative" according to one study (Yooseph et al., 2010).

Heterotrophic microbes are also far from sources of organic material they need to support growth. The abundance of detritus and phytoplankton varies greatly but is on the order of 10^3–10^4 particles or cells per milliliter. At these densities and again assuming even distributions, a heterotrophic bacterium would be >100 μm away from these potential organic carbon sources. Even the number of many dissolved molecules is low in the immediate neighborhood of a microbe. As has already been mentioned and will be emphasized again and again, concentrations of all but a few essential compounds are very low in natural environments. These low concentrations mean that only a few molecules are nearby an average microbial cell. For example, consider a dissolved amino acid with a concentration of 100 nmol liter^{-1}. Taking advantage of Avogadro's number, we can calculate that only about 30 molecules of this amino acid occur in a

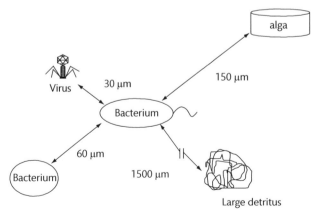

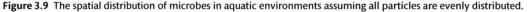

Figure 3.9 The spatial distribution of microbes in aquatic environments assuming all particles are evenly distributed.

Box 3.5 Physics of microbial motility

The problem of how microbes move through a viscous liquid has attracted some of the best minds in science. Perhaps most noteworthy is E.M. Purcell (Nobel Prize in 1952 for discovering nuclear magnetic resonance) who wrote a classic paper (Purcell 1977) on the physical environment of microbes. Albert Einstein did not mention microbes in his *annus mirablis* (1905) paper on Brownian motion but he did calculate how far a microbe-sized particle would move in 1 minute (about 6 μm).

0.5 μm^3 sphere of water surrounding a cell. The number of molecules for many compounds would be even lower. Even the total concentration of all 20 protein amino acids is usually much less than 100 nmol liter^{-1}. The same is the case for many other organic and inorganic compounds in natural microbial environments.

Motility and taxis

The calculations just discussed assumed an even distribution of cells and other particles, but this is not the case in actuality. Many microbes are motile. Those protists that swim by using flagella are referred to as "flagellates" although they may not be phylogenetically closely related. Protists using cilia are called ciliates and belong to the Ciliophora phylum. These eukaryotic microbes swim through water to increase their chances of encountering prey. Many bacteria also use flagella for propulsion, although the structure of a bacterial flagellum differs greatly from its eukaryotic counterpart (Chapter 2). Some attached bacteria and diatoms glide along surfaces, propelled by the secretion of polymers. One marine *Synechococcus* strain, which is a free-living coccoid cyanobacterium, swims, albeit slowly, without the benefit of flagella and other visible locomotion apparatuses (McCarren and Brahamsha, 2005). Microbes can swim incredibly fast, with speeds ranging from 1–1000 μm per second. This speed sounds more impressive when scaled up to our size. If a one meter-tall person could swim as many body lengths as a microbe can (e.g. 100 μm per second), her speed would be over 300 km per hour.

Analogous to the use of motility by microbial predators, other motile microbes swim to increase uptake of inorganic and organic nutrients. If the microbe is big enough, the mere act of moving helps to break down limits to uptake set by diffusion. Some motile microbes may use chemotaxis to swim toward sources of essential dissolved compounds. To do so, microbes have to sense and follow up the concentration gradient toward the nutrient source. One mechanism for sensing this gradient is by the "tumble and run" strategy (Fig. 3.10). The duration of runs, or swimming in a straight line, increases when the microbe swims against the concentration gradient. If concentrations decrease, implying that the microbe is going in the wrong direction, then it tumbles and heads randomly in another direction. The net result is chemotaxis towards the source of the desired compound. Alternatively, the opposite may happen for negative chemotaxis if a microbe is to avoid an inhibitory compound, perhaps its own waste products.

Microbes are attracted to or repelled by other environmental clues besides dissolved compounds. Phototropic microbes may sense light and use phototaxis, while other microbes rely on aerotaxis to search for

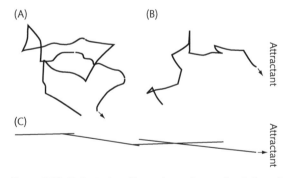

Figure 3.10 Trajectories of bacteria moving randomly (panel A) and moving by positive chemotaxis toward an attractant (B and C). The classic "run and tumble" (biased random walk) is shown in panel B. The bias towards the attractant is only about 1% and has been exaggerated here to show the effect. Another mechanism for chemotaxis, "run and reverse", is common in aquatic habitats (panel C). In this mechanism, long runs towards the attractant are interrupted by short reversals. The angle between the paths is only a few degrees and has been exaggerated in panel C for clarity. Figures provided by Jim G. Mitchell and used with permission.

oxygen. Magnetotactic bacteria use intracellular magnets, composed of magnetite (Fe_3O_4) or greigite (Fe_3S_4), to align along the earth's magnetic field. This mechanism in combination with aerotaxis enables these bacteria to find the depth in sediments with optimal concentrations of oxygen and other dissolved compounds (Chapter 13).

Given the many advantages of motility and moving to more suitable micro-habitats, it may be surprising to learn that not all microbes are motile. The estimates for the oceanic bacteria, for example, range from 5 to 70%, depending on the season and location (Mitchell and Kogure, 2006). Some microbes may not be active enough to afford the energetic expense of motility, and others may be just too small. Theory suggests that microbes have to be at least 3.7 or 8.5 µm (depending on the calculations) in order to use motility to escape limitations set by diffusion (Dusenbery, 1997). There are many advantages to being small, but there are downsides as well.

Submicron- and micron-scale patchiness in aqueous environments

While not all microbes are motile, certainly enough are for us to revisit the picture sketched before about the spatial distribution of microbes and detrital particles suspended in a milliliter of water from a natural aquatic habitat. Remember the calculations using typical abundances and concentrations gave the impression of everything being isolated and widely dispersed. But we now know that some microbes swim at top speed to track down nutrient-rich sources. These sources, such as an algal cell or decaying detrital particle, may be surrounded by swarms of chemotactic bacteria, which in turn attract grazers in search of easy prey.

Even the distribution of organic material in a milliliter of natural water is complex and is far from being a well-mixed, homogenous soup. Because of how dissolved and particulate material is separated (see Box 3.6), the dissolved pool contains many types of particles that are closer to being particulate than to being dissolved. These particles include colloids (any 1–500 nm particle), inorganic and organic aggregates, and gels. Gels spontaneously form by coagulation of smaller organic and inorganic components and can range in size from less than one micron to several microns. Related to gels are

---

Box 3.6 Dividing line between dissolved and particulate

Filters made of glass fibers are used to examine dissolved and many particulate components from natural environments. These filters can be cleaned easily of contamination, most drastically by burning ("combusting") the filter at about 500 °C to remove all organic compounds. These filters are needed to analyze particular organic carbon (POC) because the most common method for measuring POC is to combust the material collected on the clean filter and then to measure the resulting CO_2. Glass fiber filters that retain the smallest particles are Whatman GF/F filters ("**G**lass **F**iber **F**ine"). Anything that passes through a GF/F filter is, by definition, dissolved, while anything retained is particulate. The GF/F filters are advertised to retain particles of about 0.6 µm, but things smaller than this size can be trapped or stick to the glass fibers while particles larger than 0.6 µm may slip through and appear to be components of the dissolved pool. Delicate cells and detrital particles may be broken up during the filtration process and the pieces may pass through the filter. Many approaches, collectively referred to as "size fractionation", in aquatic microbial ecology rely on filters composed of different materials, such as polycarbonate or cellulose nitrate, and with various pore sizes, ranging from 0.1 to 10 µm.

---

transparent extracellular polymers (TEP), which become visible after staining with Alcian Blue, and are probably composed of polysaccharides from various microbes. Patchiness of microbes at larger scales is perhaps even greater than at smaller scales. It is clear that the microbe community attached to large particles (tens of microns) differs from the "free-living" community, or those microbes not on large particles. Microbial ecologists have looked at patchiness by sampling water samples with small pipettes separated by millimeters (Long and Azam, 2001). Both total abundance and the type of microbes differed in these samples.

Microbial life in soils

The immediate physical environment of a microbe in soil is similar in many respects to the aqueous environment just described. Active soil organisms are found in a film of water covering soil particles. These organisms have been called "terrestrial plankton" by one soil ecologist (Coleman, 2008) who borrowed the term "plankton" used by aquatic ecologists to describe free-floating biota. Redox state, pH, and temperature all affect soil microbes as much as aquatic microbes, and soil microbes also live in a low Reynolds-number world. But of course soils differ from the water column of aquatic habitats in several important ways. The soil environment is defined by inorganic and organic particles separated by open channels (pore space) through which air and water pass. The physical environment for microbes, including key properties such as oxygen concentrations and redox state, can differ drastically between locations separated by microns, much more so than seen in the water column of aquatic environments.

Total pore space and the size, shape, and connections between pores are all important for understanding microbial life in soils (Voroney, 2007). These properties vary with the type of soil (Table 3.5). Mineral soils are 35–55% pore space by volume while organic soils are 80–90%. In pores larger than about 10 μm (macropores), air and water readily move by diffusion and drainage. Macropores can be created by plant roots or movement of earthworms and other non-microbial soil organisms. Pores smaller than about 10 μm (micropores) retain water and can limit the movement of soil organisms.

Water content of soils

The extent to which pore space is filled with water has a huge impact on soil chemistry and microbial life. Diffusion of gases is much slower in water than in air. For example, the diffusion coefficients at 20 °C for oxygen are 0.205 cm^2 s^{-1} in air and 0.0000210 cm^2 s^{-1} in water. Oxygen penetrates into soils mainly by diffusion at rates that differ greatly depending on water content and soil type and texture. Clay has more micropores than sand, so soils rich in clay are often more poorly aerated than sandy soils. Waterlogged soils quickly become anoxic if microbial activity is high.

Water in soils can be described by two fundamental properties: water content and water potential. Water content is simply the amount or volume of water per amount or volume of soil. It can be measured by weighing soils before and after drying. Water potential is the potential energy or the amount of work potentially done by water moving without a change in temperature. Water potential is the sum of four components: matrix, osmotic, gravitational, and atmospheric pressure. The matrix component consists of adsorption of water to soil constituents, leading to a negative water potential. Osmotic effects, which are also negative, are due to the solutes dissolved in water. Both contribute to the retention of water in soils while gravitational and atmospheric pressures, which are usually positive, will pull and push water out of soils. The units of water potential are the same as for pressure, pascals (Pa) or more commonly, kilopascals (kPa).

To see how these different components of water potential interact, consider a field that has been satu-

Table 3.5 Some properties of the three major inorganic constituents of soils. Particles are assumed to be spherical. Data from Hartel (1998).

Property	Sand	Silt	Clay
Porosity	large pores	small pores	small pores
Particle size (mm)	0.05–2	0.02–0.05	<0.002
Permeability	rapid	low to moderate	slow
Number of particles per gram	10^2–10^3	6×10^6	9×10^{10}
Water holding capacity	limited	medium	very large
Soil particle surface (cm^2 g^{-1})	10–200	450	8×10^6
Cation exchange capacity	low	low	high (but varies with mineral)

rated with water such as after a heavy rain or the spring thaw. At first, the negative matrix and osmotic effects on water potential are balanced by positive pressure effects, giving a water potential of 0 kPa. Once the rains have stopped or all of the snow has melted, gravity takes over, draining the soils of water until the matrix and osmotic forces are large enough to retain water in the soil. At this point, the soil is said to be at field capacity. The amount of water left, the field capacity, varies with soil type. Loam soils have a soil water potential of –33 kPa while in sandy soils it is –10 kPa. The water potential of –50 kPa corresponds to water contents of 10% in sandy soil and 45% in clay soils.

Water potential can be used to describe how much water is needed for microbial activity. Even terrestrial microbes need water to grow and be active. In order for microbes to move, soils have to have a water potential of about –30 to –50 kPa, which corresponds to a water film of 0.5 to 4.0 μm thick on soil particles. Bacterial activity becomes limited in soils with a water potential of –4000 kPa or a <3.0 nm film of water, although experimentally dried-out soils still have some metabolic activity (respiration) even at -1.0×10^5 kPa (Fig. 3.11). There is some evidence that different types of bacteria respond differently to water content, and fungi are thought to grow better in dry soils than bacteria. Fungal hyphae can

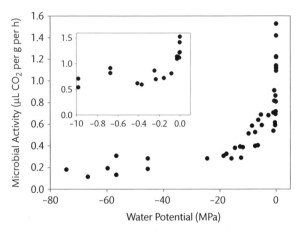

Figure 3.11 Microbial activity (here respiration) as a function of water potential, which was manipulated experimentally. The insert uses the same data but only for high water potential. Data from Orchard and Cook (1983).

traverse dry spaces in soils better than even filamentous bacteria. Some soil microbes may form resting stages, spores, to survive periods of desiccation.

Interactions between temperature and water content in soils

Temperature affects aquatic and soil microbes equally with both groups having a Q_{10} of about 2 over the typical temperature range in nature. However, unlike aquatic habitats, temperature can have an additional impact on microbial activity in soils via its effect on water content. Warmer temperatures due to climate change would increase microbial activity, but would also lead to drier soils and eventually lower microbial activity. Extensive field work has demonstrated that temperature is an excellent predictor of soil respiration until a threshold above which soil moisture comes into play.

Water potential and other factors can greatly complicate attempts to look at the relationship between microbial activity and temperatures. The apparent Q_{10} of soils often is substantially higher than the canonical value of 2 and is more variable when estimated from the change in microbial rates over the seasons as temperatures naturally rise and fall. An analogous problem is seen in aquatic habitats. In water and soils, temperature often is an excellent predictor of microbial activity but the implied Q_{10} is higher than that measured in controlled experi-

---

Box 3.7 Tough nuts

Some Gram-positive soil bacteria, including *Bacillus* and *Clostridium*, form spores, said to be the most resilient biological structure in the biosphere. The resilience of spores is in part due to the number and nature of protein coats surrounding the core protoplast consisting of the bacterium's DNA. The coats contain unusual compounds not found in "regular", vegetative cells, including dipicolinic acid in complexes with calcium ions. Spores can remain viable for at least decades, and there are controversial reports of spore-forming bacteria being recovered from 24–40 million year old bees trapped in amber and from 250 million year old salt crystals (Vreeland et al., 2000).

ments. The difference is in part caused by the response of microbes to factors co-varying with temperature. For example, temperature increases in the spring in temperate environments, but so does the organic carbon supply which also affects heterotrophic microbes. These other factors may not depend much on temperature directly (organic carbon supply increases because of higher primary production), whereas in other cases, the temperature effect is intertwined with the other factor. Soil moisture is an example of the latter (Fig. 3.12), as discussed above. Understanding how microbes in soils and aquatic environments respond to temperature is critical for predicting the feedback of the biosphere to global warming.

The biofilm environment

Biofilms are complex communities of microbes attached to surfaces. The term is usually used for communities on large surfaces of at least millimeters in length and width and that are inorganic, such as rocks or stones in a stream, a ship hull in the ocean, or teeth in an animal's mouth. But biofilms also grow on living tissue of plants and animals. As these examples suggest, biofilms can cause problems in many industrial and biomedical settings, in addition to being important in natural environments. Biofilms can be part of the solution, such as in removing dissolved compounds from waste water in sewage treatment plants. Entire research institutes are focused on biofilms. Microbes other than bacteria such as diatoms and other algae can be important in some biofilms growing attached to submerged surfaces exposed to sunlight. However, most of the work has focused on bacteria.

Biofilms form any time and every time a surface is immersed in water or moist soil. They start with the colonization of a surface by planktonic bacteria, perhaps attracted to organic compounds at the surface or as a way to escape predation (Fig. 3.13). These initial colonizers divide and are joined by other free-living bacteria, such that several layers of microbes form over time. The timescale of this process differs depending on the environment, but the initial colonization phase may last hours to a couple of days. More complex biofilm structures may take weeks to months to form. Along with the addition of new cells by colonization and growth, biofilm bacteria and other microbes secrete extracellular polymers, mainly polysaccharides, as mentioned in Chapter 2. These polymers help anchor cells to the surface and they store carbon, protect against predators, or perhaps to fend off competitors, and keep extracellular

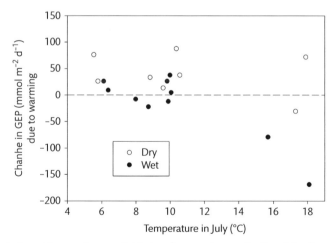

Figure 3.12 Interactions between two environmental factors (temperature and soil moisture) on a biological process. In this example, gross ecosystem production (GEP) was measured after tundra soils were experimentally warmed by 1–2 °C, the low end of the expected increase due to climate change. In most cases, production increased significantly with temperature (rates are positive and the points are above the dashed line), but the effect depended on soil moisture and the natural, in situ temperature (here given as the average for July). This example illustrates the complexity of predicting the effects of global warming and other climate changes on biological processes. Data from Oberbauer et al. (2007).

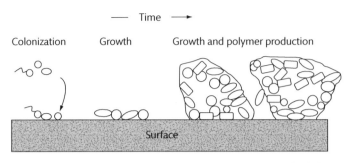

Figure 3.13 Development of a biofilm over time, starting with the colonization phase and culminating with extracellular polymer production.

enzymes close to the cell (Flemming and Wingender, 2010). The polymers often constitute a larger fraction of a mature biofilm's total mass than do the cells and often determine the role of a biofilm in applied and basic environmental problems.

Bacteria in biofilms differ from their free-living counterparts in several respects. First, the species composition of biofilms differs from the free-living community, just as the species make-up of microbes attached to detritus differs from its surrounding planktonic counterpart. Second, the metabolism of an initially planktonic bacterium changes after it colonizes a surface and after the biofilm matures with time. In contrast to the isolated existence of a planktonic cell, a biofilm bacterium is sur-

rounded by other microbes which may or may not be daughter cells, with limited exchange with the outside environment. A microbe at the biofilm's outer boundary may experience the same dissolved compounds as if it were in the bulk fluid, but concentrations of metabolic by-products from the biofilm would still be high. In contrast, a microbe buried deep within a biofilm may never see some compounds from the bulk phase.

A microbe embedded deep in a biofilm may have little contact or exchange with the outside world, but not because the biofilm restricts diffusion. Since a biofilm is mostly water, diffusion within a biofilm is still 60% of that in the bulk fluid (Stewart and Franklin, 2008). The problem facing a deeply buried microbe is consumption of compounds by other biofilm microbes. Oxygen is an important, well-studied example. One study of a mature biofilm found that oxygen concentrations were completely depleted within a 175 μm layer of a 220 μm-thick biofilm. In this example, concentrations even at the biofilm outer surface were only 40% of the bulk fluid. So, anoxic niches and anaerobic microbes and processes can occur in biofilms immersed in an oxic environment.

Microbial ecologists once thought that a mature biofilm resembled tiramisu and consisted of layers upon layers of microbes evenly covering a surface. Confocal microscopic studies demonstrated that rather than being two-dimensional, biofilms were complicated three-dimensional structures with channels of fluid flowing over bare surface between towers of microbes. This complex three-dimensional structure helps to explain variability in chemical and other biofilm properties.

Box 3.8 Microbial life in 3-D

Confocal microscopy has been instrumental in understanding biofilm microbes and structure. In regular fluorescence microscopy, a single plane of focus is excited by light and the resulting emitted light is analyzed, giving only a two-dimensional view of the sample. Anything not in the plane of focus is not seen. This approach is adequate for many applications in microbial ecology. In contrast, confocal microscopes are capable of taking images at several planes of focus which are then compiled to reconstruct a three-dimensional image of the sample. These 3-D images have provided new insights into biofilm structure and function.

Summary

1. Different types of microbes are able to survive and grow in environments that vary greatly in temperature, pH, salt, pressure, and other physical properties. Terms building on the suffix "philic" are used to describe these microbes, including psychrophile, thermophilic, acidophilic, halophilic, and piezophilic.

2. Reaction rates in microbes often increase by twofold when temperatures increase by 10 °C (Q_{10} = 2). Temperature affects several aspects of a microbe's environment, which complicates efforts to understand the impact of global warming.

3. Microbes live in a low Reynolds number environment in which the movement of compounds is affected by diffusion more so than turbulence. Diffusion limits large cells more so than small cells in the uptake of dissolved compounds.

4. The physical structure of aquatic habitats is potentially quite sparse for microbes because of relatively low numbers of cells and of other particles and because of very low concentrations of many dissolved compounds. However, chemotaxis and the presence of various sized particles create a patchy environment at the microbial scale.

5. Soil microbes live in pores of various dimensions in between soil particles. Pore sizes vary with soil type and determine water content, which in turn affects many soil properties and microbial activity.

6. Biofilms are complex structures of microbes living so close together that the availability of oxygen and other dissolved compounds is restricted. In addition to cells, many properties and practical roles of biofilms are determined by extracellular polysaccharides and other polymers.

CHAPTER 4

Microbial primary production and phototrophy

Previous chapters mentioned a few processes carried out by microbes in nature, but this chapter is the first to be devoted entirely to a process, the most important one in the biosphere. Primary production is important because it is the first step in the flow of energy and materials in ecosystems. The organic material synthesized by primary producers supports all food chains in the biosphere, setting the stage for the cycle of carbon and of all other elements used by microbes and larger organisms. Food web dynamics and biogeochemical cycles depend on the identity of the primary producers, their biomass, and rates of carbon dioxide use and biomass production.

Primary production by microbes is very important on both global and local scales. Mainly because of their abundance in the oceans, microbes account for about half of all global primary production while the other half is by terrestrial higher plants. This means that microbes also account for about half of the oxygen in the atmosphere. In contrast to life on land, in most aquatic habitats, primary production is mainly by microbes, the eukaryotic algae and cyanobacteria. Even though higher plants dominate primary production in terrestrial ecosystems, there are some places on land where higher plants cannot live but a few microbes can. Photosynthetic microbes can grow on rocks (epilithic) or even within rocks (endolithic) in deserts and in Antarctica (Walker and Pace, 2007). Another common rock-dweller, lichen, is a symbiosis between photosynthetic microbes (green algae or cyanobacteria) and fungi. Photosynthetic microbes are found in the top surface layer of soils if enough light is available.

The main primary producers are photoautotrophs that carry out oxygenic photosynthesis, meaning they evolve oxygen during photosynthesis (Table 4.1). Oxygenic photosynthesis is used by a diverse array of eukaryotes and cyanobacteria. There are no known photoautotrophic archaea, although some hyperhalophilic archaea use light energy to synthesize ATP by a mechanism quite different from that used by photosynthetic microbes and higher plants. All photoautotrophic microbes, whether eukaryotic or prokaryotic, are called algae (alga is the singular), and those free-floating in aquatic habitats are phytoplankton. Some algae, the macroalgae, such as kelp and other brown algae, are quite visible to the naked eye and are not counted here as microbes. Other microbes carry out anoxygenic photosynthesis and do not evolve oxygen (Chapter 11). This chapter is focused on primary production carried out by oxygenic photosynthetic microbes and will end with some discussion of other microbes (photoheterotrophs) that use light for ATP synthesis (phototrophy) and organic material for both ATP synthesis and as a carbon source.

Basics of primary production and photosynthesis

Light-driven primary production is based on photosynthesis (Fig. 4.1). The first part of photosynthesis, the light reaction, generates reducing power (NADPH), energy (ATP), and a useful by-product, oxygen. Oxygenic photosynthetic organisms use light energy to "split" or oxidize water:

Table 4.1 Use of light energy by microbes. The main pigment used in energy production is given here, although these microbes may have other pigments for light harvesting. Chlorophyll a = Chl a, and bacteriochlorophyll a = BChl a.

Metabolism	Purpose of light	Pigment	C source	Role of O_2	Organisms
Oxygenic photosynthesis	ATP and NADPH production	Chl a	CO_2	Produces O_2	Higher plants, eukaryotic algae, and cyanobacteria
Anaerobic anoxygenic photosynthesis	ATP and NADPH production	BChl a	CO_2 or organic C	O_2 inhibits photosynthesis	Bacteria
Photoheterotrophy	ATP production	Chl a, BChl a or rhodopsin	Organic C	Consumes O_2	Protists, archaea, and bacteria
Mixotrophy	ATP and NADPH production	Chl a	CO_2 or organic C	Produces and consumes O_2	Protists
Heterotrophy	Sensing	Rhodopsin	Organic C	Consumes O_2	Eukaryotes, bacteria, and archaea

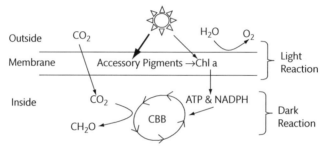

Figure 4.1 Summary of oxygenic photosynthesis. Light energy harvested by "accessory pigments" is transferred to chlorophyll a in the reaction center where water is "split" in order to synthesize ATP and NADPH, evolving oxygen in the process during the light reaction. The ATP and NADPH are then used to fix CO_2 and synthesize organic material ("CH_2O") in the dark reaction by the Calvin-Benson-Bassham (CBB) cycle.

$$2H_2O + light \rightarrow 4H^+ + 4e^- + O_2 \qquad (4.1).$$

The four electrons ($4e^-$) produced by the light reaction are used to reduce $NADP^+$ to NADPH and to add a high energy phosphate bond to ADP to produce ATP. Photosynthetic organisms have to make ATP and NADPH in order to synthesize organic carbon (Falkowski and Raven, 2007):

$$CO_2 + 2NADPH + 2H^+ + 3ATP \rightarrow$$
$$CH_2O + O_2 + 2NADP^+ + 3ADP + 3Pi \qquad (4.2).$$

"CH_2O" refers to organic material, not a specific compound, and "Pi" is inorganic phosphate. Equation 4.2 is the second half of photosynthesis, the dark reaction. This process is also called carbon fixation, because the C in the gas CO_2 is added or "fixed" to a nongaseous form of C, an organic compound.

Light and algal pigments

A key step in photosynthesis is the absorption of light by various pigments in the photoautotroph, a process that is sometimes called light harvesting. For terrestrial plants and green algae, the light-harvesting pigments are chlorophylls a and b and "accessory pigments", but chlorophyll a is the dominant one. Anoxygenic phototrophic bacteria have bacteriochlorophyll a, which is structurally similar to chlorophyll a. More than 99% of the chlorophyll a molecules in phytoplankton are used for light harvesting. This light energy is transferred to special chlorophyll a (or bacteriochlorophyll a) molecules that lie at the heart of the reaction centers of photosynthesis. It is in the reaction centers that light energy is converted into chemical energy. Because the reaction center chlorophyll a is essential, all oxygen-evolving, photosynthesizing organisms have it (but they do not necessarily have chlorophyll *b*).

Phototrophic microbes synthesize a greater diversity of accessory pigments than seen in higher plants on land. These accessory pigments enable aquatic phototrophs to harvest the wavelengths of light found in lakes and the oceans, especially green light (roughly 450–550 nm), the main wavelengths penetrating to deep waters. The wavelengths absorbed by chlorophyll *a* and *b* (>650 nm) are simply not available in water deeper than a few meters. Some of the main accessory pigments include fucoxanthin (a carotenoid), which is found in diatoms and some other eukaryotic algae, peridinin (another carotenoid), which is found in dinoflagellates, and phycoerythrin, which is made by cyanobacteria and by eukaryotic red algae and cryptomonads. These pigments absorb in the green part of the light spectrum (Fig. 4.2). Because these pigments are more abundant than chlorophyll *a*, phototrophic microbes are colored yellow, red, or brownish hues, or shades of green not seen in higher plants.

Ecologists use pigment data to address two questions about the ecology of phototrophic microbes in natural habitats. The most common use is in estimating algal biomass or "standing stock" from chlorophyll *a*. This pigment is easily measured in acetone extracts by fluorometry or spectrophotometry, with the former being much more sensitive than the latter. Algal biomass (µgC per sample) is then estimated by multiplying chlorophyll *a* concentrations (µg chlorophyll per sample) by an assumed ratio of algal biomass per chlorophyll *a*. A commonly used algal C-to-chlorophyll ratio is 50:1. However, the chlorophyll-based approach is a very crude way to estimate algal biomass because the C to chlorophyll ratio can vary substantially, up to tenfold, mainly due to differences in light intensity and temperature (Wang et al., 2009, Geider, 1987). Pigments are also used to identify phototrophic microbes (Table 4.2). Aquatic ecologists use high performance liquid chromatography (Chapter 5 describes this instrument) to measure pigment concentrations in order to determine the taxonomic composition of the phytoplankton community. Other methods include direct microscopic analysis and molecular approaches (Chapter 9).

Transport of inorganic carbon

The light energy harvested by pigments in photoautotrophic organisms is used to synthesize ATP and

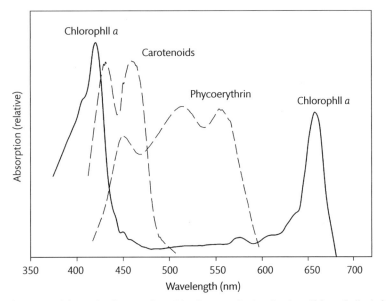

Figure 4.2 Absorption by some of the main pigments found in photosynthetic microbes. Chlorophyll *a* is found in all oxygenic photosynthetic organisms, ranging from higher plants to cyanobacteria. Phycoerythrin is found in some cyanobacteria and red algae. Several different carotenoids, such as fucoxanthin, are found in photosynthetic microbes.

Table 4.2 Some important eukaryotic algal groups and their pigments. Chl is chlorophyll. Based on Dawes (1981) and other sources.

Division	Common name	Characteristic pigments	% Marine	Comments
Chlorophyta	Green algae	chl b	13	Predecessor to vascular (land) plants
Phaeophyta	Brown algae	chl c and fucoxanthin	99	Includes kelp
Rhodophyta	Red algae	phycobilins	98	Few microbial representatives
Chrysophyta (Bacillariophyceae)*	Diatoms	chl c and fucoxanthin	50	Diatoms often dominate spring blooms
Chrysophyta (Coccolithophoridales)*	Coccolithophorids	chl c and fucoxanthin	90	Outer covering made of $CaCO_3$
Chrysophyta (Raphidophytes)*		chl c and fucoxanthin	?	Brown tide algae
Cryptophyta		chl c; xanthophylls; phycobilins	60	Motility driven by flagella
Pyrrhophyta	Dinoflagellates	chl c and peridinin	93	Some heterotrophic; red-tide organisms

* There are other members of Chrysophyta besides those listed here.

NADPH which goes on to support carbon dioxide fixation. Equation 4.2 indicates that photosynthesis uses CO_2, and in fact the actual fixation step in the Calvin-Benson-Bassham (CBB) cycle involves this form of inorganic carbon. However, CO_2 is just one of four forms of inorganic carbon found in nature which vary in relative concentrations depending on pH. The exchanges between these compounds are described by:

$$H_2O + CO_2 \Leftrightarrow H_2CO_3$$
$$\Leftrightarrow H^+ + HCO_3^- \Leftrightarrow 2H^+ + CO_3^{2-}$$

(4.3).

H_2CO_3, HCO_3^-, and CO_3^{2-} are called carbonic acid, bicarbonate, and carbonate, respectively.

The pH governs the relative concentrations of these inorganic carbon compounds. The pK_a of the first deprotonation step ($H_2CO_3 \Leftrightarrow H^+ + HCO_3^-$) is about 6 and the second one ($HCO_3^- \Leftrightarrow H^+ + CO_3^{2-}$) is about 9. Consequently, the main inorganic carbon form in most natural waters is bicarbonate (HCO_3^-), given that the pH usually is between 7 and 8 (Fig. 4.3). The pH of lakes and soils can vary greatly, however, both naturally and due to pollution, such as acid rain and run off from coal mining. In contrast, the pH of the oceans is much more uniform and constant at about 8.2, so again bicarbonate is the dominant species of dissolved inorganic carbon (DIC). But even the oceans are being threatened by human activity.

The increase in atmospheric CO_2 concentrations has led to more acidity in the oceans and already a drop of about 0.1 pH units, with another 0.3 to 0.6 pH unit decrease projected for the next century at current trends (Doney et al., 2009). Since pH is a logarithmic scale, a drop of 0.3 pH units implies a 100% increase in acidity. This huge increase in acidity has several potential impacts on oceanic biota.

Even with increasing atmospheric CO_2 and more CO_2 dissolved in natural waters, concentrations are still low. For example, in the oceans the total DIC concentration is about 2 mM of which dissolved CO_2 makes up only about 10 µM. So, except for acidic lakes, concentrations of HCO_3^- and CO_3^{2-} are much higher than of dissolved CO_2 in natural waters. This can be a problem for aquatic algae because the charged forms of DIC cannot readily cross membranes. H_2CO_3 and HCO_3^- tend to deprotonate, yielding CO_2, but this conversion is actually rather slow compared to photosynthetic reactions.

Algae have come up with various solutions to the problem of acquiring inorganic carbon (Fig. 4.4), in addition to relying on passive diffusion of CO_2 into cells. There is evidence for active transport of HCO_3^- (Chen et al., 2006). Another mechanism is to convert HCO_3^- to CO_2 using the enzyme carbonic anhydrase and then again rely on passive diffusion of CO_2. In this case, carbonic anhydrase can be

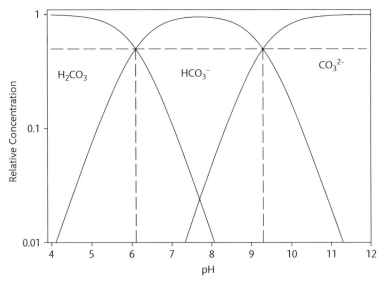

Figure 4.3 Concentrations of the three major inorganic carbon compounds in natural waters as a function of pH. These curves were calculated assuming pK_1 = 6.1 (first vertical line) and pK_2 = 9.3 (second vertical line) when concentrations of H_2CO_3 and HCO_3^- are equal (pH = 6.1) or HCO_3^- equals CO_3^{2-} (pH = 9.3) indicated by the horizontal dashed line (relative concentration = 0.5).

associated with the outer membrane or with the plasmale-mma of chloroplasts in eukaryotic autotrophs. Differences in CO_2 concentrating mechanisms help to explain the variation in phytoplankton groups over geological timescales, such as the appearance of dinoflagellates in the early Devonian period (about 400 million years ago) when atmospheric CO_2 was eightfold higher than today (Beardall and Raven, 2004).

Carbonic anhydrase is one of the most common enzymes in biology. In addition to its occurrence in algae and higher plants, animals have a form (alpha-carbonic anhydrase) to maintain a balanced pH and to facilitate CO_2 transport. Interestingly, carbonic anhydrase normally requires zinc. Since concentrations of this metal are often very low in the oceans, some marine phytoplankton have replaced zinc with cadmium, one of the few biological uses of an otherwise toxic metal (Lane and Morel, 2000).

The carbon dioxide-fixing enzyme
Once inside the cell or chloroplast, carbon dioxide is fixed by reduction to form organic compounds that are used for biomass production. The main pathway for carbon dioxide fixation in oxic surface environments is the Calvin-Benson-Bassham (CBB) cycle, which is found in the biosphere's main primary producers: higher plants, eukaryotic algae, and cyanobacteria. It is also used by many chemolithoautotrophic microbes (Chapters 11 and 12). Other microbes use other carbon dioxide fixation pathways with different enzymes (Table 4.3) and different requirements for ATP and NADPH (Hanson et al., 2012). These alternative carbon dioxide fixation pathways may have been more important during the early evolution of photosynthesis on the planet (Fuchs, 2011).

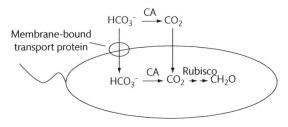

Figure 4.4 Uptake of dissolved inorganic carbon by photosynthetic organisms. CA is carbonic anhydrase and Rubisco is ribulose-1,5-bisphosphate carboxylase-oxygenase.

Table 4.3 Summary of CO_2 fixation pathways used by autotrophic organisms. "Phototrophs" includes eukaryotic algae, cyanobacteria, and anaerobic anoxygenic photosynthetic bacteria. Modified from Hanson et al. (2012).

Pathway*	Key Enzymes	Pathway found in		
		Phototrophs	Chemoautotrophic Bacteria	Chemoautotrophic Archaea
CBB	RubisCO Phosphoribulokinase	Yes	Yes	No
rTCA	Pyruvate synthase ATP:citrate lyase	Yes**	Yes	Yes
3-HP	Malonyl-CoA reductase Propionyl-CoA synthase	Yes**	Yes	No
Acetyl-CoA	CO dehydrogenase: Acetyl-CoA synthase Pyruvate synthase	No	Yes	Yes
3-HPP: 4-HB	Succinate semialdehyde reductase 4-Hydroxybutyryl-CoA synthetase	No	No	Yes
Dicarboxylate/4-HB	Succinate semialdehyde reductase 4-Hydroxybutyryl-CoA synthetase	No	No	Yes

* Abbreviations: CBB = Calvin-Benson-Bassham (CBB), rTCA = reductive TCA cycle, and 3-HP = 3-hydroxypropionate; 4-HB = 4-hydroxybutyrate.
** Present in anaerobic anoxygenic photosynthetic bacteria, but not eukaryotic phototrophs or cyanobacteria.

In today's oxic surface environments, however, the CBB cycle is the most common physiological basis of primary production.

A CBB enzyme examined by microbial ecologists in natural environments is ribulose-bisphosphate carboxylase/oxygenase (Rubisco). This enzyme catalyzes the following reaction:

$$\text{Ribulose 1,5- bisphosphate} + CO_2 \rightarrow 2 \text{ 3-phosphoglycerate} \qquad (4.4).$$

This enzyme is so important to autotrophs that it sometimes makes up 50% of cellular protein, making it one of the most abundant proteins in nature (Tabita et al., 2007). The enzyme consists of a large subunit (about 55 000 Da) containing the catalytic site of the enzyme and a small subunit (about 15 000 Da) involved in enzyme regulation. A common version (Form I) of the holoenzyme has eight copies each of the large and small subunits, resulting in a very large molecule (about 550 000 Da). This version of Rubisco is found in higher plants, cyanobacteria, and chemoautolithotrophic and photoautotrophic

bacteria (Tabita et al., 2008). Forms II and III have only the large subunit. Form II is found in some abundant types of eukaryotic algae (dinoflagellates), and chemoautolithotrophic and photoautotrophic bacteria whereas Form III has been found only in archaea to-date. Form IV Rubiscos, also called Rubisco-like proteins (RLP), have sequences somewhat similar to bona fide Rubisco, but are not involved in CO_2 assimilation. The RLP proteins have a variety of other functions in microbes.

The differences among Rubiscos have been used by microbial ecologists to explore the contribution of various autotrophs to primary production (Bhadury and Ward, 2009). The presence of the Rubisco gene indicates the potential for primary production by specific types of autotrophs carrying that gene. It is one example of a classic problem in microbial ecology: to establish the ecological and biogeochemical function of microbes that cannot be cultivated and grown in the lab. While genes for small subunit ribosomal RNAs (rRNA) are used to identify uncultivated microbes in nature (Chapter 9), it is difficult to connect a specific function with an rRNA sequence. Pigments give some clues about photoautotrophic communities (Table 4.2), but only at broad phylogenetic levels. The synthesis of mRNA for Rubisco, that is, its "expression", is more indicative than pigments or the Rubisco gene of actual carbon fixation activity in nature (Wawrik et al., 2002).

Primary production, gross production, and net production

The rate of primary production is perhaps the most important parameter for describing an ecosystem and for understanding microbial and biogeochemical processes. How we measure this rate affects our interpretation of the data and the implications of those data for understanding other processes.

In the light-driven ecosystems discussed here, primary production can be defined as the sum of Equations 4.1 and 4.2, which is

$$CO_2 + H_2O \rightarrow O_2 + CH_2O \qquad (4.5).$$

This simple equation summarizes the basis of most of life in the biosphere. It suggests that to estimate primary production, we could measure the movement of two elements (C or O) or changes in concentrations of O_2, CO_2, or CH_2O. All of these possibilities are used for various purposes by ecologists, except for following changes in CH_2O; that is too imprecise to be useful. A common method is to add $^{14}CO_2$ (actually $NaH^{14}CO_3$) to a sample and trace the ^{14}C into organic material (CH_2O). The advantages of this method are that it is easy and quick, and the instrument to measure radioactivity (liquid scintillation counter) is relatively inexpensive and common.

Changes in dissolved O_2 concentrations are also relatively easy to measure with the modified Winkler method or with O_2 electrodes. One of the first approaches, the "light-dark bottle method", for estimating production was to measure changes in O_2 concentrations over time in light and dark bottles. Oxygen decreases in the dark bottle due to respiration but it increases in the light bottle if net production is above zero. The change in oxygen in the light bottle is a measure of net community production (NCP) whereas respiration (R) is the decrease in oxygen in the dark bottle (Fig. 4.5). Gross production (GP) then is:

$$GP = NCP + R \qquad (4.6).$$

In words, gross production is the production of O_2 before respiration takes its toll while net community production

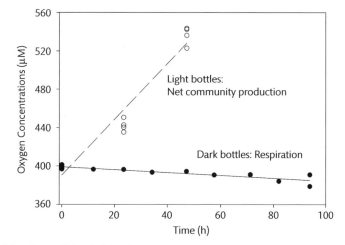

Figure 4.5 An example of data from a light-dark bottle experiment to measure net community production and respiration. Net community production is the change in oxygen concentration in the light bottles over time while respiration is the decrease in the dark. Each point is an individual bottle. The data are from the Arctic Ocean where oxygen concentrations in surface waters are high because of high primary production, relatively low respiration, and cold water temperatures. Data from Cottrell et al. (2006a).

is gross production minus respiration. Terms related to net community production include "net primary production" and "net ecosystem production". These terms emphasize the organisms and scale of processes contributing to respiration, ranging from just primary producers and closely connected organisms (net primary production) to the entire ecosystem (net ecosystem production); net community production is in between the two extremes. The light-dark bottle approach traditionally examines oxygen but the rates can be converted to carbon units to estimate biomass production and release of carbon dioxide. The CO_2:O_2 ratio, also called the respiratory quotient, is usually assumed to be about 0.9 (Williams and del Giorgio, 2005). The hidden assumption with the light-dark bottle approach is that respiration is the same in the light and dark bottle. Stable isotope studies with ^{18}O can help test this assumption and in measuring production.

Investigators who have compared ^{14}C-based and ^{18}O-based measurements of primary production have concluded that the ^{14}C method is measuring something between net and gross production. The ^{14}C method gives a rate that is smaller than gross production because of loss of ^{14}C due to several processes. Any respiration of ^{14}C-organic carbon back to $^{14}CO_2$ during the ^{14}C incubation would go unnoticed and would lead to a rate lower than the gross production estimate. Another problem is excretion or release of dissolved ^{14}C-organic material during the incubation. When this ^{14}C loss is not measured, the fixed C is not included in the primary production estimate. Some of the dissolved ^{14}C organic material can also be taken up by small heterotrophic microbes that are not sampled by standard approaches. Even given these problems, the ^{14}C method is still a powerful and frequently used tool for estimating primary production.

The magnitude and even the sign of net community production have several important implications for carbon fluxes, other biogeochemical processes, and the biota in an ecosystem. Net community production can be negative (respiration exceeds gross production) for short periods of time or in regions supplied by organic carbon from more productive waters or from terrestrial inputs. Many lakes are heterotrophic with negative net production because of organic inputs from terrestrial primary production (Fig. 4.6). In these systems, the partial pressure of CO_2 (pCO_2) is higher in the water than in the atmosphere, leading to the release of CO_2, or outgassing, to the atmosphere, a non-trivial flux in global budgets (Tranvik et al., 2009). These heterotrophic lakes may still have large build-ups of algal biomass due to high nutrient inputs. In spite of the algal bloom, oxygen production by the algae may still be lower than respiration

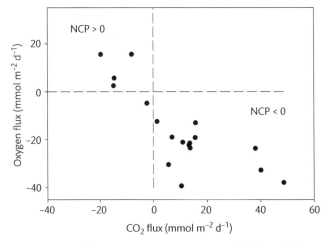

Figure 4.6 Net ecosystem production in four lakes over four years in northern Michigan (USA), as determined by oxygen and carbon dioxide fluxes. Positive values indicate net fluxes from the lake to the atmosphere. Net ecosystem production (NEP) was usually negative in these lakes, as indicated by the negative oxygen fluxes and positive CO_2 fluxes. Data provided by J.J. Cole (Cole et al., 2000).

fueled by inputs of terrestrial organic material. More controversial are studies showing negative net production (net heterotrophy) in large regions of the oceans. These studies are controversial because the source of organic carbon for these regions is unclear.

Ecosystem production has an impact on oxygen as well as on carbon. Our atmosphere has oxygen because photosynthesis exceeds respiration and the consumption of oxygen. The burning of fossil fuels has led to a slight decline in atmospheric oxygen as well as the more widely known increase in atmospheric CO_2 (Prentice et al., 2001). Notably, positive net ecosystem production draws down CO_2 concentrations and lowers the pCO_2 in aquatic systems. When dissolved pCO_2 is less than atmospheric pCO_2, there is a net flux of atmospheric CO_2 into the water. An important example of this flux is one into the ocean. Many regions of the world's oceans are a net sink for atmospheric CO_2. In fact, about half of all CO_2 released by the burning of fossil fuel ends up in the oceans (Houghton, 2007). There would be even more CO_2 in the atmosphere if it were not for this flux into the oceans.

Primary production by terrestrial higher plants and aquatic microbes

Using the methods for examining biomass and primary production just discussed, ecologists and biogeochemists found that higher plants and aquatic photoautotrophic microbes each account for about half of all global primary production, as mentioned before. But a closer look at the data reveals profound differences in how these photoautotrophs contribute their 50% of the total. Of the many differences between terrestrial and aquatic primary producers, size is the most obvious and probably the most important. The difference in size ends up having huge consequences for how terrestrial and aquatic ecosystems are organized and structured. It also helps to explain levels of biomass and growth rates in the two systems. Finally, this comparison of aquatic and terrestrial environments illustrates important principles about how per capita rates and standing stocks contribute to fluxes.

Although biomass and rates of primary production vary greatly, the generic averages given in Table 4.4 still accurately illustrate the huge differences in biomass and primary production rates among terrestrial and aquatic ecosystems. The biomass of even the most nutrient-rich ("eutrophic") lake or small pond is very small compared with an equal area on land, with the exception of barren deserts. The nutrient-poor ("oligotrophic") open oceans have even less biomass, as is obvious from their clear blue waters. Yet rates of primary production in aquatic ecosystems can rival those found on land. The oceans account for a large fraction (about 50%) of global primary production not only because of large surface area but

Table 4.4 Photoautotrophic biomass and growth in the major biomes of the planet. "NPP" is net primary production. Turnover time was calculated by biomass/NPP. Data from Valiela (1995).

	Location	Area (10^6 km^2)	Biomass (kg C m^{-2})	NPP (gC m^{-2} y^{-1})	Turnover time (y)
Aquatic					
	Open oceans	332	0.003	125	0.02
	Upwellings	0.4	0.02	500	0.04
	Continental shelves	27	0.001	300	0.00
	Estuaries	1.4	1	1500	0.67
	Other wetlands	2	15	3000	5.00
	Lakes	2	0.02	400	0.05
Terrestrial					
	Tropics	43	8	623	12.6
	Temperate	24	5.5	485	11.3
	Desert	18	0.3	80	3.8
	Tundra	11	0.8	130	6.2
	Agriculture	16	1.4	760	1.8

also because rates per square meter can be high for some oceanic regions.

The reason why low biomass systems can have high primary production rates is because of high growth rates. To see why, note that the relationship between production (P), growth rate (μ) and biomass (B) is:

$$P = \mu \cdot B \qquad (4.7)$$

where B has units of mass per unit area or volume (gC m^{-2}, for example) and μ has units of per time (e.g. d^{-1}). Consequently, production has units of mass per unit area or volume per time (gC m^{-2} d^{-1}, for example). The growth rate has units of per time, such as d^{-1}. Although growth rates can be measured directly for photoautotrophic microbes (one method is to measure ^{14}C incorporation into algal pigments), here we estimate it simply by using Equation 4.7 and dividing P by B. This method has many experimental and theoretical problems, but it does give a rough idea of how fast algae grow. To compare with terrestrial ecosystems, the inverse of μ is calculated to give the turnover time (Chapter 6).

This calculation indicates that algae have growth rates of about 0.1 to 0.2 d^{-1} and turn over in 4–7 days. More accurate measurements indicate growth rates of about 1 d^{-1}. Even the slow growth rates are one hundred to one thousandfold faster than growth rates of land plants (Table 4.4). Consequently, although biomass per square meter is much less in aquatic systems than on land, the difference is nearly cancelled out by the much higher growth rates in freshwaters and the oceans. More generally, growth rates increase as the size of organisms becomes smaller. The result is similar rates of primary production per square meter in aquatic systems as on land. This is the first of several examples presented in this book that illustrate the contribution of turnover (with units of time) and standing stocks (units of mass per unit volume or area) to determining a rate or flux (mass per unit volume or area per unit time).

If growth by primary producers in aquatic habitats is so fast, why do biomass levels remain so low? The short answer is that aquatic primary producers die off nearly as fast as they grow. Some of the larger phytoplankton sink to deep waters and die because of the lack of light. Most, however, are eaten by various herbivores (Chapter 7) while others may be killed off by viruses (Chapter 8).

The spring bloom and controls of phytoplankton growth

The numbers given in Table 4.4 give some hints about the variation in rates and standing stocks of primary producers among various ecosystems, but these numbers also vary over time, on temporal scales ranging from hours to years. Primary production varies from zero every night when the sun goes down to high rates on bright sunny days. It and algal biomass levels increase from year to year in lakes and coastal oceans receiving more and more inorganic nutrients from land run off and processes causing eutrophication. The importance of variation with season, which will be the focus here, is seen in the change in atmospheric CO_2 concentrations over a year. Concentrations are low in the summer and high in the winter due to changes in net primary production.

These changes are due to the seasonal procession in higher plants and algae in both terrestrial and aquatic systems. In temperate lakes and oceans, phytoplankton abundance and biomass increase from very low levels in winter to high levels in spring (Fig. 4.7). This large increase is called an algal or phytoplankton bloom. During blooms, net production is high and growth of phytoplankton exceeds mortality due to viral lysis and grazing. Blooms are large biogeochemical events that affect many ecosystem processes while they occur and long after their demise. If we understand blooms and how they end, we have gone a long way to understanding the controls of the phytoplankton community and of algal growth in aquatic ecosystems. The factors affecting phytoplankton or any group of organisms can be divided into "bottom-up factors", such as light and nutrient concentrations for algae, that affect growth and "top-down factors", such as grazing, that affect biomass levels. Here we'll concentrate on the bottom-up factors and leave the top-down factors for later chapters.

It is worthwhile pointing out that blooms do not occur everywhere, and these exceptions are especially interesting to aquatic ecologists. They do not occur in lakes in the tropics that do not have the strong seasonal cycles in light and temperature found in the temperate zone. In marine systems, phytoplankton blooms also do not occur in high nutrient-low chlorophyll (HNLC) oceans, such as the subarctic Pacific Ocean, the equatorial Pacific

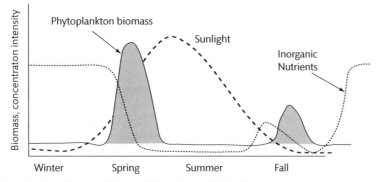

Figure 4.7 Peaks in phytoplankton biomass ("blooms") in spring and in fall in temperate aquatic ecosystems. The primary inorganic nutrients include nitrate, phosphate and, for diatoms, silicate.

Ocean, and the largest HNLC regime, the Southern Ocean. In these waters, concentrations of iron are extremely low and limit rates of primary production (Boyd et al., 2007).

Most temperate lakes and some oceanic regions do experience algal blooms, especially in spring. Why is algal biomass so high in the spring? Likewise, why is it so low in the winter and in the summer, although less so? Temperature may be one factor that comes to mind. Indeed, many biological processes are affected by temperature (Chapter 3), but phytoplankton can form blooms even in very cold waters, such as seen in the Arctic Ocean and Antarctic seas where water temperatures hover near freezing. So temperature is not the complete answer.

The key in explaining low algal biomass in the winter is mixing and how it affects the amount of light available for phytoplankton. Both the quality and quantity of light have a large impact on the taxonomic composition of the phototrophic community and rates of biomass production. Quality is the wavelength of light and quantity the light intensity. These two parameters of light vary greatly in aquatic ecosystems and with water depth. Light intensity declines exponentially with depth in lakes and the oceans. Light intensity (I_z) as a function of depth (z) is described by:

$$I_z = I_0 e^{-kz} \qquad (4.8)$$

where I_0 is light intensity at the surface (z = 0) and k is the attenuation coefficient. This coefficient is small for open ocean water and large for a murky pond. There is not a simple linear relationship between photosynthesis (P) and light intensity (I), as illustrated in Figure 4.8. Of the many equations proposed to describe the general curve given in Figure 4.8, one of the simplest is:

$$P = P_{max} \tanh(\alpha \cdot I/P_{max}) \qquad (4.9)$$

where P_{max} is the maximum rate of photosynthesis and α (or α^{chl} when primary production is normalized to chlorophyll) is the slope of the initial part of the curve in Figure 4.8, the "P versus I" curve. One complication not included in Equation 4.9 is inhibition of photosynthesis at high light intensities, which often occurs in surface waters. Because of photo-inhibition, primary production is often highest not at the surface but deeper in the euphotic zone; the bottom depth of the euphotic zone is set where light is roughly 1% of the surface intensity.

The effect of mixing through the water column on light availability and phytoplankton growth is encapsulated in the critical depth theory. First developed for explaining spring blooms in the oceans (Sverdrup, 1953), the theory gets its name for the depth at which availability of light and nutrients is high enough for net phytoplankton growth. This depth is set by mixing. Mixing by winds brings up nutrients to the surface layer from deep waters where concentrations are high. Blooms are not possible in the winter, however, even though nutrient concentrations are high because phytoplankton are limited by light. Winter mixing sends phytoplankton deep into the water column where they spend too much time in poorly lit deep waters to grow much, and thus blooms cannot form. The mixed layer is too deep—it is below the

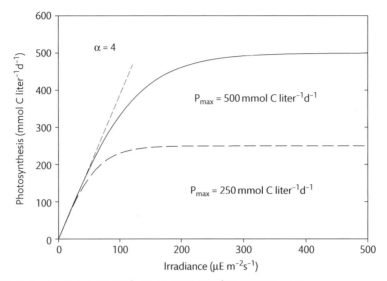

Figure 4.8 Relationship between photosynthesis (primary production) and light intensity.

critical depth, and phytoplankton growth is less than losses due to respiration. As winter gives way to spring, the upper layer of the water column warms up and becomes less dense. This warm, less dense surface layer sits on top of colder, denser deep waters. The water column overall becomes more stable and mixing is not as deep. Consequently, phytoplankton are kept in the light, enabling them to grow faster and to produce more biomass than any losses due to respiration. High net growth of the phytoplankton leads to a phytoplankton bloom. Although not perfect (Behrenfeld, 2010), the theory still provides a useful framework for thinking about the regulation of phytoplankton growth in lakes as well as the oceans.

Major groups of bloom-forming phytoplankton

The impact of the bloom can vary with the type of phytoplankton most abundant in the bloom. The major groups of phytoplankton can be distinguished by their pigments (Table 4.2), but they can also be divided into five functional groups that differ in cell wall composition, cell size, and the production of specific compounds (Table 4.5). Variation in these properties explains the different impacts of phytoplankton groups on aquatic ecosystems. Many of the algae in these groups contribute to blooms and often occur in large numbers and high biomass levels.

Table 4.5 Functional groups of phytoplankton. Diatoms, diazotrophs (except *Trichodesmium*), and picophytoplankton are found in both freshwaters and marine waters whereas coccolithophorids are only in the oceans.

Functional group	Function	Example
1. Diatoms	Silicate use. Blooms in lakes and coastal waters	*Thalassiosira, Asterionella*
2. Coccolithophorids	$CaCO_3$ production	*Emiliania huxleyi*
3. *Phaeocystis*	DMS production	*Phaeocystis*
4. Diazotrophs	N_2 fixation	*Trichodesmium*
5. Picophytoplankton	Accounts for large fraction of biomass and production in oligotrophic waters	*Synechococcus, Prochlorococcus*

Diatoms

This algal group, the class Bacillariophyceae, is often the major photoautotroph making up spring phytoplankton blooms in lakes and coastal oceans. Diatoms do well in the spring probably because they are better than other algae at using high concentrations of nitrate (the dominant form of inorganic nitrogen source in spring) and phosphate, because they grow faster than other algae in the low water temperatures of spring, and because they are able to cope with the huge variation in light that is typical of the spring. Diatoms are successful in spite of the fact that they require a nutrient, silicate, not needed by other algae; silicate is used for the synthesis of a cell wall, the frustule, found only in diatoms (Fig. 4.9A). Silicate concentrations are high in the spring, but decrease as the bloom progresses. Low silicate concentrations eventually limit diatom growth at the end of spring. The depletion of silicate, among other factors, allows other algal groups to dominate the phytoplankton community as the seasons progress.

Spring blooms of diatoms and other algae, such as green algae in lakes, are critical for ensuring the success of higher trophic levels (larger organisms) and in some sense of the entire aquatic ecosystem. These blooms fuel the growth of herbivorous zooplankton, and the zooplankton in turn are prey for larvae of invertebrates and fish. The spawning of larvae is timed to coincide with spring blooms. Playing on a line from the poem *Leaves of Grass* by Walt Whitman (who in turn borrowed it from Isaiah 40:6), the oceanographer Alfred Bigelow opined that "all fish is diatoms". Herbivores graze on diatoms probably simply because they are the dominant algal group. In fact, diatoms may inhibit reproduction of some types of zooplankton (Miralto et al., 1999, Ianora et al., 2004).

Coccolithophorids and the biological pump

In the oceans, another group of phytoplankton, the coccolithophorids, can form dense blooms sometimes after diatom blooms are finished. Quite unlike other algae, coccolithophorides are covered by calcified scales (coccoliths) made of calcium carbonate ($CaCO_3$) (Fig. 4.9B). This algal group is not abundant in freshwaters where calcium concentrations are too low. Coccolithophorides contribute to the carbon cycle by affecting dissolved CO_2 via primary production and $CaCO_3$ formation (Chapter 13) and by the export of $CaCO_3$ to deep waters and sediments. The sinking of $CaCO_3$-rich coccolithophorids and other calcified microbes to deep waters leads to loss of CO_2 from the upper layer of the oceans and burial in sediments. Calcium carbonate and other inorganic carbon buried in oceanic sediments make up the largest reservoir of carbon on the planet (Chapter 13).

(A)

(B)

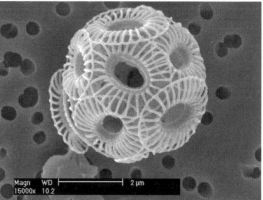

Figure 4.9 Electron micrographs of two common eukaryotic algae, both about 50 µm in diameter. Centric diatom (A). Coccolithophorid (B). Pictures used with permission from Ken Bart, University of Hamilton, and Jeremy Young, Natural History Museum, London.

Box 4.1 Microbial fossil beds

Because their cell walls are preserved in sediments, diatoms and coccolithophorids are studied by paleo-oceanographers and limnologists as a proxy for primary production and calcium carbonate fluxes over geological times. The famous White Cliffs of Dover, England are made of coccoliths ("chalk") that were deposited about 140 million years ago during the Cretaceous period when southern England was submerged under a tropical sea. The sediments then became exposed as the sea retreated during the ice ages. After the ice ages, the rising sea cut through these soft sediments, leaving behind the English Channel and exposing the cliffs of coccoliths.

Coccoliths make up one component of the "biological pump" in the ocean, which is the sinking of organisms and detritus from the surface layer to the deep ocean. $CaCO_3$-containing structures from coccolithophorides and other organisms make up the "hard" part of the pump. The "soft" part consists of organic material originally synthesized by phytoplankton which can be repackaged into fecal pellets from zooplankton and into other organic detritus that sinks to the deep ocean. The sinking of both hard and soft parts "pumps" carbon from the surface layer into the deep ocean and some as far as the sediments.

The amount of carbon exported as $CaCO_3$ is generally <10% of total export (Sarmiento and Gruber, 2006). Albeit small, it is an important percentage because of the contribution of $CaCO_3$ to carbon storage in sediments for millennia or more (Chapter 13). $CaCO_3$-containing structures also act as ballast, speeding up the sinking of organic carbon. Most of the carbon never makes it to the sediments, however, because heterotrophic organisms oxidize the organic carbon back to CO_2, and $CaCO_3$ structures dissolve back to soluble ions. Even so, the CO_2 regenerated in the deep ocean does not reach the surface layer and remains out of contact with the atmosphere for hundreds of years. In the absence of the biological pump, atmospheric CO_2 concentrations would be 425–550 ppm (depending on various assump-tions) or 70 to nearly 200 ppm higher than current levels. At the other extreme, if the biological pump were operating at 100% efficiency, atmospheric CO_2 would only be 140–160 ppm (Sarmiento and Toggweiler, 1984).

Phaeocystis *and dimethylsulfide*

A genus of phytoplankton, *Phaeocystis*, which belongs to the Prymnesiophyceae, is another bloom-forming photoautotroph which has its own functional group. It also has an unusual life cycle. Although it can occur as a solitary, flagellated cell, unknown environmental factors trigger the formation of a colonial form. Some species of the *Phaeocystis* family form blooms in coastal waters and cause serious water-quality problems. *Phaeocystis* can excrete large quantities of extracellular polymers, so large that unsightly foam several meters deep builds-up on the beaches of northern Europe and in the Adriatic Sea. Other species are abundant in Antarctic seas, especially in the Ross Sea where *Phaeocystis* colonies are not grazed on by zooplankton and can sink rapidly from the water column.

The main reason why *Phaeocystis* has its own functional group is due to its production of dimethylsulfide (DMS), the major sulfur gas in the oceans. DMS is produced during the degradation of another organic sulfur compound, dimethylsulfoniopropionate (DMSP) (Fig. 4.10), which *Phaeocystis* and some other algae make as an osmolyte, as mentioned in Chapter 3. Other roles of DMSP in algal physiology, such as an antioxidant (Sunda et al., 2002), have been suggested. Initially, oceanographers thought that algae directly make DMS from DMSP but later work showed that DMSP is released from algae, perhaps during zooplankton grazing, and then is degraded by heterotrophic bacteria. Select heterotrophic bacteria are known to produce DMS during cleavage of DMSP whereas other bacteria demethylate DMSP to produce 3-methiolpropionate (González et al., 1999).

These organic sulfur compounds are important in the sulfur cycle. DMSP, for example, can supply nearly all of the sulfur used by heterotrophic bacteria (Kiene and Linn, 2000). Of even more significance is the possible role of DMS in affecting the world's heat budget. Because it is supersaturated in the upper ocean, DMS outgasses to the atmosphere where it is oxidized to sulfate and contributes to aerosol formation. These aerosols can

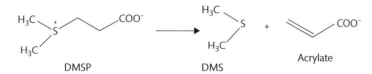

Figure 4.10 Production of dimethyl sulfide (DMS) and acrylate from dimethylsulfonic propionate (DMSP).

scatter sunlight and act as cloud nucleation sites, hot spots of cloud formation. The extent of cloud cover has impacts on the amount of light and heat reaching the earth's surface. This interaction between the plankton and global climate via DMS production is one example often mentioned in discussion of the "Gaia hypothesis", originally proposed by James Lovelock (Charlson et al., 1987, Kleidon, 2004). According to the Gaia hypothesis, negative feedbacks between the biosphere and the rest of the planet maintain the earth's climate and biogeochemical processes at homeostasis.

Diazotrophic filamentous cyanobacteria

Lakes and reservoirs sometimes experience large blooms of cyanobacteria in summer. The cyanobacteria making up these blooms often occur as cells strung together in long filaments reaching several millimeters in length. Although these microbes have some unique features and thus unique roles in the environment, cyanobacteria carry out the same main function (primary production) as do eukaryotic algae. In fact, cyanobacteria share many physiological traits with eukaryotic algae and higher plants. All of these organisms use the same mechanism to convert light energy to chemical energy, and they fix inorganic carbon into organic carbon by the same pathway, the CBB cycle. One key enzyme, Rubisco, is also nearly the same in all of these autotrophs.

But in every other respect, cyanobacteria are firmly in the Bacteria kingdom. Cyanobacteria do not have chloroplasts or any organelles, their genome is usually a single circular piece of DNA, and they have cell walls like Gram-negative heterotrophic bacteria with components such as muramic acid and lipopolysaccharides that are found only in bacteria. The composition and organization of the light-harvesting pigments of cyanobacteria also differ from eukaryotic algae and higher plants. These distinct pigments enable them to flourish in low light

environments and occasionally outcompete eukaryotic algae. The pigments are also useful taxonomic markers. Cyanobacteria used to be called "blue-green algae", and the chromatic terms of the old name still accurately describe the color of some cyanobacteria; the green is due to chlorophyll *a* while one phycobilin, phycocyanin, gives these microbes a blue tinge. Isolated phycocyanin is indeed a brilliant blue. In contrast, another type of cyanobacterium common in the oceans, *Synechococcus*, has large amounts of a blood-red pigment, phycoerythrin. Dense liquid cultures of this cyanobacterium are pink even though the microbe also contains some phycocyanin, other phycobiliproteins and chlorophyll *a*. Table 4.6 lists some cyanobacteria found in nature.

Summer blooms of filamentous cyanobacteria in freshwaters can be caused by high water temperatures which favor these prokaryotes over eukaryotic microalgae. Of even more importance is the supply of phosphate (Levine and Schindler, 1999). Since phosphate is often the nutrient limiting primary production in freshwater ecosystems and some oceans, eukaryotic algal groups with superior uptake systems for phosphate outcompete cyanobacteria when phosphate concentrations are low. However, phosphate pollution of freshwaters removes the competitive edge of eukaryotic algae and allows cyanobacteria to bloom. These massive outbreaks negatively affect water quality and the general "health" of the ecosystem. There is evidence that harmful cyanobacterial blooms have become more common over the years (Paerl and Huisman, 2009).

The success of some filamentous cyanobacteria in freshwaters with high phosphate concentrations is largely due to their ability to fix nitrogen gas to ammonium (NH_4^+). Organisms capable of fixing N_2 are called diazotrophs. Some diazotrophs have heterocysts, which are specialized cells where N_2-fixation is carried out. When phosphate is plentiful, the supply of nitrogen nutrients, mainly NH_4^+ and nitrate (NO_3^-), become important

Table 4.6 Some important cyanobacteria found in natural habitats.

Genus	Morphology	Habitat	Noteworthy ecology
Anabaena	Filament	Freshwaters	N_2 fixation
Microcoleus	Filament	Soils, desert crust	Tolerates harsh conditions
Microcystis	Filament	Freshwaters	Produces toxins
Trichodesmium	Filament	Marine	N_2 fixation
Richelia	Endosymbiont	Marine	N_2 fixation
Synechococcus	Coccus	Marine and freshwater	Primary production*
Prochlorococcus	Coccus	Marine	Primary production*
Unknown	Coccus	Marine	N_2 fixation

* Although all of these cyanobacteria carry out oxygenic photosynthesis like eukaryotic algae, *Synechococcus* and *Prochlorococcus* are especially important as primary producers and account for a large fraction of primary production and biomass in oligotrophic oceans.

factors in controlling primary production and potentially allowing N_2-fixing cyanobacteria to proliferate. While nitrogen fixation is carried out by many prokaryotes, not eukaryotes, only some cyanobacteria contribute substantially to both nitrogen fixation and primary production. In freshwater ecosystems, filamentous cyanobacteria with heterocysts are the dominant diazotrophs, whereas in the oceans, other cyanobacteria without heterocysts are more important (Chapter 12).

In addition to N_2-fixation, several other traits of filamentous cyanobacteria contribute to their success in freshwater systems. Some of these traits help to minimize losses due to grazing by zooplankton. One strategy is to be too big for a grazer to eat. Individual filaments of some cyanobacteria are often too large to be grazed on effectively by filter-feeding zooplankton. To make matters worse for zooplankton, some filamentous cyanobacteria form large colonies or aggregates that are even harder for zooplankton to graze on. These aggregates may be visible to the naked eye as green scum on the surface of lakes and reservoirs. Some cyanobacteria float to the surface by regulating buoyancy with gas vacuoles. The end result of high growth fueled by N_2 fixation, assisted by warm summer temperatures, plus low losses due to zooplankton grazing, is a dense bloom of filamentous cyanobacteria.

Some cyanobacteria also have chemical defenses against grazing. We know the most about toxins produced by freshwater cyanobacteria such as *Anabaena*, *Aphanizomenon*, *Microcystis*, and *Nodularia* species.

Microcystins, a suite of toxins produced by toxic strains of *Microcystis*, cause liver damage in humans (Carmichael, 2001) and affect the heart and other muscles of zooplankton (Fig. 4.11). These toxins also deter, if not kill off, zooplankton and herbivorous fish grazing on cyanobacterial mats (Nagle and Paul, 1999). Because of these toxins and other secondary metabolites, cyanobacterial blooms can lead to a decrease in water quality; drinking water from reservoirs with dense cyanobacterial populations may taste poor. Worse, cyanobacteria-tainted water can be toxic to humans, domestic pets, livestock, birds, and fish (Pitois et al., 2000).

Toxin production is just one of several negative impacts of cyanobacterial blooms. The switch at the base of aquatic food webs, from primary production by eukaryotic algae to that by filamentous cyanobacteria, can negatively affect the rest of the food web. Herbivorous zooplankton suffer if their only prey are inedible cyanobacteria. Carnivorous zooplankton and larvae feeding on the herbivores are affected next.

In addition to being ugly, cyanobacterial scum floating on the surface of lakes and reservoirs shades other phytoplankton, leading to a reduction in oxygen production below the surface. To make matters worse, consumption of oxygen (respiration) is high in scum-filled water, fueled by the release of organic compounds from living and dying cyanobacteria. The end result is anoxia or hypoxia just underneath the luxuriant floating mat of cyanobacteria, which still may be actively photosynthesizing and producing oxygen. Among many negative effects of

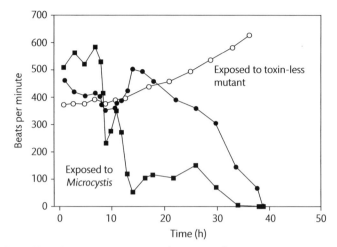

Figure 4.11 An example of the effect of a cyanobacterial toxin (microcystin) on the physiology of a freshwater zooplankton, *Daphnia*. Individual animals were glued in place and fed either a toxic (solid symbols) or a toxin-less mutant (open circles) of *Microcystis*. Heart beat (circles), leg motion (squares), and other physiological responses were recorded over time. Data from Rohrlack et al. (2005).

anoxia and hypoxia, the noxious compounds produced by microbes in low oxygen waters add to the toxins produced directly by cyanobacteria.

In addition to blooms of noxious cyanobacteria, several eukaryotic algae can cause problems when their numbers are high. Many of these algae are dinoflagellates (Table 4.7).

After the bloom: picoplankton and nanoplankton

As the spring bloom progresses and biomass levels build up, microbes strip nutrients from the water column, reducing concentrations from micromolar to nanomolar or lower levels. The result is that algal growth becomes limited by nutrients, thus explaining why growth slows down and blooms stop, but it does not explain the disappearance of phytoplankton and the crash of the bloom. Various types of protists and zooplankton eat the phytoplankton and viruses infect phytoplankton, both processes leading to the loss of phytoplankton biomass from the water column. If big enough and of the right shape, phytoplankton cells can also simply sink out of the euphotic zone, sometimes all the way to the bottom. Intact phytoplankton cells have been observed in sediments 2000 m below the ocean surface after a spring bloom.

Competition for limiting nutrients

As the spring bloom declines, other algal groups start to dominate the phytoplankton community due to the lack of inorganic nutrients. Large diatoms and other large algae decline and are taken over by a complex community of small algae in the nanoplankton (cell diameter 2–20 µm) and picoplankton (0.2–2 µm) size classes. The nanoplankton include dinoflagellates, cryptophytes, and other, poorly characterized algal groups. The picoplankton include coccoid cyanobacteria and small eukaryotic algae. These small cells are more abundant and account for more primary production than large phytoplankton in the oligotrophic habitats such as the open ocean where concentrations of nutrients such as ammonium and phosphate are extremely low (<10 nM). One implication of these observations is that small cells somehow outcompete large cells for the limiting nutrient when concentrations of that nutrient are low. Why is that?

The first problem faced by large cells is physics. In the microbial world, diffusion is the main process by which nutrients are brought to the cell surface. A cell cannot

Table 4.7 Some toxic or harmful algae in aquatic habitats. In addition to impact on the organisms listed in the table, these algae can affect the health of humans in contact with water containing high densities of these algae or by eating contaminated fish or shellfish. Based on http://www.whoi.edu/redtide/ and other sources.

Algal type	Genera	Problem	Affected organisms
Dinoflagellates	*Gambierdiscus*	Ciguatera fish poisoning	Some tropical fish
Dinoflagellates	*Dinophysis, Prorocentrum*	Diarrhetic shellfish poisoning	Mussels, oysters, scallops
Dinoflagellates	*Karenia brevis*	Neurotoxic shellfish poisoning	Manatees, bottlenose dolphins, oysters, fish, clams, birds
Dinoflagellates	*Alexandrium, Gymnodinium, Pyrodinium*	Paralytic shellfish poisoning	Mussels, clams, crabs, oysters, scallops, herring, sardines, marine mammals, birds
Diatoms	*Pseudo-nitzschia*	Amnesic shellfish poisoning	Razor clams, dungeness crabs, scallops, mussels, anchovies, sea lions, brown pelicans, cormorants
Chrysophytes	*Aureoumbra*	Brown tide	Reduced grazing by zooplankton and bivalves, higher bivalve mortality, reduced light penetration, and a decline in seagrass beds
Chrysophytes	*Heterosigma*	Fish poisoning	Fish

take up more nutrients than are supplied to it by diffusion. Equation 3.12 in Chapter 3 indicated that the upper limit set, which is set by diffusion, of the flux per unit surface area is a function of diffusion coefficient (D), the nutrient concentration away from the cell (C), and the cell radius (r). To calculate the flux per unit volume or biomass of the cell, Equation 3.12 can be divided by the volume of a spherical cell ($V = 4/3 \pi r^3$) to yield:

$$\text{Flux per biomass} = 3D{\cdot}C/r^2 \qquad (4.10).$$

Equation 4.10 indicates that the upper limit for uptake set by diffusion increases as cell size decreases. As cell size and the radius increase, nutrient concentrations must increase proportionally to achieve the same flux to the cell. For example, a 1.0 μm cell could grow at 1 d^{-1} with about 15 nM of nitrate while a 5.0 μm cell would require >100 nM to grow at the same rate (Chisholm, 1992). The story is complicated by cell shape and motility, but the basic physics remains the same. Big cells are more likely to be diffusion-limited in oligotrophic waters than small cells.

In addition to physics, biochemistry and physiology work against large cells in oligotrophic waters. To understand this, it is necessary to look at how uptake varies as a function of nutrient concentrations. This relation is described by the Michaelis-Menten equation, the same used to describe enzyme kinetics:

$$V = V_{max}{\cdot}S/(K_s+S) \qquad (4.11).$$

Here uptake (V) is a function of the nutrient concentration (S) and two parameters of the uptake system, the maximum uptake rate (V_{max}) and the half-saturation constant (K_s), which is equivalent to the concentration at which the uptake rate is half of V_{max}. Over a full range of substrate concentrations, V increases until it reaches V_{max}, resulting in the curve illustrated in Figure 4.12. Note, however, that when S is very low, especially low relative to K_s, then $K_s + S \approx S$ and Equation 4.10 reduces to

$$V = (V_{max}/K_s)S \qquad (4.12).$$

Equation 4.12 says that at very low nutrient concentrations, such as after the spring bloom, the uptake rate depends on the ratio of V_{max} to K_s, termed the affinity constant. So, the prediction is that small cells must have either a lower K_s or higher V_{max}, or both, if they are to outcompete large cells for limiting nutrients. In fact, there is evidence of small cells having a lower K_s, but small cells may also have a higher V_{max} normalized per unit of biomass.

To see this, consider two cells with equal number of total transport proteins per unit of membrane area, but one cell is larger than the other and has more biomass (assuming equal density for each cell). The larger cell will also have more surface area and a higher total number of transport proteins, but it will have a lower ratio of surface area to volume. Since surface area (SA) is a function of the radius squared (r^2) whereas the volume varies as r^3, the ratio of surface area to volume is $1/r$ and thus will decrease as r increases. So, the number of transport proteins per biomass will be smaller for a large cell than for a small cell. This may affect V_{max} if it is determined by the number of transport proteins. In this case, then V_{max} per cell is larger for a large cell, but V_{max} normalized per unit of biomass and the number of transport protein per unit biomass will be larger and greater for the small cell.

So, size explains why small phytoplankton, both cyanobacteria and eukaryotes, dominate systems with very low nutrient concentrations. Size is also very important in thinking about the interactions between heterotrophic bacteria and phytoplankton in using dissolved compounds. Again, the prediction is that the smaller heterotrophic bacteria would outcompete larger phytoplankton for inorganic nutrients, which is often the case. Size also has a big impact on grazing and top-down control (Chapter 7), another reason why small phytoplankton take over after blooms end.

Primary production by coccoid cyanobacteria

One phytoplankton group that is common after blooms is coccoid cyanobacteria. These microbes are especially important in the surface waters of oligotrophic oceans where nutrient concentrations are extremely low (<10 nmol liter^{-1}). Biological oceanographers first recognized the importance of coccoid cyanobacteria when they used filters with different pore sizes to examine the size distribution of primary producers ($^{14}CO_2$ uptake) and of phytoplankton biomass (chlorophyll a concentrations). An example is given in Figure 4.13. These studies demonstrated that in the open oceans, such as the North Pacific Gyre, as much as 90% of all $^{14}CO_2$ uptake and of chlorophyll a is associated with organisms in the <1 µm size fraction. Subsequent microscopic examination indicated that many of the algae among the picoplankton are coccoid cyanobacteria. Given the vast coverage of the open oceans, these data

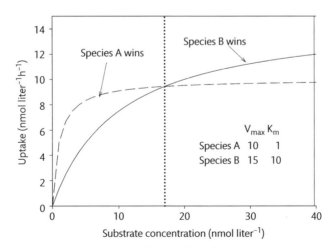

Figure 4.12 Uptake of a dissolved compound as described by the Michaelis-Menten equation. The example has two species competing for the same dissolved compound. The winner is the one with higher uptake rates for a given range of substrate concentrations.

imply that roughly 25% of global primary production is by cyanobacteria.

The phycoerythrin-rich cyanobacteria, *Synechococcus*, were first examined by marine microbial ecologists in part because these microbes could be counted by epifluorescence microscopy (Chapter 1). But this method missed another type of cyanobacteria, *Prochlorococcus*. These microbes are very difficult to see by epifluorescence microscopy, so they were not observed by microbial ecologists until another technique, flow cytometry, was used on samples from the oceans (Chisholm et al., 1988). Both *Synechococcus* and *Prochlorococcus* are most abundant in low latitude open oceans while freshwater species of *Synechococcus* also occur in oligotrophic lakes. Table 4.8 summarizes the main traits of these two types of cyanobacteria.

The distribution of *Synechococcus* and *Prochlorococcus* varies differently with depth. Though both cyanobacterial groups are adapted to the intensity and quality of light found deeper in the water column, *Synechococcus* is generally found higher in the water column than *Prochlorococcus*. *Prochlorococcus* is divided into two types ("ecotypes") which are found at different depths: a high-light ecotype with a low ratio of chlorophyll *a* to chlorophyll *b* (chl *a*:chl *b*), and a low-light ecotype with a high chl *a* to chl *b* ratio. As expected, the low-light ecotype is usually found deeper in the water column than the high-light ecotype. The 16S rRNA genes from isolates of these two ecotypes differ only slightly (average similarity of >97%), less than that usually used to separate species (Chapter 9). However, whole genome sequencing revealed many differences between the two ecotypes (Rocap et al., 2003).

Photoheterotrophy in the oceans

The organisms discussed so far obtain energy from light and carbon from carbon dioxide, but even these photoautotrophic microbes respire and thus use O_2 and produce CO_2. Respiration by photoautotrophs ranges from 20 to 50% of primary production, depending on what group dominates the algal community (Langdon, 1993). Another type of respiration by algae, photorespiration, occurs when rates of photosynthesis and consequently O_2 concentrations are high. Even with photorespiration, however, the eukaryotic algae and cyanobacteria are still classified as photoautotrophs because their primary source of energy is light and CO_2 supplies the carbon needed for biomass synthesis.

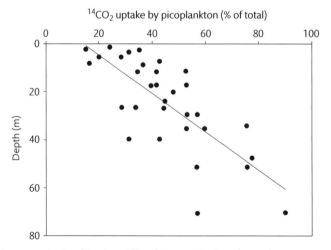

Figure 4.13 Example of primary production (fixation of $^{14}CO_2$) by picoplankton (<1 µm) in an oligotrophic ocean, the eastern tropical Pacific Ocean. Primary production by picoplankton increases with depth because of faster growth and higher biomass in deeper waters. Depth profiles from several locations were combined. Data from Li et al. (1983).

Table 4.8 Comparison of the two major coccoid cyanobacterial genera found in lakes and oceans.

Property	*Synechococcus*	*Prochlorococcus*
Size (diameter)	1.0 μm	0.7 μm
Chlorophyll a	Yes	Modified
Chlorophyll b	No	Yes
Phycobilins	Yes	Variable
Distribution	Cosmopolitan	Oceanic gyres
Nitrate use	Yes	No*
N_2 fixation	Some strains	No

* So far, no cultured strains of *Prochlorococcus* can use nitrate (Coleman and Chisholm 2007).

Photoheterotrophs, in contrast, use both light and organic material. Photoheterotrophs include mixotrophic protists (Chapter 7), which carry out both photosynthesis and phagotrophy, and bacteria that obtain energy from light and the oxidation of organic material. Different photoheterotrophs rely on phototrophy and heterotrophy for energy to different extents, putting them somewhere in the middle of a continuum between "pure" photoautotrophic and heterotrophic organisms (Fig. 4.14).

Uptake of organic material by algae

Heterotrophic bacteria dominate the uptake and degradation of dissolved organic material (DOM) in most aquatic habitats, most of the time (Chapter 5). But this does not rule out use of DOM by cyanobacteria and eukaryotic algae. Laboratory studies have shown that some species of both groups of phototrophic microbes are capable of using DOM components, most notably compounds such as amino acids. In these cases, algae take up organic compounds not necessarily for the carbon and energy, but rather for inorganic nutrients, such as P and N. One of the best-studied examples of DOM uptake by algae is that of *Prochlorococcus* (Zubkov, 2009).

Aerobic anoxygenic phototrophic bacteria

The next two groups of photoheterotrophic bacteria are quite different from cyanobacteria and eukaryotic algae. The first of these are aerobic anoxygenic phototrophic (AAP) bacteria. Anoxygenic photosynthesis was first

observed being carried out by anaerobic bacteria in anoxic environments (Chapter 11). Long after this discovery, obligate AAP bacteria were found in oxic environments (Yurkov and Beatty, 1998). As implied by the name, AAP bacteria require oxygen for growth (they are aerobic), do not evolve oxygen during phototrophy (they are anoxygenic), but can use light energy to augment the energy gained from heterotrophy. Unlike anaerobic

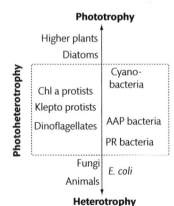

Figure 4.14 Photoheterotrophy, indicated by the box, along a continuum from heterotrophy to phototrophy. "Chl *a* protists" refer to those microbes with chlorophyll but which also are capable of phagocytosis and grazing (Chapter 7). These protists include dinoflagellates, which deserve to be singled out because some feed solely by phagocytosis, while others also photosynthesize. "Klepto protists" partially rely on photosynthesis by chloroplasts stolen from ingested phytoplankton (Chapter 7). AAP bacteria carry out aerobic anoxygenic phototrophy, and PR bacteria have proteorhodopsin.

anoxygenic photosynthetic bacteria, AAP bacteria do not have Rubisco and require organic carbon for growth. At first, AAP bacteria were not considered to be ecologically important, but their discovery in the oceans changed that picture.

Oceanic AAP bacteria were first detected by the distinctive fluorescence characteristics of bacteriochlorophyll *a* found in these bacteria as well as in anaerobic anoxygenic phototrophic bacteria (Kolber et al., 2000). This fluorescence characteristic is due to the main absorbance peak in bacteriochlorophyll being at 770 nm in the infrared (IR) region of the light spectrum (Fig. 4.15). To detect AAP bacteria, cells are excited by green light, which is absorbed by carotenoids, the main light-harvesting pigments of AAP bacteria. These cells then fluoresce in the IR region after the light energy is transferred from the carotenoids to bacteriochlorophyll. By taking advantage of the unique IR fluorescence, a direct count approach has shown that the abundance of AAP bacteria can be quite high in some marine environments. Some of the highest values have been observed in the South Pacific (about 20% of total bacterial abundance), one of the most oligotrophic environments on the planet (Lami et al., 2007), but AAP bacteria are nearly as abundant in eutrophic estuaries (Waidner and Kirchman, 2007). It is not clear why AAP bacterial abundance varies so much.

Pigment concentrations give some hints about the importance of phototrophy to AAP bacteria. Bacteriochlorophyll concentrations in AAP bacteria are only about 0.1 fg per cell, 100–200-fold lower than divinyl chlorophyll *a* levels in *Prochlorococcus* (Cottrell

et al., 2006b). While *Prochlorococcus* can use DOM, it cannot grow on it exclusively without light, implying that it is closer to the strict phototroph end of the spectrum. In contrast, the AAP bacteria in culture can grow without light using the carbon from DOM. These experiments and pigment concentrations put AAP bacteria closer to the other, strict heterotrophic end of the spectrum (Fig. 4.14).

Rhodopsin in photoheterotrophic bacteria

Rhodopsin consists of the protein opsin bound to a light-absorbing molecule, retinal. Rhodopsins are found in all three kingdoms of life and have similar structures but different functions (Spudich et al., 2000). In metazoans such as mammals, rhodopsins are in photoreceptor cells of the retina where they detect different colors of light. Among prokaryotes, rhodopsin was first found in extreme halophilic archaea, such as the well-studied *Halobacterium salinariu* (Mukohata et al., 1999). Unfortunately, to distinguish them from rhodopsins in metazoans, rhodopsins in archaea are often called "bacteriorhodopsins" even though archaea are not bacteria.

Some of these rhodopsins (the sensory rhodopsin I and II) act as light sensors similar to the rhodopsins in our eye, and are part of the phototactic system of some halophilic archaea. Other rhodopsins (halorhodopsin) are light-driven chloride pumps, which help halophilic archaea survive and grow in high salt concentrations. The fourth type of rhodopsin found in these archaea is a light-driven proton pump that synthesizes ATP via the proton-motive force. So, this last type of rhodopsin allows halophilic archaea to make energy from light and to grow phototrophically. Halophilic archaea like *H. salinarium* can grow autotrophically, if no organic carbon is available.

Rhodopsin was not known to occur in bacteria until a culture-independent method discovered it in the Pacific Ocean (Béjà et al., 2000). Since the bacterium first discovered to carry rhodopsin belonged to the SAR86 cluster, a group of Gammaproteobacteria, the term "proteorhodopsin" was used to distinguish it from bacterial rhodopsin of archaea. After its discovery in SAR86 bacteria, rhodopsin genes were found in bacteria belonging to other groups, including an entirely different phylum or division of bacteria, the

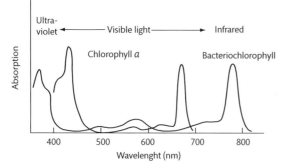

Figure 4.15 Absorption spectra of chlorophyll *a* and bacteriochlorophyll *a*.

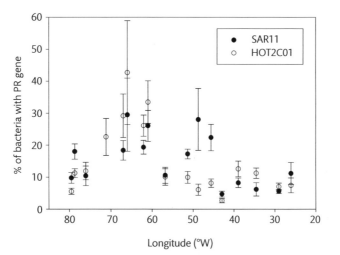

Figure 4.16 Relative abundance of two types of bacteria containing proteorhodopsin (PR) along a transect from Florida to the Azores. The number of proteorhodopsin genes was estimated by quantitative PCR and then normalized to the number of bacteria determined by quantitative PCR of 16S rRNA genes (Chapter 9), corrected for the fact that the average oceanic bacterium has about two 16S rRNA genes. Both SAR11 and HOT2C01 are groups in the Alphaproteobacteria. Data from Campbell et al. (2008).

Bacteroidetes (Venter et al., 2004). We now know that about half of all bacteria in the North Atlantic Ocean have proteorhodopsin (Fig. 4.16), and proteorhodopsin genes have been found in many lakes and brackish waters. Some species of Actinobacteria, a bacterial phylum common in lakes and soils (Chapter 9), have rhodopsin (Sharma et al., 2008). The freshwater proteorhodopsins are quite different from those found in the oceans.

In spite of proteorhodopsin being so widespread, it has been difficult to show that it confers any advantage to bacteria bearing it. Only one of the four proteorhodopsin-bearing bacteria examined so far grows faster with light than without it in laboratory experiments (Gómez-Consarnau et al., 2007). A proteorhodopsin-bearing marine bacterium survives starvation better in the light than in the dark, while a mutant strain of this bacterium without proteorhodopsin is not helped by light (Fig. 4.17), proving the role of proteorhodopsin in the starvation response (Gómez-Consarnau et al., 2010). There are probably other, yet-to-be discovered roles of proteorhodopsin in actively growing bacteria in nature.

Ecological and biogeochemical impacts of photoheterotrophy

Photoheterotrophy and other forms of mixotrophy potentially complicate simple models of how energy and elements are transferred among organisms in natural environments. It was relatively easy to think about these models with just two compartments: photoautotrophs depending on light for energy, and heterotrophs depending on the organic material synthesized by photoautotrophs. With photoheterotrophy, we have to consider that some of these photoautotrophic microbes may have heterotrophic capacities, and likewise organic material degradation may be directly affected by light. In particular, a photoheterotrophic microbe may use organic material with higher efficiency than a strict heterotrophic microbe because less of the organic carbon would need to be oxidized to yield energy if energy is harvested from light. Mixotrophic protists can gather critical elements, such as N, P, and Fe, from their prey, allowing them to survive and potentially outcompete strict photoautotrophs in oligotrophic waters. Most biogeochemical models still need to be modified to include mixotrophic and photoheterotrophic microbes.

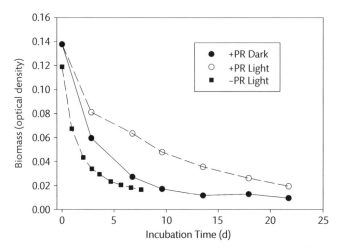

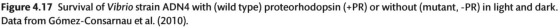

Figure 4.17 Survival of *Vibrio* strain ADN4 with (wild type) proteorhodopsin (+PR) or without (mutant, -PR) in light and dark. Data from Gómez-Consarnau et al. (2010).

Summary

1. Microbes account for about 50% of global primary production and the other half is by land plants. Cyanobacteria alone are responsible for 50% of aquatic primary production or about 25% of total global primary production.

2. Rates of primary production are governed by light intensity and quality, along with nutrients such as ammonium, nitrate, phosphate, and iron. One important type of photoautotrophic microbe, diatoms, also requires silicate.

3. Several eukaryotic microbes can dominate the algal community with various effects on the ecosystem. These microbes include diatoms, green algae, coccolithophorids ($CaCO_3$ formation), and *Phaeocystis* (DMS producer).

4. Cyanobacteria are functionally similar to eukaryotic algae, including having the same ecological role in primary production and the same mechanism for photosynthesis (both have chlorophyll *a* and the CBB cycle), but phylogenetically cyanobacteria are members of the Bacteria domain.

5. Filamentous N_2-fixing cyanobacteria are common in freshwaters, whereas coccoid cyanobacteria (*Synechococcus* and *Prochlorococcus*) are abundant in marine waters, particularly oligotrophic open oceans.

6. Cyanobacterial blooms can cause problems with the quality of water in freshwater reservoirs and ecosystem function in small ponds and lakes. Other algae are harmful to aquatic life and humans.

7. Photoheterotrophic microbes harvest light energy while also grazing on other microbes (in the case of mixotrophic protists) or using DOM.

CHAPTER 5

Degradation of organic material

The previous chapter discussed the synthesis of organic material by autotrophic microbes, the primary producers. This chapter will discuss the degradation of that organic material by heterotrophic microbes. These two processes are large parts of the natural carbon cycle. Nearly all of the 120 gigatons of carbon dioxide fixed each year into organic material by primary producers is returned to the atmosphere by heterotrophic microbes, macroscopic animals, and even autotrophic organisms (Chapter 4). Note the "nearly" in the last sentence. While primary production is mostly balanced by degradation, imbalances occur, affecting many aspects of the ecosystem. It is these imbalances that set whether the biota is a net producer or consumer of atmospheric carbon dioxide. Since these imbalances depend on degradation as well as primary production, do both primary production and organic matter degradation determine the net contribution of the biota to fluxes of carbon dioxide to and from the atmosphere.

As with all biogeochemical cycles, the carbon cycle consists of reservoirs (concentrations or amounts of material) connected by fluxes (time-dependent rates) made of both natural and anthropogenic processes (Fig. 5.1). The natural rates of exchange between carbon reservoirs are much larger than the anthropogenetic ones. In particular, the natural production of carbon dioxide by heterotrophs is much higher than the anthropogenic production due to the burning of fossil fuels and other human activities. The problem is that because the anthropogenetic production of carbon dioxide is not balanced by carbon dioxide consumption, concentrations in the atmosphere are increasing and our planet is warming up (Chapter 1). Many of the natural processes

in the carbon cycle are huge and variable. This complicates the efforts of biogeochemists to understand how human activity is affecting the carbon cycle and to determine the implications for climate change. One example is the missing carbon problem. Of the eight petagrams of carbon burned by human activity every year, "only" about three stay in the atmosphere. Some of the remaining five petagrams goes into the ocean or is taken up by plants on land, but about three petagrams per year were missing until recently (Stephens et al., 2007). While the missing carbon problem may be solved, many parts of the carbon cycle remain mysteries, greatly complicating predictions of how the biosphere will respond to climate change over the coming decades.

The carbon cycle has several reservoirs of both inorganic and organic material (Fig. 5.1). The largest reservoirs are dissolved inorganic carbon (DIC), mostly bicarbonate, in the ocean, and calcium carbonate (a major mineral in limestone) on land and in oceanic sediments. Compared to the dissolved pools, the amount of carbon in organisms and in non-living particulate organic is small. Aquatic ecologists call this dead material detritus, while terrestrial ecologists also use the terms "plant litter" or simply "litter" when discussing material that is still recognizable as coming from plants. Another large dissolved pool is dissolved organic carbon (DOC), and there is also much organic carbon in sediments of the oceans. The largest reservoir of organic carbon, however, is in soils and in other terrestrial compartments. These organic reservoirs are as large (oceanic DOC) or larger (soil organic material) than the atmospheric reservoir of CO_2 which was 391 parts per million in January 2011, or over 760 gigatons for the entire atmosphere.

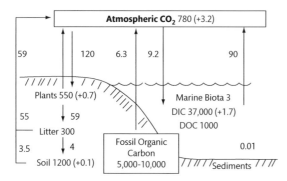

Figure 5.1 Global carbon cycle. The units for the numbers next to the reservoir names are Pg of carbon (Pg = 10^{15} g) and next to the arrows are PgC y^{-1}. The numbers in parentheses are the yearly changes. Some features not shown here include the CO_2 produced by land-use change (2 PgC y^{-1}) and the biggest carbon reservoir, carbonate rocks (Chapter 13). Some budgets use higher fluxes into and out of the oceans, closer to the rates seen for terrestrial systems (Sarmiento and Gruber 2006). Based on data presented in Houghton (2007).

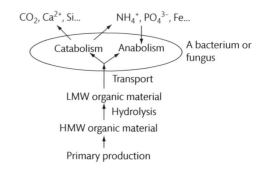

Figure 5.2 Mineralization of organic material by heterotrophic bacteria and fungi. LMW and HMW refer to low and high molecular weight material, respectively. Catabolism is the energy-producing parts of microbial metabolism whereas anabolic reactions lead to synthesis of cellular components and eventually growth. Some inorganic ("mineral") compounds are potentially used by heterotrophic microbes (NH_4^+, PO_4^{3-}, and Fe) whereas others (CO_2, Ca^{2+}, and Si) are not used substantially for energy production or biomass synthesis.

Microbes are very important in setting the fluxes between these large carbon reservoirs.

In this chapter, we discuss aerobic respiration and degradation of particulate detritus, litter, and dissolved organic material (DOM) in oxic environments, leaving anaerobic degradation in anoxic environments to later chapters.

A simple equation for aerobic respiration is:

$$CH_2O + O_2 \rightarrow CO_2 + H_2O \tag{5.1}$$

where as before CH_2O symbolizes generic organic material, not a specific compound. In oxic environments, the complete degradation of organic matter is due to aerobic respiration which consumes oxygen and produces carbon dioxide and water. But degradation involves more than just carbon because organic material always has several other elements. Consequently, organic matter degradation releases several other inorganic or mineral nutrients, such as ammonium and phosphate, in addition to CO_2 (Fig. 5.2). Some authors use "remineralization" to highlight the never-ending cycle of uptake and release of compounds containing essential elements like N and P. The degradation and mineralization of detritus is the traditional role assigned to heterotrophic microbes.

Mineralization of organic material in various ecosystems

Before discussing mineralization at the microbial scale, let us take a global view and examine where mineralization and respiration are the highest. In the previous chapter, we saw that roughly half of all primary production was by land plants and the other half by microbes in the ocean. As a first approximation, respiration is also split evenly between land and the sea. As with primary production, respiration rates for the oceans, especially the open oceans, are rather low when expressed per unit volume (m^{-3}) or per unit area (m^{-2}), but they add up to a large number for the entire ecosystem because the oceans cover so much area and are so deep. Likewise, respiration rates for lakes and rivers are low when summed over the entire ecosystem, but per cubic meter rates are actually quite high. Finally, in spite of covering only about 30% of the earth, soils account for nearly as much respiration as the oceans because of very high per square meter rates.

As a general rule, degradation rates follow primary production, and the same ecosystems with high primary production also have high rates of respiration (Fig. 5.3). Overall, there is an excellent correlation between the two processes and the regression analysis indicates that

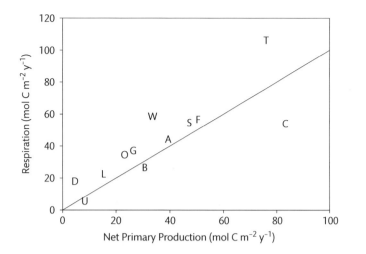

Figure 5.3 Respiration rates (R) and net primary production (NPP) in the major biomes of the world. The line is the 1:1 line. The slope and intercept of the regression line are not significantly different from one and zero, respectively (R = 0.868NPP+11.6; r^2 = 0.68), indicating that respiration follows net primary production. D = desert; U = tundra, G = temperate grassland, B = boreal forest, W =Mediterranean woodland, A = agriculture, S = tropical savannahs, F = temperate forest, T = moist tropical forest, O = open ocean, C = continental shelf, L = lakes. Data from Pace and Prairie (2005), Raich and Schlesinger (1992), Bond-Lamberty and Thomson (2010), and Field et al. (1998).

respiration and net primary production are in balance overall. However, there are important exceptions when the two rate processes are not in balance. When primary production exceeds respiration, the system is said to be net autotrophic, one example being spring blooms in aquatic habitats (Chapter 4). Subsurface environments are net heterotrophic because primary production is zero where light cannot reach. More intriguing to microbial ecologists and biogeochemists are net heterotrophic aquatic systems in which respiration exceeds primary production, as mentioned in Chapter 4. It is not clear why respiration is higher than net primary production for nine of the 12 biomes in Figure 5.3. The difference between the two processes may not be significant considering the huge spatial and temporal variation in these numbers.

Who does most of the respiration on the planet?

In Chapter 4, we saw that microbes were responsible for about half of all global primary production due to photosynthesis by eukaryotic phytoplankton and cyanobacteria in the oceans. Microbes account for much more than half of total global respiration, although the precise percentage is difficult to estimate. The global estimate may be less important than the percentages for individual ecosystems. These percentages indicate the importance of microbes in structuring the flow of carbon and other elements in these ecosystems.

In aquatic environments, respiration by microbes can be estimated by incubations in which large organisms are removed by filtration, leaving only microbes in the water. The consumption of oxygen is then measured over time in the dark (to stop photosynthesis and oxygen production), sometimes along with estimates of DOM consumption. Simultaneously, respiration by all organisms is estimated from changes in oxygen in other, dark incubations with unfiltered water.

This experiment has shown that nearly all of the respiration is by organisms < 200 μm in size (Fig. 5.4), which would include many zooplankton and most phytoplankton. Of more interest here, nearly half of total respiration is by organisms passing through a filter with 0.8 μm pores. The exact percentage varies with the environment, but usually it is very high, 50% or greater. Other analyses show that these organisms are

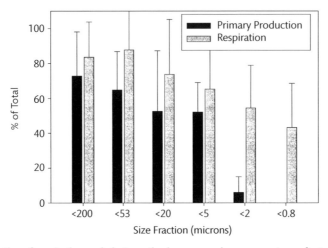

Figure 5.4 The size distribution of respiration and photosynthesis, expressed as a percentage of rates in unfiltered samples. Data from Williams (2000).

mostly bacteria. Several other methods and approaches support the conclusion that over half of the total respiration in aquatic ecosystems is by bacteria.

It is much harder to compare respiration by microbes versus macro-organisms in sediments and by bacteria versus fungi in soils. For macro-organisms in sediments the only way devised so far has been to combine data on abundance and on rates per organism determined in laboratory experiments. These studies indicate that macro-organisms account for 5–30% of total respiration in freshwater and coastal marine sediments (Canfield et al., 2005). For soils, respiration has been measured before and after removing plant roots. These studies found that roughly half of total respiration is by roots (called autotrophic respiration) and associated microbes in the rhizosphere (Andrews et al., 1999, Raich and Mora, 2005), and nearly all of the rest is by other microbes. Little respiration in soils is by large organisms, such as nematodes, earthworms, and insect larvae, as their biomass is a small fraction (<5%) of total biomass (Fierer et al., 2009). Large soil organisms have a much more important role in breaking up large pieces of plant litter and detritus, in the process creating more surface area for microbes to grow and to degrade detrital organic material.

In contrast to aquatic ecosystems, on land, fungi contribute substantially to soil respiration. Here we focus on fungi living on dead organic material (saprophytic fungi),

and leave discussion of root-associated fungi (mycorrhizal fungi) to Chapter 14. According to experiments using antibiotics and other inhibitors, bacteria and fungi account for about 35 and 65% of microbial respiration in soils, respectively (Joergensen and Wichern, 2008). These percentages may be inaccurate due to inefficiencies in stopping activity with inhibitors, and the contributions by fungi and bacteria certainly vary among soils, depending on environmental factors such as water content and temperature. Fungi do better than bacteria in dry soils, and may also contribute more to respiration than bacteria at low temperatures (Pietikåinen et al., 2005). More so than bacteria, fungi degrade dead plants still standing above soil or water. In any case, both bacteria and fungi are important in contributing to total respiration and organic matter decomposition in soils.

Like their contribution to respiration, the biomass of fungi as a fraction of total biomass is much higher in soils than in aquatic habitats (Table 5.1). Again, the exact percentage varies with the soil type, geographical location, and method for estimating microbial biomass. It is possible to examine both bacteria and fungi by epifluorescence microscopy (Chapter 1), yielding direct counts of individual bacterial cells and estimates of total length of fungal hyphae. Both are then converted to common units of grams of cellular carbon per gram of soil or sediment or per milliliter of water. Other methods for estimating bacterial and fungal biomass also rely on conversion factors

Table 5.1 Abundance and biomass of bacteria and fungi in various habitats. "ND" is not detectable. The values depend on the location and time of sampling, varying as much as tenfold. Data from Frey et al. (1999), Busse et al. (2009) and Whitman et al. (1998).

Habitat	Bacterial abundance (10^6 cells ml^{-1} or g^{-1})*	Fungal length (m g^{-1})	Bacteria as % of total microbial biomass
Soil, agriculture	900	164	71
Soil, forest	300	330	35
Lakes	1	ND	100
Ocean	0.5	ND	100
Marine sediments	460	ND	100

*Abundance in lakes and oceans is expressed as cells ml^{-1} whereas in soils and sediments the units are g^{-1}.

and are imperfect. Regardless, the data indicate that fungi make up on the order of 50% of microbial biomass in soils (Joergensen and Wichern, 2008), but they are hard to detect at all in lakes and oceans. Some fungi are indigenous to aquatic systems, and they may be abundant on large particles and fresh detritus (Findlay et al., 2002). Still, their overall biomass is low compared to bacteria in aqueous environments.

Bacteria and saprophytic fungi appear to have the same ecological role in nature, but their abundance and contribution to total degradation are quite different in aquatic habitats versus in soils. Why? Bacteria win out in the water column of aquatic habitats because their small size makes them superior competitors for dissolved compounds. This competitive edge is less important in soils, unless they are waterlogged. In terrestrial environments, the hyphae life form taken on by many fungi allows them to cross dry gaps between moist microhabitats and to access organic material not available to waterbound bacteria. Some bacteria also grow as filaments, but the resemblance to fungal hyphae is superficial. Unlike bacteria, the cytoplasm of fungi moves within the rigid hyphae to take advantage of favorable growth conditions. A microbial ecologist in the nineteenth century thought of fungi as tube-dwelling amoeba (Klein and Paschke, 2004). The hyphal body form goes a long way to explaining the success of fungi in soils.

Slow and fast carbon cycling pathways

The amount of respiration, biomass, and biomass production (Chapter 6) by bacteria versus fungi has several

important implications for understanding soil ecosystems (Moore et al., 2005). One is that bacteria are thought to mediate a fast carbon cycling pathway while fungi are responsible for a slow pathway, reflecting the types of organic matter used by the two microbial groups (Rinnan and Bååth, 2009). In soils, bacteria use labile organic compounds, while fungi degrade refractory material, the most important being ligno-cellulose, as discussed below. These differences in organic carbon use have effects on growth rates; as discussed in Chapter 6 in more detail, bacteria appear to grow more quickly than fungi in soils, another reason to think that bacteria mediate the fast pathway while fungi do the same for the slow pathway. As with all generalizations, there are exceptions. But the slow-fast pathway model is a useful simplification for thinking about the implications of microbial growth.

High mineralization and respiration rates by microbes, whether bacteria or fungi, have many implications for the flow of carbon in ecosystems, and radically transform the view of a world with just plants, herbivores, and carnivores. High respiration usually means high degradation of organic material (Equation 5.1). Unless the organic material fueling respiration is old and was synthesized by primary producers in the distant past, respiration by microbes represents primary production not being used by herbivores whether on land or in water (Fig. 5.5). So, when microbes account for most of total respiration, it implies that most primary production is routed through them and not through larger organisms. Given such high microbial activity, it is sometimes amazing that large herbivores and carnivores exist on the planet. They do exist because some take advantage of microbe-based food webs (Chapter 7). Even those organisms that consume only other large organisms still depend on microbes for digestion and for other facets of their existence (Chapter 14).

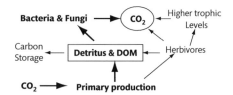

Figure 5.5 The fate of primary production in natural ecosystems, illustrating the central and often dominant role of the detritus pathway as indicated by the thick arrows.

Chemical characterization of detrital organic material

Microbial ecologists have to know something about the chemical make-up of detrital organic material to understand mineralization, respiration, and growth of heterotrophic microbes because all of these microbial processes depend greatly on what compounds and elements are in organic material. As with the composition of microbial cells, there are two complementary approaches for thinking about the composition of non-living organic material. One examines the relative amounts of the major biochemicals, and the other elemental ratios. Both the biochemical composition and the elemental ratios differ the most for organic material in soils versus in water, because of differences in which primary producer dominates these environments.

While they share with phytoplankton many traits necessary for carrying out photosynthesis, higher plants had to evolve several additional structures to succeed on land. Terrestrial plants need these structures in order to grow up and out into air away from soil and to fend off attack by herbivores. Plants partially solve both problems by having lots of cellulose, related complex carbohydrates, and lignin, the latter being especially abundant in wood. Cellulose is a polymer of glucose linked by β 1,4 bonds whereas lignin is a very complex, ill-defined structure consisting of several phenolic or aromatic groups (Fig. 5.6). Lignin is the major component of wood and its strength explains why trees can grow so high. Lignin also explains why wood is so hard for herbivores to eat. Although some phytoplankton and other aquatic primary producers have cellulose in their cell walls, they do not make lignin. Suspended by water, phytoplankton

and macroalgae do not need lignin and woody structures to survive.

Consequently, phytoplankton are rich in protein, much more so than terrestrial plants, because they lack many of the carbohydrates and all of the lignin required for life on land (Table 5.2). Some macroalgae have more carbohydrates, such as alginate, than phytoplankton, but still not as much as terrestrial plants. Likewise, the particulate detritus in aquatic environments is protein-rich whereas detritus on land reflects the carbohydrate make-up of terrestrial plants. The chemical properties of carbohydrates and lignin that give terrestrial plants structural strength also make them hard to degrade by microbes.

Other than the main biochemicals just listed, many components of detritus cannot be assigned a chemical name. Unidentified components make up 90% or much more of detrital mass, depending on the age of the detritus and stage of decomposition. The unidentified fraction is low in fresh detritus and plant litter, but then

Table 5.2 Biochemical composition of plant detritus and organisms in terrestrial and aquatic ecosystems. Data from Canfield et al. (2005) and Randlett et al. (1996).

	% of Total				
	Lignin	Carbohydrate	Protein	Lipid	C:N ratio
Terrestrial					
Straw	14	81	1	2	80
Tree leaves	12	77	7	12	50
Pine wood	27	72	0	1	640
Aquatic					
Kelp	0	91	7	<1	50
Diatom	0	32	58	7	6.7
Zooplankton	0	14	46	<1	6.7

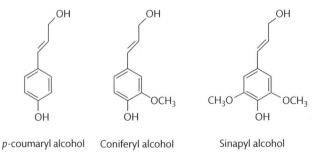

p-coumaryl alcohol   Coniferyl alcohol   Sinapyl alcohol

Figure 5.6 The structure of common subunits of lignin, the main structural element of wood. The amounts of these and other subunits vary with the type of lignin.

increases with detritus age and as degradation proceeds. Characterizing these unidentified organic compounds and determining how they are formed are major topics in organic geochemistry.

Biochemical composition drives the relative abundance of crucial elements making up detritus and plant litter. Examining elements (usually C, N, and P) is the second approach for studying DOM and particulate detritus. Aquatic organisms and detritus are rich in nitrogen because of their high protein content whereas the opposite is true for terrestrial plants and detritus. Another crucial element, phosphorus, is also more abundant in aquatic material relative to its total mass. There is some nitrogen and phosphorus in plant detritus because, of course, terrestrial plants have proteins, nucleic acids, and lipids, but these N- and P-rich compounds are diluted by the high amounts of carbon in carbohydrates and lignin. Consequently, C:N and C:P ratios are very high for terrestrial organic material, in contrast to the much lower ratios for detritus in aquatic environments (Table 5.2).

Dissolved organic material
As seen in Figure 5.1, the reservoir of DOC in the biosphere is very large, much larger than that of particulate detritus, plant litter or of the biota. DOC is a large component of DOM. In aquatic habitats, DOM is defined as whatever passes through a filter with pore sizes about 0.5 μm (Chapter 3). As with particulate detritus, concentrations of DOM are usually expressed in terms of key elements, mostly C, N, and P. We know the most about DOC and less about dissolved organic nitrogen (DON) and phosphorus (DOP). Soil ecologists focus on plant litter and particulate detritus, but DOC and other DOM compounds are present in pore water of soils and in aquifers. In soil ecology, the term "soluble organic material" is sometimes used instead of DOM.

Concentrations of DOC generally follow phytoplankton biomass levels (chlorophyll) and primary production in aquatic habitats (Fig. 5.7). Surface waters of the open ocean have much lower DOC concentrations than in euphotic reservoirs and lakes. Concentrations range from about 50 μM-C in the winter of the Ross Sea (Antarctica) to over 500 μM-C in some eutrophic lakes and reservoirs. Some of the DOC found in freshwaters comes from land, which explains why concentrations are higher in freshwater than in marine habitats with similar phytoplankton biomass levels. Terrestrial organic carbon makes its way to the oceans as well, but that input is small compared to the marine DOC pool. Concentrations are usually higher in the euphotic zones of both lakes and the oceans and then decrease with depth. In the deep ocean, DOC is present at a minimum of about

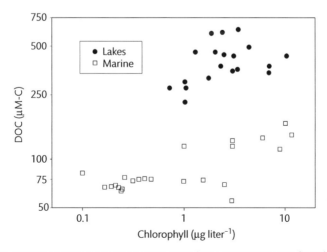

Figure 5.7 DOC in the surface layer of some lakes and oceans. Data from del Giorgio et al. (1999) and Kirchman et al. (2009).

35 µM-C. Most of the DOC in the biosphere is in the deep ocean because its volume is so large.

Only about 10% of the DOM reservoir can be identified chemically. Some of the largest components with known, defined structures include polysaccharides and proteins. The concentrations of these two DOM components are usually estimated by measuring the monomers resulting from acid hydrolysis of a DOM sample. The acid breaks up, for example, protein and any amino acids complexed with other material, yielding "free" amino acids, which can be measured by high performance liquid chromatography (described below). The difference then in amino acid concentrations before and after acid hydrolysis gives an estimate of the dissolved combined amino acid concentrations. The same procedure is used to estimate free and combined carbohydrates. Great progress is being made in characterizing DOM using techniques such as mass spectrometry-Fourier transform-ion cyclotron resonance (FT-ICR-MS) (Dittmar and Paeng, 2009).

The concentrations of simple monomers like free amino acids and sugars are usually very low, about tenfold lower than combined forms. In aquatic habitats, the concentration of each free amino acid may range from <1 to 20 nM and the total concentrations are usually <100 nM. To put this concentration in perspective, there are more amino acids on your fingertips than there are in a liter of water. Concentrations in soils are much higher, in the micromolar range (Jones et al., 2009a), perhaps because some of the measured monomers are released during the extraction of soil pore waters. Even in soils, however, concentrations of simple monomers are much lower than concentrations of the polymers in which they occur. That is, concentrations of free glucose, other sugars, and amino acids are much lower than concentrations of polysaccharides, peptides, and proteins.

Much of soil organic material and DOM in aquatic habitats is said to be humic material. This and related terms came from soil chemists examining fractions of soil organic material isolated by acid and base extractions and other simple procedures. The fractions are defined by the isolation procedure, resulting in material with predictable bulk characteristics (Fig. 5.8). The structure of humic material is often depicted as being incredibly complicated with many aromatic rings, studded with phenolic and organic acid (-COOH) moieties

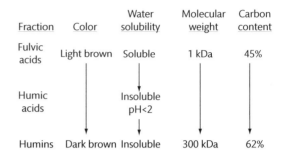

Figure 5.8 Classic definitions of organic material fractions isolated from soils. The terms, especially humic acids, are often used to describe DOM in aquatic habitats. Based on Stevenson (1994).

(Stevenson, 1994). The ligno-cellulose detritus from terrestrial plants has humic-like properties and some classic humic moieties. This detritus can be abundant in small lakes and rivers receiving large inputs of terrestrial organic material. But the classic model of humic substances probably does not accurately reflect the chemical composition of organic material in soils and aquatic habitats (Kleber and Johnson, 2010). It is difficult for any single model to capture the complexity of natural organic material in soils and aquatic habitats.

Detrital food webs

Detritus and plant litter are produced when phytoplankton, higher plants, and animals senesce and die. Detritus also can be a by-product of herbivore grazing or of lysis by viruses. One type of detritus is the fecal material from metazoan grazers. Fecal pellets vary in size depending on the grazer and the prey concentration. Even protists can produce submicron particles ("picopellets"), although many of these particles would be included in the DOM reservoir. Dissolved compounds can also stick together, "coagulate", and form particulate detritus in aquatic systems. The detritus produced by these different mechanisms differs in composition and rates of degradation.

Detritus supports a complicated food web of bacteria, fungi, protists, and metazoans, all living directly or indirectly on particulate dead organic material rather than on live plants or algae. Detrital food webs are especially important in detritus-rich habitats, such as salt marshes, many estuaries, bogs, and all soils. In addition to large reservoir sizes, the flux of detritus and plant litter is also

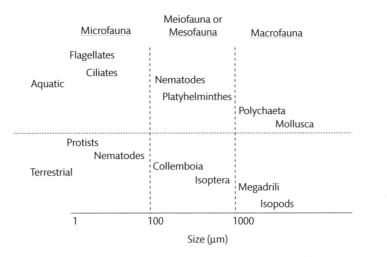

Figure 5.9 Organisms grouped by size set by the nets and sieves used for collection. Only a few of many possible organisms are given here as examples. "Meiofauna" is used by sediment ecologists while soil ecologists prefer "mesofauna". Many of these organisms are capable of feeding on detritus and detritus-associated microbes. See also Table 5.3 for more examples.

very high. Nearly all primary production from trees is routed through detrital food webs while roughly half is, in the case of grasslands (Cebrian, 1999). The percentage may be equally high in detritus-rich aquatic habitats, such as small ponds and salt marshes, but it is low (<10%) in the open ocean and large lakes without high amounts of particulate detritus.

Many types of organisms are able to ingest detritus and potentially obtain some carbon, other elements, and energy, if not use it as a sole food source (Table 5.3). These organisms are called detritivores. Marine benthic ecologists use the term deposit-feeders, reflecting the fact that detritus is deposited onto sediments from plankton production in overlying surface waters. In the water column of aquatic habitats, relatively few metazoans seem to specialize on detritus, as the grazers there are more selective and ingest individual food items, although some filter-feeding zooplankton do appear to ingest all particles of the certain size. In contrast, detritivores feed more indiscriminately in sediments and soils where detritus, plant litter, and inorganic particles are much more abundant. These organisms are grouped together by size (Fig. 5.9).

In all cases, microbes, which otherwise are too small to be grazed on by these animals, are included with the detritus as it is ingested. Which is more important nutritionally to the animal, the detritus or the attached

microbes? With few exceptions, the detritus has more organic carbon than the microbes in terms of sheer mass.

Table 5.3 Some examples of detritivores, which are eukaryotes able to consume detritus and use it for carbon, other elements, and energy. Bacteria and other microbes associated with the detritus may be as important as or more so than the detrital carbon itself to these organisms.

Habitat	Organism	Comments
Aquatic water column	Zooplankton	Most are mainly herbivores and carnivores
Aquatic sediments	Nematodes	
Aquatic sediments	Harpacticoid copepods	
Aquatic sediments	Polychaetes	Mainly marine
Soils	Enchytraeids	Microdrile annelids, commonly known as "potworms"
Soils	Oligochaetes	Megadrile annelids, including earthworms
Soils	Nematodes	
Soils	Collembola	Small arthropods (<5 mm)

However, microbes may be more nutritious because of their high protein content, whereas detritus consists largely of structural polysaccharides, such as ligno-cellulose, depending on its age and source. Even with the help of symbiotic bacteria (Chapter 14), these polysaccharides are difficult for metazoans to digest and are low in nitrogen. So, there is no simple answer to the microbes versus detritus question. The relative contribution of each to animal nutrition depends on the detritus and the detritivore.

Detritivores are very important in effecting the degradation of particulate detritus in both terrestrial and aquatic ecosystems, even though their direct contribution to detritus mineralization is small. Rather than accounting for much respiration, the more important role of detritivores is to break up detritus and plant litter physically, which decreases the size of detrital particles and as a result increases the surface area where microbes can adhere and degrade the exposed organic compounds (Fig. 5.10). Macrofauna, which are organisms larger than 2 mm, can have additional effects on the microbial environment. In soils, these large animals, large in the microbial world at least, break up aggregates and increase aeration and water flow. Earthworms in particular constitute a "geomorphic force" tenfold stronger than other, purely physical processes (Chapin

et al., 2002). Likewise, in sediments, burrows of macrofauna allow penetration of oxygen into otherwise anoxic environments, greatly affecting sediment chemistry. In both soils and sediments, macrofauna disrupt the orderly layers, horizons, and gradients in geochemical properties that would otherwise form in a world without animals. The end result is that detritivores and other macroscopic organisms help to speed up the degradation of detritus even though most of the actual mineralization is done by bacteria and fungi.

DOM and the microbial loop

In addition to particulate detritus, plant and algal organic material becomes available to microbes when it is transferred from cells and particulate detritus to dissolved reservoirs. In soils and sediments with rooted plants, this release is part of "below-ground production", in contrast to the more visible "above-ground production". Although difficult to estimate, below-ground production can be a very large fraction, as high as 50%, of total primary production by higher plants (Högberg and Read, 2006). Dissolved or soluble organic material released by roots fuels soil microbial activity while bypassing herbivores.

Like excretion by plant roots, DOM is released directly by phytoplankton in aquatic ecosystems, but it is also

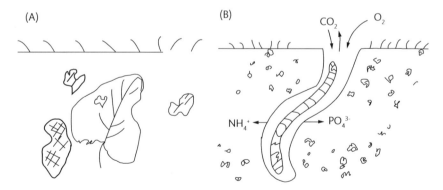

Figure 5.10 Effect of macrofauna-like worms on the degradation of organic material and on the structure of soils and sediments. Panel A illustrates a worm-less world in which large pieces of detritus are not broken down. In contrast, in the environment depicted by Panel B, worms and other macrofauna help to break up detritus and facilitate the mineralization of the organic material to inorganic nutrients like ammonium and phosphate. The burrows of these large organisms also allow faster diffusion of gases in soils and of dissolved compounds in aquatic sediments.

produced by many heterotrophic organisms. The release of DOC by phytoplankton can be measured by tracing $^{14}CO_2$ into phytoplankton cells and eventually into the DOM reservoir. Although these experiments indicated that as much as 50% of primary production can be released as DOM, the overall average is closer to 10%. Some of the ^{14}C-labeled DOC comes directly from phytoplankton cells, while other components may be released inadvertently by herbivores trying to eat phytoplankton cells, a process sometimes called "sloppy grazing". Still other DOM is released during excretion by herbivores and carnivores in aquatic ecosystems. The internal content of cells lysed by viruses also adds to the DOM reservoir (Chapter 8). Since bacterial respiration amounts to about 50% of primary production but only 10% of it comes from direct phytoplankton excretion, most of the DOM production must be by mechanisms involving organisms other than phytoplankton. This complicates efforts to compare bacterial production and respiration with primary production (Chapter 6).

In addition to containing large amounts of carbon and other elements (large reservoir size), fluxes through the DOM reservoir are also quite large and can support much microbial growth and respiration. In soils, it is difficult to compare the relative importance of DOM with that of particulate detritus in supporting microbial activity, but DOM accounts for at least 50%, on average, of soil respiration because that is the percentage attributable to root exudation and below-ground production. This is a high percentage even if we assume that the other 50% is from particulate detritus. In aquatic habitats, it is easier to compare the activity of "free-living" and particle-associated bacteria. ^3H or ^{14}C-labeled dissolved compounds, such as glucose or amino acids, are added to a water sample and incubated for an hour or so. Then the radioactivity in attached microbes is collected by filtration using large pore-sized filters (1 or 3 μm) and compared to the radioactivity going through these filters.

This type of experiment demonstrates that usually >75% of total bacterial activity is by free-living cells rather than the particle-associated ones. Any uptake by non-bacterial microbes, such as fungi, large phytoplankton, and other protists, which are in the large-size fraction, would lead to even higher estimates for the free-living bacteria relative to the attached bacteria. The percentage may be lower if some of the particles are broken up

by the filtration process. Also, the apparently free cells may in fact be associated with small particles that pass through GF/F filters (Chapter 3) used to separate the dissolved and particulate pools.

Regardless of the precise form of DOM, this organic material is not readily available to larger organisms and the non-microbial parts of food webs. Once taken up by microbes, however, the carbon, nitrogen, and other elements now potentially can be used by other organisms and transferred up the food chain. The DOM-based pathway, more precisely, primary production → DOM → microbes → grazers, is called the microbial loop (Fig. 5.11). The term was coined by aquatic microbial ecologists (Azam et al., 1983), but the concept is applicable to terrestrial ecosystems as well (Bonkowski, 2004). Bacteria and fungi turn indigestible organic material, such as ligno-cellulose, into food for soil metazoans. This connection and the microbial loop concept emphasize that in addition to being mineralizers, bacteria and fungi can be large components of food webs in natural ecosystems.

However, not all of the carbon taken up by microbes is available for grazers and higher trophic levels. Some of it may be respired as CO_2 and thus is lost from the system until it is fixed again by primary production. The rest of the carbon taken up by microbes would be used for biomass production and would be available as food for

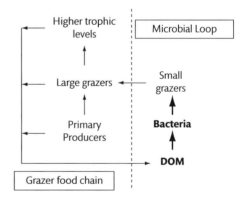

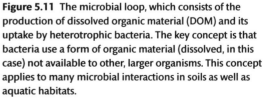

Figure 5.11 The microbial loop, which consists of the production of dissolved organic material (DOM) and its uptake by heterotrophic bacteria. The key concept is that bacteria use a form of organic material (dissolved, in this case) not available to other, larger organisms. This concept applies to many microbial interactions in soils as well as aquatic habitats.

grazers. Figuring out which of these two fates of carbon—respiration or biomass production—is most important has been called the "sink or link" question (Pomeroy, 1974). Is the microbial loop a sink in which the carbon is mostly respired and lost from the system? Or is it a link, meaning that organic carbon taken up by microbes is passed on to higher trophic levels? Which fate dominates?

For aquatic ecosystems, the sink-link question was answered experimentally by examining the use of ^{14}C-glucose by bacteria and the rest of the plankton. This radioactive form of glucose was added to large mesocosms (big bags containing 10 to >1000 liters of water) and the radioactivity was then followed in organisms of various sizes over several days. Microbial ecologists found that little of the radioactivity appeared in large organisms, implying that little of the glucose taken up by bacteria was transferred to other food chains (Fig. 5.12). The link between the microbial loop and larger organisms was weak. Most of the radioactivity was simply respired to CO_2, indicating that the microbial loop is mainly a sink.

This conclusion was later supported by data on the bacterial growth efficiency (BGE). This parameter is the ratio of biomass production (P) to the sum of production and respiration (R):

$$BGE = P / (P + R) \bullet 100 \qquad (5.2).$$

When the sink-link question was first posed, microbial ecologists thought that the growth efficiency of bacteria was high, on the order of 50%. Growth efficiencies of fungi were also thought to be high. However, results from new experiments with natural microbial communities indicated that the growth efficiency was much less than 50%, ranging from 15% in the oceans to 35% in estuaries (Fig. 5.13). Even less is known about growth efficiencies of microbes in soils, aside from respiration of simple compounds like glucose and acetic acid (Six et al. 2006; Herron et al., 2009). The few studies found higher efficiencies for soil microbes than for aquatic bacteria. There is little reason to believe that fungi and bacteria differ in growth efficiency, if they use the same organic material, since heterotrophic metabolic pathways are the same in both. In any case, growth efficiencies of less than 50% mean that most of the carbon is released as CO_2 and little remains in biomass available to be eaten and passed on to higher trophic levels. So, the low growth efficiency estimates indicate that the microbial loop is mainly a sink.

Still, the microbial loop is also a link, transferring otherwise unavailable material and energy, starting as DOM or complex detritus, to larger organisms and higher

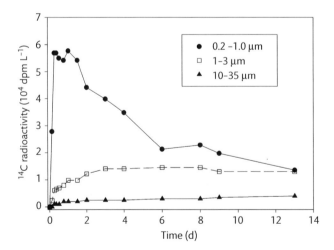

Figure 5.12 The fate of ^{14}C-glucose added to a mesocosm. Note the large amount of ^{14}C radioactivity (given here as dpm) in the small size fraction (0.2 to 1.0 μm) and the small amount in the larger size fractions, indicating that most of the organic carbon was assimilated by bacteria and then respired rather being transferred to larger organisms and higher trophic levels. This experiment indicated that the microbial loop is a sink. Data from Ducklow et al. (1986).

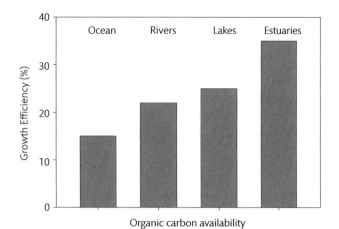
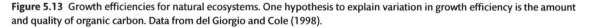

Figure 5.13 Growth efficiencies for natural ecosystems. One hypothesis to explain variation in growth efficiency is the amount and quality of organic carbon. Data from del Giorgio and Cole (1998).

trophic levels. The link percentage is similar to the percentage of C transferred by other, traditional food webs. This question has been examined by adding $^{14}CO_2$ or ^{14}C-glucose to separate incubations of lake water and then tracing the ^{14}C into large zooplankton (Wylie and Currie, 1991). The labeled glucose traces transfers by the microbial loop as just discussed while $^{14}CO_2$ is used to follow carbon fixed by primary producers and then transferred by a traditional grazer food chain. When normalized to the initial ^{14}C uptake, roughly equal amounts of ^{14}C ended up in the large zooplankton, suggesting transfer of bacterial carbon to large organisms was similar to the transfer of phytoplankton carbon. The key is the number of trophic levels and transfer steps before the top of the food chain is reached (Berglund et al., 2007). It does not matter whether those steps are taken by bacteria and other microbial loop components or by metazoans. The effect of the number of trophic levels on transfer up food chains is discussed again in Chapter 7.

Hydrolysis of high molecular weight organic compounds

Even after detritus and plant litter is broken up by metazoans, organic compounds may need to be reduced in size even further before use by microbes. Organic compounds larger than about 500 Da must be transformed somehow to smaller compounds which can then be transported across cell membranes. This transformation usually consists of hydrolysis of polymers to monomers; hydrolysis, which literally means "lysis by water", is the breaking of bonds that link monomers together into a polymer. For example, hydrolysis of protein releases amino acids and oligopeptides, but not CO_2 nor NH_4^+. Hydrolysis is often said to be the rate-limiting step or the slowest reaction in the degradation pathway, one piece of evidence being that concentrations of polymers are higher than that of monomers in natural environments.

The 500 Da cut off for transport is largely set by the capacity of membrane proteins to ferry substrates from the environment across the membrane and into the cell. The 500 Da limit applies to other organisms, not just bacteria. Although metazoans and protists can capture and retain food in a digestive tract or food vacuole, they still must use enzymes to hydrolyze high molecular weight (HMW) compounds. In metazoans and protists, these enzymes are excreted into digestive tracts and food vacuoles where they work on the HMW compounds making up the bulk of their food. The released <500 Da compounds are then transported into the protist cell or into cells lining the metazoan digestive tract.

Several types of enzymes, collectively called hydrolases, are needed to hydrolyze polymers found in the HMW pool (Table 5.4). Specific enzymes are necessary for each biopolymer, with the enzyme name usually containing the polymer name, such as cellulases for cellulose and proteases for protein. For most polymers, effective hydrolysis requires enzymes that work on different parts of the polymer chain. The breakdown of protein is a good example (Fig. 5.14). Protein must first be hydrolyzed by exoproteases, which cleave off amino acids or dipeptides (two amino acids) at the ends of the polypeptide, and endoproteases, which cleave the peptide chain far from the ends. Exoproteases can be further divided into those that work at the N terminus (aminopeptidases) or at the C terminus (carboxypeptidases) of the peptide chain. Any oligopeptides must then be hydrolyzed further by peptidases, although this hydrolysis step may be inside the cell if the oligomer is <500 Da, roughly a pentapeptide. Finally, the monomers can be used to synthesize new polymers or they are catabolized to provide energy. Note that only during catabolism of monomers, the final step in biopolymer degradation, is organic carbon oxidized to CO_2 and nitrogen mineralized to NH_4^+.

Enzymes that catalyze the initial hydrolysis of biopolymers into low molecular weight (LMW) by-products must be located outside the outer cell membrane, hence their name "extracellular enzymes". Another term is ectoenzymes. Analogous to the food vacuole of protists and the digestive system of metazoans, releasing extra-

cellular enzymes to the outer environment is an effective strategy for bacteria and fungi in biofilms or in particulate detritus. Likewise for bacteria and fungi in a soil aggregate. In these cases, the released enzyme has a good chance of reaching the targeted biopolymer and in turn, the LMW by-products cannot diffuse away before uptake by the cell originally releasing the enzyme.

Free-living bacteria in aquatic environments must use a different mechanism. In these environments, any LMW by-products resulting from biopolymer hydrolysis would diffuse away from the cell synthesizing and releasing the hydrolase. Other cells may "cheat" and utilize the LMW by-products without incurring the cost of enzyme synthesis. The released enzyme itself would be a good carbon and nitrogen source for other microbes. Rather than releasing enzymes, free-living bacteria in natural environments seem to have these enzymes somehow attached or tethered to the outer membrane. Nearly all biopolymer-hydrolyzing enzymes are cell-associated and little enzyme activity is found in the dissolved reservoir in aquatic environments. There are times, however, when activity of some enzymes in the dissolved phase is high for unknown reasons.

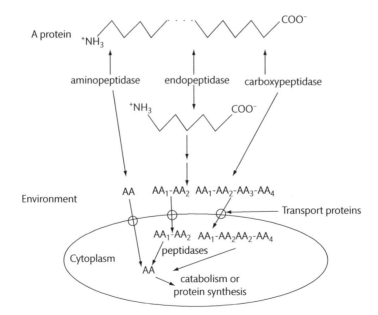

Figure 5.14 Example of the enzymes needed to degrade high molecular weight organic material. AA is a free amino acid.

Table 5.4 Some polymers, associated hydrolases, and fluorogenic analogues used to study hydrolyase activity.

Biopolymer	Hydrolyase	Analog*
Proteins	Leucine aminopeptidase,	Leu-MCA
Chitin, glycoproteins	N-acetyl-β-D-glucosaminidase	MUF-N-acetylglucosamine
Peptidoglycan	Lysozyme	MUF-N- tri-N-acetyl-β-chitotrioside
Chitin	Chitinase	MUF-N- tri-N-acetyl-β-chitotrioside
Organic phosphate	Phosphatase	MUF-phosphate
Cellulose	Cellulase	MUF-β-D-cellobioside
Polysaccharides with alpha-linkage	α-D-glucosidase	MUF- α-D-glucoside
Lipids	Lipases	Various

*MCA= methylcoumaryl; MUF= methylumbelliferyl

Lignin degradation

One of the most abundant types of HMW organic material is lignin. Although composed of characteristic compounds (Fig. 5.8), lignin is not a polymer with regular, repeating bonds like a protein or carbohydrate. Consequently, it is broken down by a mechanism quite different from how other biopolymers are degraded. One key to lignin degradation is the production of hydrogen peroxide (H_2O_2) by a variety of mechanisms, such as the excretion of aldehydes which are oxidized by extracellular enzymes to hydrogen peroxide. This highly reactive compound then serves as a co-substrate for several enzymes, such as lignin peroxidase, manganese-dependent peroxidase, and copper-dependent laccase, to attack lignin. The exact details of lignin degradation remain unclear.

In soils, white rot fungi are the main degraders of lignin, with *Phanerochaete chrysosporium*, belonging to the homobasidiomycetes, being the best-studied example (Cullen and Kersten, 2004). The name includes "white" because degradation of the brown, lignin-rich parts has the net effect of bleaching the wood. In contrast, brown rot fungi focus mainly on the white parts rich in cellulose and hemicelluloses, leaving behind the darker, lignin-rich components. Studies using both radioactive ^{14}C and stable ^{13}C indicate that lignin carbon is not used for biosynthesis by fungi, nor is it likely to be broken down to generate energy, given that lignin degradation occurs extracellularly. Rather, white rot fungi appear to degrade lignin to gain access to more easily degraded cellulose and hemicelluloses in wood detritus.

Bacteria are not important in degrading lignin in soils, and no bacterium has been isolated so far that completely degrades wood (Li et al., 2009). Fungi are probably superior degraders of wood and lignin specifically because of their enzymes and hyphal-growth form. However, bacteria may be more important than fungi in aquatic environments where their sheer numbers give them an advantage. This question has been examined by following the fate of lignin or cellulose of ligno-cellulose complexes labeled with ^{14}C in incubations with added inhibitors that act against either bacteria or fungi (Benner et al., 1986). Bacteria account for what little lignin degradation occurs in anoxic environments where fungi and most other eukaryotes cannot survive.

Uptake of low molecular weight organic compounds: turnover versus reservoir size

After hydrolysis or the breakdown of large compounds, the next step in organic material degradation is the assimilation of monomers and other LMW compounds. These compounds could come from microbial hydrolysis of biopolymers, but monomers and other LMW compounds can also be released by plant roots in soils and by phytoplankton and zooplankton in aquatic environments. We know the most about the fate of free amino acids and glucose. These compounds have been examined extensively because proteins and polysaccharides are large components of cells and of the known fraction of organic material. In addition, the use of amino acids and some sugars can be followed easily because their concentrations can be measured by high pressure liquid chromatography (Fig. 5.15) and they are available labeled with ^{13}C, ^{14}C, or [3]H.

If judged by concentrations alone, LMW compounds would not seem very important in fueling microbial

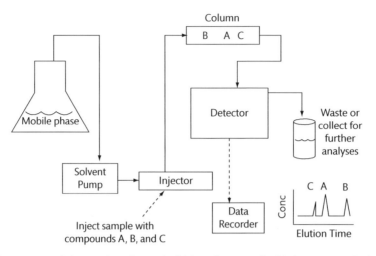

Figure 5.15 Quantifying compounds in complex mixtures by high performance liquid chromatography (HPLC). Microbial ecologists use HPLC to estimate concentrations of one or more compounds in mixtures of several compounds. As with all types of chromatography, the basic principle is that compounds differ in their affinity for the solid material in the column versus the solvent or mobile phase carrying the compounds. This difference in affinity results in differences in the time (elution time) that the compound is retained in the column. Because the small bead size results in high pressure, sometimes the "p" in HPLC means "pressure".

growth and in overall degradation of organic material. However, in spite of low concentrations, the flux of amino acids and other monomers can be quite high. Flux refers to both production and uptake, which are equal at steady state ($dS/dt = 0$). The change in a compound (or substrate, S) over time is:

$$dS/dt = P - \lambda \bullet S \qquad (5.3)$$

where P is the production rate and λ the turnover rate constant. The units of flux combine the units of both concentration (mass per unit area or volume, such as nanomol liter^{-1}) and of the turnover rate constant (per time, such as per day). In spite of very low concentrations, turnover is fast enough to result in very high fluxes (Fig. 5.16). Microbial ecologists often use the inverse of turnover rate constants, the turnover time, to quantify the relationship between fluxes and reservoir size. Geochemists use "residence time" for the same concept.

The turnover time of LMW compounds like amino acids can range from minutes to hours, even at the high concentrations found in soils (Jones et al., 2009a). The end result is that the flux of free amino acids or of glucose alone can support a high fraction, sometimes all of bacterial growth in natural environments (Kirchman, 2003). More generally, low concentrations of a compound may

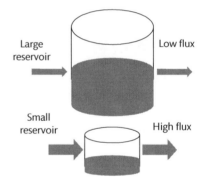

Figure 5.16 Relationship between reservoir size and fluxes. Not shown are cases in which a low flux is due to a small reservoir and a high flux due to a large reservoir.

result from low production, but they may also result from high production and rapid use by microbes.

Chemical composition and organic material degradation

So, decomposition seems to be a process of cutting compounds down in size from large to small. Molecular size is certainly important but it is not the only property that determines degradation rates by microbes. Chemical

Box 5.1 How long is the turnover time?

Most microbial ecologists think that the turnover time is the time required for reservoir contents to be completely used, to be turned over once. This definition is close but not quite correct. To obtain a more accurate picture of turnover time (τ), consider the fate of a tracer (R) added to a reservoir, in this case a dissolved compound. If flux of this tracer follows first-order kinetics, then

$$dR / dt = -\lambda \bullet R$$

where λ is the first-order rate constant with units of per time and is equal to the inverse of the turnover time ($\lambda = 1/\tau$). The solution to this equation is:

$$R_t = R_0 e^{-\lambda t}$$

where R_0 is the initial amount of added tracer. Note that R never goes completely to zero until t → ∞. So the reservoir is never "turned over" completely. It can be shown that the turnover time is the time for about 63% of the reservoir contents to be removed.

composition has a large effect. We probably know the most about how degradation varies with the type of molecules found in the detritus from higher plants (Fig. 5.17). The LMW compounds quickly leach from plant litter and are easily degraded, resulting in fast turnover times, as discussed above. Next to go are simple carbohydrates such as starch, a major storage compound in plants, consisting of α 1,4 glucose. Most proteins are also easily degraded, although some, such as keratin found in hair, are not. Cellulose is another glucose-containing polymer but with β 1,3 glycosidic bonds, making it harder to degrade than starch. Still, it is used more quickly than those compounds making up wood, primarily lignin. Lignin slows the degradation of cellulose and other biopolymers in wood by impeding access by hydrolytic enzymes.

We can draw some generalizations from studies of the plant litter degradation and of organic pollutants about how chemical structure affects degradation rates. In general, the bonds of large, naturally occurring polymers with many branches are difficult for microbes to hydrolyze. Also difficult to degrade are compounds with many aromatic and heterocyclic rings, the prime example being again lignin. Many organic pollutants in natural environments also contain aromatic components, making them persistent and potentially toxic to larger organisms.

One example is polycyclic aromatic hydrocarbons (PAH), which are produced when petroleum is not completely burnt

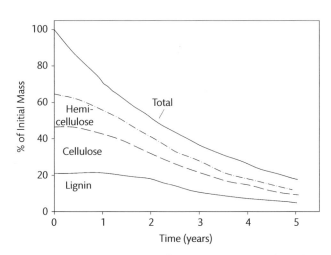

Figure 5.17 Decomposition of various chemical components of litter. The example is of litter from Scots pine needles, but the general trends apply to other types of litter. Modifed from Berg and Laskowski (2006).

and oxidized to carbon dioxide. Two factors affecting PAH degradation are worth mentioning here. First, the addition of moieties such as -Cl, $-NH_2$, or -OH, often leads to slower degradation rates and less bacterial growth. Second, experimental work has shown that the degree of aromaticity has an impact on PAH degradation. For example, naphthalene with only two aromatic rings is used rather easily by microbes, whereas chrysene with four rings is not. There is much concern about contamination by HMW PAHs that can persist in the environment in spite of microbial degradation and photochemistry (see below).

Other than those few generalizations, microbial ecologists and geochemists know surprisingly little about the relationships between chemical structure and degradation rates. Part of the problem is the lack of information about the chemical make-up of naturally occurring organic material and the complexity of microbial communities. Rather than detailed information about chemical structures, geochemists often look at gross properties, such as lignin amounts and the C:N and C:H of the organic material. Degradation tends to be faster, for example, with low C:N and C:H, the latter being an index of the oxidation state of a compound. But there are many exceptions to these generalizations.

Release of inorganic nutrients and its control

To complete the degradation and mineralization of organic material, LMW compounds are transported across cell membranes by specific transport proteins. Once inside the cell, the compounds are fed into various parts of central metabolism and used either for biomass synthesis or energy production via respiration, depending on the growth efficiency. If used for energy production, the carbon is eventually oxidized to CO_2, and other elements can be released. Excretion of ammonium, phosphate, and other inorganic compounds during organic matter degradation is the traditional role assigned to bacteria and fungi in natural ecosystems. As pointed out before, some compounds, such as ammonium and phosphate, can also be assimilated and used for biomass synthesis. Whether microbes release or take up compounds like ammonium is governed by elemental ratios and the bacterial growth efficiency. The specific case for ammonium is discussed in Chapter 12. While uptake of ammonium and of other inorganic nutrients occurs, the net

effect of mineralization is the release of these inorganic nutrients, making them available to primary producers in aquatic and terrestrial ecosystems. Primary production runs on mineralization by heterotrophic microbes.

Consequently, it is important to understand the factors controlling rates of mineralization. Biogeochemists have explored how factors such as temperature and inorganic nutrient concentrations affect various indices of organic matter mineralization, such as oxygen consumption, carbon dioxide production, and the release of ammonium. Microbial ecologists take a different view of the same problem by examining how these factors affect microbial growth, as discussed in Chapter 6. The two approaches usually give the same answer, if the bacterial growth efficiency is constant. Suffice it to say that the concentration and quality of organic material and temperature have large impacts on mineralization and growth rates. Oxygen concentration is another important factor. Oxygen is most the important electron acceptor for organic matter mineralization as long as concentrations remain above about 5 µM (Stolper et al., 2010), below which other electron acceptors take over, if they are available (Chapter 11).

Photo-oxidation of organic material

Microbial ecologists usually assume that detrital organic material is degraded by biotic processes mediated by microbes. However, one abiotic factor—light—can contribute substantially to degradation. In addition to direct effects on microbes (Chapters 3 and 4), light can affect detritus and DOM itself. Light affects non-living organic material by the same biophysical mechanisms affecting organic compounds in microbes and other organisms. DOM that absorbs light is called colored or chromophoric DOM (CDOM). Like the rest of the DOM reservoir, the composition of CDOM is not entirely known, but it is thought to be dominated by aromatic compounds and other compounds with alternating double-bonds. These types of compounds are common in terrestrial organic material, and waters receiving high inputs of terrestrial material, such as tea-colored ponds and small lakes, have high CDOM concentrations. Some CDOM is also produced by phytoplankton-based food webs.

Regardless of its source, CDOM is studied intensely by oceanographers using data from satellites to estimate phytoplankton biomass, primary production, and other properties of the oceans that can be deduced from ocean color. Optical oceanographers and limnologists are interested in CDOM because it can account for a very large fraction of the attenuation of all light in water.

Microbes undoubtedly contribute to the degradation of CDOM, but light appears to be more important. In the example given in Figure 5.18, lake DOM was incubated in the dark or with natural sunlight for over two months, and DOC and CDOM concentrations were measured periodically. In this experiment, CDOM was rapidly degraded in the light but hardly at all in the dark. By the end of the experiment, the CDOM exposed to sunlight was bleached out and was not measurable. Total DOC concentrations also decreased more so in the light than in the dark; about 40% was degraded in the light versus 10% in the dark. The light effect on total DOC degradation is very large in this experiment, because of the large amount of terrestrial DOM and other CDOM susceptible to degradation by light. The drop in total concentrations in this experiment implies that DOC is photo-oxidized to CO_2, and indeed this is the main by-product of photo-oxidation. Another gas released by photo-oxidation is carbon monoxide (CO), which is used by microbes, even though it is nearly as oxidized as carbon dioxide. Photochemistry can also lead to the production of labile compounds that are quickly used by microbes. These include carbonyl compounds, mainly small fatty acids and keto-acids, as well as ammonium and free amino acids from DON (Bushaw et al., 1996).

Refractory organic matter

Microbes are amazingly effective at degrading organic compounds, including exotic ones made by industrial processes. Yet a very small amount of primary production does in fact escape immediate degradation. This small fraction has built up over geological time, resulting in soils and oceans now having large reservoirs of organic carbon compounds that are hundreds to thousands of years old. Studies using ^{14}C-dating found that about 50% of the DOC in the surface layer and nearly all of it in the deep ocean is ancient, with some components dated at being 12 000 years old (Hansell et al., 2009). Also according to ^{14}C-dating, the estimated age of refractory organic carbon in soils ranges from about 300 to over 15 000 years, depending on the extraction method and geological setting (Falloon and Smith, 2000, Trumbore, 2009). The mechanisms preserving organic material are not completely understood. Microbes may be involved in helping to produce refractory organic material in both soils and aquatic systems (King, 2011, Ogawa et al., 2001). Adsorption onto clay

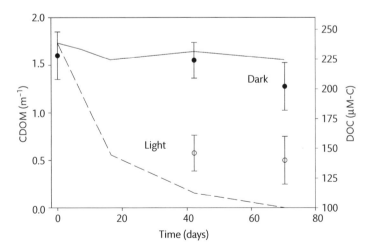

Figure 5.18 Degradation of lake DOC and CDOM in the light and dark. The solid and dashed lines are CDOM concentrations measured by absorbance (300–700 nm). The filled-in and open points are DOC concentrations. Data from Vahatalo and Wetzel (2004).

particles in soils and lakes or onto diatom frustules in all aquatic systems can protect otherwise labile compounds from degradation by microbes. But other mechanisms are needed to explain how compounds can survive for millennia.

Regardless of how it is formed, refractory organic carbon is a large and important component of the carbon cycle (Fig. 5.1). Even a small change in this large reservoir has huge effects on levels of atmospheric carbon dioxide, with equally large implications for climate change.

Summary

1. Bacteria and fungi account for 50% or more of total respiration in the biosphere.

2. Bacteria are much more abundant than fungi in aquatic habitats, whereas in soils both are about as abundant and important in degrading organic material.

3. Detritus food webs consist of many organisms that feed on detritus and associated microbes as carbon and energy sources. Although contributing little directly to carbon mineralization, detritivores and other eukaryotes break up detritus and increase surface area for attack by microbes.

4. Biopolymers and large organic compounds must be broken down by hydrolysis to compounds smaller than about 500 Da. Hydrolysis is carried out by a complex suite of enzymes (hydrolyases) specific for the biopolymer and sometimes location within the biopolymer.

5. Compounds like lignin with many different types of chemical bonds are difficult to degrade whereas polysaccharides and most protein are easily degraded by microbes.

6. A very small fraction of the organic material synthesized by primary producers is not degraded by heterotrophic microbes, resulting in the storage of carbon away from atmospheric CO_2 and allowing for the net production of oxygen.

CHAPTER 6

Microbial growth, biomass production, and controls

In the previous chapter we learned about the degradation and the mineralization of organic material back to its original inorganic constituents, most notably carbon dioxide. The chapter pointed out the importance of heterotrophic microbes, such as fungi in soils and bacteria in all environments, in carrying out this degradation. In many aquatic ecosystems and soils, heterotrophic microbes are responsible for a large fraction (50% or more) of this degradation and thus they consume an equally large fraction of primary production. Microbes degrade organic material to support their survival and growth with the evolutionary goal of passing on their genes to future generations. So, to understand organic material degradation, we need to understand microbial growth and what controls it. Also, growth and production along with standing stock are fundamental properties of populations in nature. This chapter will discuss these properties.

Here we focus on heterotrophic bacteria and fungi in oxic environments. But many of the topics discussed here are also relevant to other microbes and anoxic environments. In anoxic systems, however, oxygen and other electron acceptors have to be considered first, as often these control growth of anaerobic microbes (Chapter 11).

Are bacteria alive or dead?

The high abundance of bacteria was an important discovery back in the 1970s when epifluorescence microscopy was first applied to natural samples. The question

then became, are these cells really active and alive? It is possible that the observed degradation of organic material is mediated by a small number of live bacteria and other microbes and that most of the cells visible by epifluorescence microscopy are dead or dormant. Questions about the metabolic state of bacteria were raised back in the 1960s in part because it was known that the number of bacteria that grew up on agar plates (the plate count method) was much smaller than the "direct count" estimate from epifluorescence microscopy (Chapter 1). Although the usual cultivation methods may miss bacteria (which we know is true), it seemed also possible that the difference could be due to dead bacteria, 99% of the total or more.

We now know that the extreme estimate—99% dead—is not correct, but the actual number of alive and dead cells for a given environmental sample is rather hard to pin down. Part of the difficulty is that microbial cells can be in different states of "activity" (Fig. 6.1), ranging from truly dead cells, which never could be resuscitated, to microbes that are actively metabolizing and dividing. These metabolic states can be explored by a variety of single cell methods, methods that examine each cell rather than bulk properties of the entire community. In the end, the estimate of the number of active or inactive cells depends on the method and what aspect of microbial activity is being examined.

Microautoradiography is one single cell method of activity first used in microbial ecology to examine the uptake of ^{3}H-thymidine by cells growing in coastal marine waters (Brock, 1967). Here is how it works. A radi-

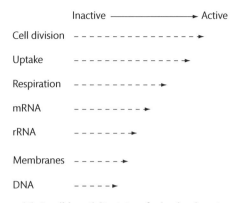

Figure 6.1 Possible activity states of microbes in nature, illustrating how the definition of "activity" depends on what is being measured. Cell properties at a particular level depend on those below it. All actively dividing heterotrophic cells take up organic compounds, respire, and so on. All cells taking up compounds may not be dividing, but are respiring and synthesizing mRNA and so on down the ladder. Inspired by a diagram in del Giorgio and Gasol (2008).

Box 6.1 **Photography before the digital age**

A photograph once was a product of chemistry, rather than of electronics. In the pre-digital age, a photograph resulted from chemical reactions driven by light hitting compounds in an emulsion that coated photographic paper. Microautoradiography uses a similar emulsion to detect radioactivity. While regular photography uses visible light energy, microautoradiography uses energy in the particles given off by the decay of unstable, radioactive elements. Nearly all of the radioisotopes used in microautoradiography and other approaches in microbial ecology are quite safe. The isotopes most commonly used by microbial ecologists include ^{3}H, ^{14}C, ^{35}S, and ^{33}P, all of which emit low energy beta particles.

---

olabeled organic compound, ^{3}H-amino acids, for example, is added to a sample, incubated for a few hours, and then filtered, or the cells collected by centrifugation. The microbes are placed into photographic film emulsion. After an exposure time ranging from hours for highly active samples to days for relatively inactive samples, the film emulsion is developed, the microbes are stained for DNA, and the sample is viewed with epifluorescence microscopy. Cells that have taken up ^{3}H-amino acids have silver grains associated with them (Fig. 6.2). These silver grains arise from the decay of ^{3}H which produces beta particles that strike compounds in the photographic emulsion.

The relative number of active bacteria detected by this method varies from <10% to 50% or even higher, depending on the environment and radiolabeled compound. This is a large fraction, given that microautoradiography probably detects only cells synthesizing new biomass because the radiolabeled compound has to be incorporated into biomass if a cell is to be scored as being active. So, a large fraction of cells are active to some extent in natural environments. Microautoradiography also has been used to examine which microbes—bacteria or phytoplankton—assimilate organic material. Studies using microautoradiography, along with other methods, demonstrated that uptake of dissolved organic material

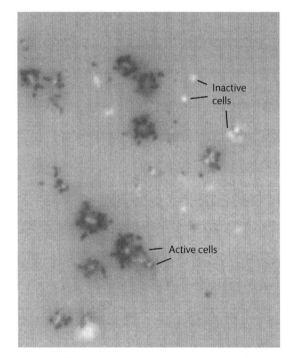

Figure 6.2 Example of microautoradiography showing cells with silver grains that have taken up ^{3}H-amino acids. The white dots are bacteria stained with the DNA stain, 4′,6-diamidino-2-phenylindole (DAPI). The dark areas around active cells are silver grains.

(DOM) is dominated by heterotrophic bacteria in aquatic habitats (Chapter 5).

Other methods have been used to examine the metabolic state of bacteria (del Giorgio and Gasol, 2008). Several approaches take advantage of the counting and sorting capabilities of flow cytometry to examine the number of active cells in a sample, even from soils (Shamir et al., 2009).

Activity state of bacteria in soils and sediments

The same questions about whether bacteria are active or not are relevant to thinking about soils and sediments. However, the physical complexity of these environments creates problems. There are many practical problems in trying to assay single cells within the complex matrix of detritus and inorganic particles, and there are conceptual problems in dealing with the range of possible activity states of bacteria inhabiting the many microhabitats in a single sample. A bacterium on one side of a particle may be quite active while another on the other side may not. In spite of these difficulties, a few generalizations can be made.

Bacteria in moist soils and sediments seem to be as active as those in the water column of aquatic habitats. For example, one study found that about 50% of bacteria were actively respiring in the rhizosphere of pine seedlings (Norton and Firestone, 1991). Likewise, over 50% of bacteria in many soils have ribosomes (Eickhorst and Tippkotter, 2008), a minimum requirement for protein synthesis and general cellular activity. In contrast, one of the few studies to use microautoradiography in sediments found that <10% of bacteria took up acetate (Carman, 1990), a key organic compound in anaerobic systems (Chapter 11). Likewise, <5% of the bacteria in an aquifer used naphthalene (Rogers et al., 2007), although that low percentage may not be surprising. As mentioned in Chapter 5, naphthalene is a pollutant produced during the incomplete burning of fossil fuels.

Bacteria and other microbes in dry soils are largely inactive, but their response after the addition of water is revealing. Respiration is negligible in dry soils, but then increases within minutes after the addition of water, a response called the "Birch effect" after H.F. Birch who first described it (Birch, 1958). Growth rates of both bacteria and fungi increase also, although not as much or as

fast as respiration, and fungi do better after re-wetting than bacteria (Bapiri et al., 2010). The implication is that microbes are completely inactive or dormant in dry soils but are not dead, as they are capable of rapidly responding to the improved environmental conditions.

Activity state of soil fungi

It may not make much sense to ask what fraction of soil fungi is active because of the hyphal growth form of many soil fungi. For this reason and because of technical difficulties in working with soils, few microautoradiographic studies have examined activity of fungi in soils (Bååth, 1988). However, microbial ecologists can gain insights into the activity state of fungi by examining the portion of hyphae that is filled with cytoplasm (Fig. 6.3). Fungi are capable of moving cytoplasm out of regions without adequate resources to support cytoplasmic metabolism in more favorable microhabitats, leaving behind inactive, empty, or evacuated hyphae. The proportion of cytoplasm-filled hyphae to total hyphae is then one index of the activity state of the fungal community, analogous to the fraction of bacterial cells that are positive in an activity assay.

The activity state of soil fungi is similar to that we have seen for bacteria. Cytoplasm-filled hyphae make up 10–50% of total hyphal length, with the exact percentage varying with many environmental properties (Klein and Paschke, 2004). For example, cytoplasm-filled hyphae are more prevalent near plant roots and other sources of

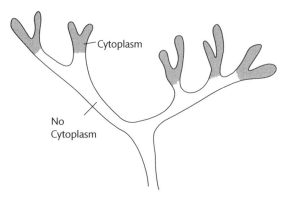

Figure 6.3 A network of filled and empty fungal hyphae. Cytoplasm has moved into the tips, leaving behind a empty sheath of the hyphae. Adapted from Klein and Paschke (2004).

organic material. The percentage of cytoplasm-filled hyphae vary with plant species and, as is the case for bacteria, with treatments that stimulate growth, such as the addition of organic carbon, ammonium, or water (Klein et al., 2006). Physical disturbances also can change the distribution and fraction of cytoplasm-filled hyphae.

Microbial growth and biomass production

As microbial ecologists were addressing the question about the number of active cells, it became clear that we did not know how fast bacteria were growing in natural environments. Even if all bacteria were alive, their growth still could be very slow. The same question applies to fungi in soils. Microbial ecologists need information about growth rates and biomass production of the entire bacterial or fungal assemblage to understand the role of these microbes in material and energy fluxes. Many metabolic processes carried out by microbes "scale" with growth rate. When growth is fast, so too is the process. Growth rate is a basic property of organisms in nature.

Before we examine growth in natural environments, let us review some basic parameters and definitions. These parameters are summarized in Table 6.1.

Growth of pure cultures in the lab: batch cultures
Microbes growing as a single species in the laboratory provide two models for growth in nature: batch cultures and continuous cultures. The simplest model, a batch culture, consists of growth in a fresh medium in a closed environment, such as a laboratory flask. When inoculated into new medium, growth usually does not begin

immediately but only after a delay of a few hours, depending on the bacterial strain and how different the environment is from that which the strain was growing in previously (Fig. 6.4). This delay is called the lag phase. Once microbes start to grow, they enter into the log or exponential phase during which abundance increases exponentially. The change in bacterial numbers (N) as a function of time (t) is:

$$dN/dt = \mu N \qquad (6.1)$$

where μ is the specific growth rate (sometimes called the instantaneous growth rate) of the bacterial population. Growth rates in pure cultures are calculated from the slope of ln(N) versus time; "ln(N)" is the natural log of N or 2.30·log(N). The change in numbers or biomass (dN/dt) is equal to biomass production. The solution to Equation 6.1 is:

$$N_t = N_0 e^{\mu t} \qquad (6.2)$$

where N_t is the number of cells at time t and N_0 is the initial abundance (t = 0). Note that the units for μ are per time; for example, for rapidly growing lab cultures, convenient units are per hour whereas they would be per day for microbial assemblages growing more slowly in nature.

Parameters related to the growth rate (μ) include the turnover time of the population ($1/\mu$) and the generation time (g), both of which having units of time (e.g. hours or days). The generation time is defined as the amount of time required for a population to double. That is,

$$2N_t = N_t e^{\mu g} \qquad (6.3)$$

which yields, after some algebra:

Table 6.1 Terms for basic parameters of microbial biomass and growth.

Parameter	Symbol	Units[a]	Method
Cell numbers	N	cells liter^{-1}	Microscopy, flow cytometry
Biomass	B	mgC liter^{-1}	Cell numbers, biomarkers
Growth rate	μ	d^{-1}	From production and biomass
Biomass production	BP	mgC liter^{-1} d^{-1}	Leucine incorporation, others
Generation time	g	Days	From the growth rate
Growth yield	Y	cells liter^{-1}	Cell numbers or biomass
Growth efficiency	BGE	Dimensionless	Various

[a] These units are most appropriate for water samples. For sediments and soils the analogous units would be per gram of dry weight. In soils, total biomass, not just carbon, may be measured. Also, for both soils and aquatic habitats, the parameters can be expressed per unit area, such as m^{-2} rather than per liter or per gram.

$$g = \ln(2)/\mu = 0.692/\mu \qquad (6.4).$$

Even in pure cultures, some resource becomes limiting, and growth slows down and eventually stops completely. At this point, the culture enters the stationary phase (Fig. 6.4). In some cultures, some cells may continue to grow, but others die and lyse. The end result is the same; abundance is constant over time in the stationary phase. The entire growth curve depicted in Figure 6.4 is often said to be sigmoid. It can be described by the logistic equation:

$$dN/dt = r \cdot N/(K-N) \qquad (6.5)$$

where r is the specific growth rate and K the maximum population size or carrying capacity of the environment. Note that when N is small relative to K, Equation 6.5 becomes similar to Equation 6.1. The symbols r and K are part of terms used to define two types of selection pressures faced by organisms: r-selection and K-selection.

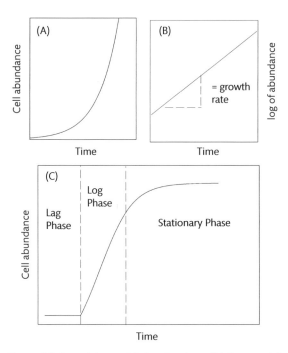

Figure 6.4 Bacterial growth in batch culture. (A) Exponential growth with no lag period or stationary phase; (B) Also exponential growth, but cell numbers are plotted on a log scale; (C) Growth phases of bacteria growing after a lag phase before the log or exponential phase. After growth-limiting substrate (usually organic carbon) is used up, the culture reaches the stationary phase when cell numbers do not change.

The terms, r-selection and K-selection, were originally derived for large eukaryotes colonizing a new habitat. The initial colonizers are r-selected and grow rapidly to take advantage of free space and new habitats. As the carrying capacity of the new habitat is reached, rapid growth is no longer favored, but rather K-selected organisms with traits for surviving crowded conditions win out. Traits of r-selected organisms allow them to flourish in unstable environments where growth conditions change rapidly, preventing the build-up of dense populations. In contrast, K-selected organisms dominate stable environments with invariant growth conditions that promote dense populations. The concepts from large organism ecology can be useful to thinking about microbes in some environments. Some bacteria, for example, are adapted to grow rapidly when organic concentrations are high, like r-selected organisms colonizing a new habitat. Another term used for these bacteria is "copiotroph". Other, K-selected bacteria are adapted to grow slowly on low concentrations in stable environments. These bacteria are referred to as being "oligotrophs".

Growth of pure cultures in the lab: continuous cultures
The key feature of a batch culture is that it is a closed system with no inputs or outputs; the inoculum is exposed to only one dose of growth substrates at the beginning, and any waste by-products excreted during growth are not removed, except for gases. In contrast to this model of microbial growth, microbes in a continuous culture are provided fresh medium continuously and the old medium—along with waste products and cells—are removed at the same rate. A chemostat is a continuous culture in which the concentrations of all chemicals are constant. Although all chemostats are continuous cultures, a continuous culture is not necessarily a chemostat.

Continuous cultures can be quite elaborate and sophisticated, but the basic design is simple (Fig. 6.5). To start off, a reaction chamber is inoculated with microbes and is allowed to operate in batch mode initially; at first there are no inputs or outputs as the microbes multiply. Then new, sterile medium is pumped into the reaction chamber at a fixed rate, and the medium in the reaction chamber is pumped out at the same rate in order to maintain a constant volume within the reaction chamber. Initially, abundance decreases when the pump is turned on, but then microbial abundance increases as they take advantage of the new media and start to grow. These oscillations continue until a steady-state is reached when abundance is constant. At this point, it can be shown that:

$$\mu = D \tag{6.6}$$

where D is the dilution rate, defined by:

$$D = f/V \tag{6.7}$$

where f is the flow rate (with units such as liters per h) and V the volume of the reaction chamber (liters). The dilution rate has the same units (e.g. per h) as the growth rate.

Equation 6.6 is a very simple but powerful statement about growth. It says that growth is set by the dilution rate, which is under control of the investigator. It also says that growth rates are independent of the supply and concentration of organic material in the continuous culture. However, the concentration of organic material, along with the growth efficiency, sets biomass levels.

Continuous cultures provide a different model of growth in nature than batch cultures. Like continuous cultures, microbial abundance is mostly constant over time in nature because growth is balanced by removal: the outflow in the case of continuous cultures, mortality

caused by grazing and viral lysis in nature. The implication is that over some space and timescales in some environments, microbial communities are in a quasi-steady-state. Growth conditions may change, but perhaps not on timescales relevant to microbes. On the other hand, if growth conditions do change on relevant timescales, a batch culture may be a more accurate description of microbial growth. It may apply to phytoplankton during the early stages of a spring bloom, for example, when nutrient concentrations are high and mortality is low (Chapter 4).

Neither batch nor continuous cultures are perfect models for growth in nature. But both provide useful terms and concepts for examining the processes controlling microbial standing stocks and production in natural environments.

Measuring growth and biomass production in nature

Measuring growth rates in the lab is usually very easy because the rate is calculated simply from the change in abundance or biomass over time in batch cultures or by knowing the dilution rate in continuous cultures. In nature, however, microbes occur in complex communities, and growth is usually balanced by mortality caused by grazing and viral lysis. Microbial abundance and biomass are usually quite constant over time and space, even though other data, like numbers of active cells, indicate that microbes must be growing. So, changes in cell numbers or biomass over time in nature, in the absence of any manipulations to minimize mortality, tell us nothing about growth rates.

Table 6.2 summarizes the methods that have been proposed over the last 30 years for measuring bacterial growth and production in aquatic ecosystems. A few of these same methods have been used to measure bacterial growth in soils. The two most commonly used methods, both in soils and in aquatic environments, are based on thymidine and leucine incorporation (Fuhrman and Azam, 1980, Kirchman et al., 1985). The two methods are quite similar. Thymidine, which is one of the four nucleotides in DNA, is used to trace DNA synthesis whereas leucine, an amino acid, is used to trace protein synthesis. Dividing cells must make more DNA and thus incorporate more thymidine as they grow. Similarly, fast-growing

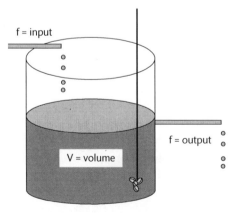

Figure 6.5 A simple continuous culture. The rate (f) at which new medium is added (input) must equal the rate at which medium from the reaction chamber flows out (output). The flow rate can be controlled by simple gravity or by pumps. Gases can be introduced to help with circulation and to provide oxygen or other compounds.

cells make more protein and thus incorporate more leucine than slow-growing cells. The same basic idea is used for estimating fungal growth, except that the starting radiolabeled compound is ^{14}C-acetate. After incubation, the common fungal sterol, ergosterol, is isolated and radioassayed for the incorporated ^{14}C (Rousk and Bååth, 2007, Newell and Fallon, 1991); this approach is called the "acetate-in-ergosterol" technique.

Bacterial biomass production in aquatic environments

The methods mentioned above have been used to estimate biomass production of heterotrophic bacteria in a great variety of aquatic environments, ranging from small ponds to entire oceans. Production rates are useful for evaluating the general importance of heterotrophic bacteria in ecosystems and for exploring what controls production and biomass levels. The most important observation is that bacterial production usually correlates with primary production; higher primary production leads to higher bacterial production (Fig. 6.6). But there is much variation in this relationship. Sometimes the correlation between bacterial and primary production is very high, indicating a tight "coupling" between the two microbial processes, while in other habitats and times, there is no significant relationship. Microbial ecologists often say that bacterial production and primary production are coupled over large spatial and temporal scales but not over small ones.

Another important observation is about the magnitude of bacterial production compared with primary production and the ratio of the two rates (BP:PP). This ratio is a measure of the importance of heterotrophic bacteria and the rest of the microbial loop in consuming primary production. The BP:PP ratio varies greatly over time and space, but usually it is low in the open oceans, about 0.1, whereas sometimes it is as high as 0.3 to 0.5 in lakes. The ratio is higher in lakes in part because of the input of terrestrial organic carbon. Consequently, small lakes which are affected more by terrestrial organic carbon often have higher BP:PP ratios than large lakes. Because of terrestrial organic carbon, the BP:PP also can be high in estuaries. At the other extreme, the BP:PP ratio for the Arctic Ocean and Antarctic seas is often low (<0.05).

A BP:PP ratio of 0.1 or less may not seem impressive, but its significance becomes clear when it is coupled with bacterial growth efficiency (BGE). Remember that BGE is:

$$BGE = BP/(BP+R) \qquad (6.8)$$

Table 6.2 Some of the methods used to estimate biomass production by bacteria and other microbes.

Method	Principle	Comments
$^{14}CO_2$ fixation	Light-dependent fixation of CO_2 into biomass	Targets autotrophs
Dark $^{14}CO_2$ fixation	Light-independent CO_2 fixation due to anaplerotic processes	Variable relationship between CO_2 fixation and total biomass production
Frequency of dividing cells (FDC)	Frequency of paired cells about to divide increases with growth rate	Variable relationship between FDC and growth rate
^3H-adenine incorporation	Adenine is used in RNA synthesis. rRNA synthesis scales with growth rate	tRNA and mRNA synthesis may not scale with growth
^{35}S-sulfate incorporation	Sulfate is used in protein synthesis which scales with growth	Hard to measure in seawater
^{14}C-acetate-in-ergosterol	Acetate is used for ergosterol synthesis which is coupled to growth	Targets fungi
Dilution or filtration	After minimizing grazing and viral lysis, increase in biomass is followed	Labor-intensive and intrusive
^3H-thymidine (TdR) incorporation	TdR is used in DNA synthesis which scales with growth	See text
^3H-leucine (Leu) incorporation	Leu is used in protein synthesis which scales with growth	See text

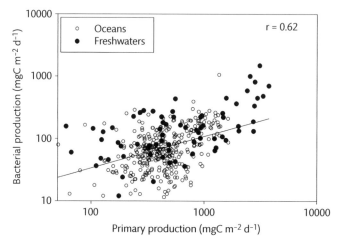

Figure 6.6 Bacterial production versus primary production in a variety of aquatic environments. The correlation coefficient (r = 0.62; n = 441; p<0.001) and least-squares line are from the entire data set of oceans and freshwaters, but here only environments with primary production greater than 50 mg C m^{-2} d^{-1} are shown for clarity. Data provided by Eric Fouilland, taken from Fouilland and Mostajir (2010).

where R is respiration. If we define total bacterial carbon demand (BCD) as the sum of both production and respiration, then:

$$BCD = BP/BGE \qquad (6.9).$$

We can now relate bacterial growth and the total use of organic carbon (BCD) by heterotrophic bacteria to primary production with data on production rates and bacterial growth efficiency. Table 6.3 gives the total impact of heterotrophic bacteria on carbon flows through aquatic environments.

These data once again indicate the importance of heterotrophic bacteria in processing primary production.

Although the open oceans tend to have lower BP:PP ratios, these are offset by low BGE, leading to the observation that about 65% of primary production is routed somehow through DOM and heterotrophic bacteria. This percentage is roughly the same as that estimated from respiration alone (Chapter 5). Studies of other marine habitats and lakes arrive at the same percentage but with higher BP:PP ratios and BGE values. A thorough analysis of these data suggested even higher BP:PP ratios and fluxes through DOM and heterotrophic bacteria than indicated in Table 6.3 (Fouilland and Mostajir, 2010). The exceptions are the Arctic Ocean and Antarctic seas. In these perennially cold environments, the extremely

Table 6.3 Average production rates for phytoplankton (primary) and heterotrophic bacteria (bacterial), the ratio of primary production to bacterial production (BP:PP), bacterial growth efficiency (BGE), and the % of primary production consumed by heterotrophic bacteria, calculated from BP/PP divided by BGE. The production data are from Figure 6.6 and the BGE values are from Figure 5.13.

Environment	Production rates		BP:PP	BGE	% of Primary Production
	Primary	Bacterial			
	(mgC m^{-2} d^{-1})				
Open ocean	1000	98	0.10	0.15	65
Arctic and Antarctica	1063	17	0.02	0.15	11
Other marine waters	780	179	0.23	0.35	66
Lakes	1385	224	0.16	0.25	65

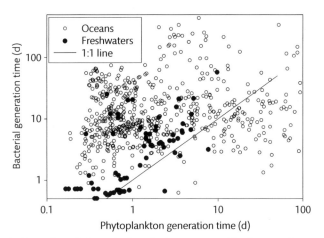

Figure 6.7 Generation times for bacteria and phytoplankton in the oceans and freshwaters. Data provided by Eric Fouilland, taken from Fouilland and Mostajir (2010).

low BP:PP ratios are not offset by equally low BGE values, leading to a low fraction of primary production routed through DOM and heterotrophic bacteria.

Bacterial growth rates in aquatic environments
It is difficult to estimate growth rates in nature for the same reasons why it is difficult to measure biomass production; microbes live in complex communities with growth usually being balanced by mortality. A few approaches, such as the frequency of dividing cells and the amount of ribosomal RNA per cell, yield estimates of growth rates directly. Other approaches use estimates of bacterial production and standing stocks (cell abundance or biomass); that is, the growth rate (μ) is bacterial production divided by cell abundance or biomass. A problem with this approach is that the calculated growth rate is a composite of all microbes in the sampled community potentially growing at quite different rates, ranging from zero (dead or dormant cells) to potentially high values. While the approach has its flaws, the estimates still give a good general picture of the timescale on which microbes grow in natural environments.

Data from aquatic habitats, which is where this approach has been used the most, illustrate a general point about microbes in most natural environments. Figure 6.7 plots generation times of heterotrophic bacteria versus generation times of the phytoplankton for various aquatic habitats. (Phytoplankton growth rates can be calculated from primary production and chlorophyll data, similar to the approach used for bacteria). These data indicate that heterotrophic bacteria grow relatively slowly in aquatic habitats, usually on the order of days, much longer than the generation time of bacteria growing in the lab where bacteria can double every 30 minutes. The record for a lab culture is less than 10 minutes, held by the marine bacterium *Vibrio natriegens*, orders of magnitude faster than the month or longer generation time of bacteria growing in polar waters in the winter. Even in warmer waters, however, bacteria grow much slower than possible under optimal conditions in laboratory experiments.

Another point to be learned from Figure 6.7 is that heterotrophic bacteria often grow more slowly than phytoplankton. According to the data given in Figure 6.7, generation times for phytoplankton range from one day in freshwaters to over 20 days in the oceans (averages for each environment), whereas the same averages for bacteria are four and 34 days. Some of these very long generation times are suspicious and should be discounted, but the conclusion is unaffected. The vast majority of points are about the 1:1 line in Figure 6.7, implying slower growth rates for bacteria. The exception may be some estuaries where bacterial generation times are quite short and bacteria seem to be growing much more quickly than phytoplankton. One explanation is that estuaries are often light-limited, preventing rapid phytoplankton growth. Similarly, growth of heterotrophic bacteria in freshwater lakes is more similar to rates for phytoplankton than seen in the oceans. Bacteria may grow quickly at the expense of organic carbon from terrestrial sources. Even so, bacterial growth is slower than phytoplankton in these lakes.

There is no simple explanation for why bacteria grow more slowly than phytoplankton in most aquatic habitats. The difference in growth rates implies differences in how these two functional groups of microbes are controlled and regulated. In waters without external inputs, bacteria depend on phytoplankton for organic carbon, so they have to grow more slowly given the high biomass levels of bacteria. In contrast, phytoplankton can take advantage of inorganic nutrients from both grazers and bacteria. Top-down controls may also differ.

Growth rates of bacteria and fungi in soils

Similar to work on bacteria in aquatic ecosystems, microbial ecologists have addressed questions about growth of bacteria and fungi in soils using the same methods introduced above: leucine or thymidine incorporation

(or both) for bacteria, and the acetate-in-ergosterol technique for fungi.

Bacteria appear to grow faster than fungi in soils and aquatic systems (Fig. 6.8). Early studies emphasize the very fast potential growth rates of bacteria with generation times on the order of an hour, tenfold faster than for fungi (Coleman, 1994). Studies using the modern methods mentioned above found much slower growth rates for both microbial groups, but still faster rates for bacteria than for fungi. In sandy loam, for example, fungi have generation times of over 100 days (Rousk and Bååth, 2007), about tenfold slower than the typical growth of bacteria in soils (Bååth, 1998). The few direct comparisons also indicate that bacteria grow faster than fungi (Buesing and Gessner, 2006). Fungal growth is also slow in aquatic environments, similar to rates in soils (Gulis et al., 2008, Pascoal and Cassio, 2004, Newell and Fallon, 1991). Some soil microbial ecologists have concluded that bacteria grow more slowly in soils than in aquatic habitats (Bååth, 1998), but this hypothesis needs more data for soils (Fig. 6.8). In any case, the available data support the general hypothesis about slow and fast carbon pathways mediated by slow-growing fungi and fast-growing bacteria, as discussed in Chapter 5.

Remember that these growth rates are for the entire community of bacteria and fungi being sampled. Studies

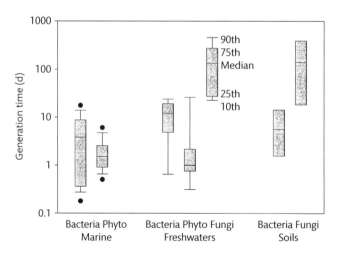

Figure 6.8 Generation times for bacteria, phytoplankton ("phyto") and fungi in soils and aquatic habitats. 10th, 25th, 75th, and 90th refer to the percentiles containing the data. The points for the marine data are the 5th and 95th percentiles. Data from Carter and Suberkropp (2004); Cole et al. (1988); Demoling et al. (2007); Gulis et al. (2008); Kirchman et al. (2009); Rousk and Bååth (2007); Rousk and Nadkarni (2009).

of growth rates in soils have not attempted to distinguish saprophytic from mycorrhizal fungi. Growth by these two fungal groups probably differs greatly because of the large differences in their environments. More generally, the physical-chemical environments of soils differ greatly for microbes over very small scales (Chapter 3), leading to huge heterogeneity in growth of both bacteria and fungi.

What sets biomass production and growth by microbes in nature?

The growth rates of bacteria and fungi are much lower than that which can be achieved in laboratory cultures. What then prevents these microbes from growing faster in nature? For phototrophic microbes, we saw in Chapter 4 that the answer is fairly simple: light and the supply of inorganic nutrients, such as compounds containing nitrogen, phosphorus, and sometimes iron. For heterotrophic microbes, the answer is more complex. Here we focus on bottom-up factors, leaving top-down factors for future chapters.

Temperature effects on growth and carbon cycling
Of all bottom-up factors, temperature is arguably the most important. Chapter 3 discussed how temperature affects all chemical reactions and rates of processes in nature, and microbial growth rates are no exception. As a general rule of thumb, the Q_{10} of growth rates is about 2, although it varies, of course. The precise value for this temperature effect is important, especially for soil ecosystems. Many studies have examined how soil respiration and organic material decomposition may respond to predicted changes in temperature due to global warming (Davidson and Janssens, 2006). The problem is very important in the Arctic where warming by only a few degrees may melt permafrost and may not only release organic carbon that can be mineralized to carbon dioxide, but also lead to higher fluxes of methane, a potent greenhouse gas, to the atmosphere (Dorrepaal et al., 2009). At the ecosystem level, respiration of the soil community has a Q_{10} of 1.4 even though controlled experiments typically lead to much higher estimates of Q_{10} (>2) (Mahecha et al., 2010). It has been argued that the experiments exclude too many important variables found in nature.

Temperature also affects the growth of bacteria in temperate aquatic environments. Often, bacterial biomass production correlates the best with temperature rather than other properties, such as dissolved organic carbon (DOC), chlorophyll, or primary production. One example is Narragansett Bay, Rhode Island (Fig. 6.9). In this environment, temperature ranges from –1 to nearly 23 °C while biomass production varies by over one hundredfold (Staroscik and Smith, 2004). The correlation between the two parameters was high during this study (r = 0.70), whereas in contrast, there was no significant correlation with chlorophyll, which is often used as a proxy for the supply of organic carbon. The investigators examining this system concluded that temperature was the most important factor controlling bacterial biomass production. However, the relationship between temperature and production varied during the year, and the Q_{10} implied by the field data was much higher than the Q_{10} estimated in experiments when only temperature is varied, suggesting that other factors also affected bacterial production and growth. Soil microbial ecologists have also concluded that high Q_{10} values indicate that factors other than temperature are at work (Davidson et al., 2006)

The Narragansett Bay study is one example of a problem often faced by microbial ecologists who need to use correlations to examine functional relationships between, in this case, microbial growth and temperature. The problem is that correlations do not necessarily imply causation. As mentioned in Chapter 3, in temperate environments, temperature varies greatly along with other ecosystem properties, such as light, primary production, and biomass, all potentially affecting growth. So, temperature may correlate significantly with bacterial production in part because temperature co-varies with another, hidden property of the ecosystem that also affects bacterial growth.

In addition to examining variation within an ecosystem, we can examine how much growth can be explained by differences in temperature among ecosystems, the biggest difference being between polar systems and low latitude waters. Growth rates of bacteria are low in the perennially cold waters of the Arctic Ocean and in Antarctica's Ross Sea and are higher in the slightly warmer subarctic Pacific Ocean and the North Atlantic Ocean (Fig. 6.10). But this increase is substantially more (about tenfold more) than

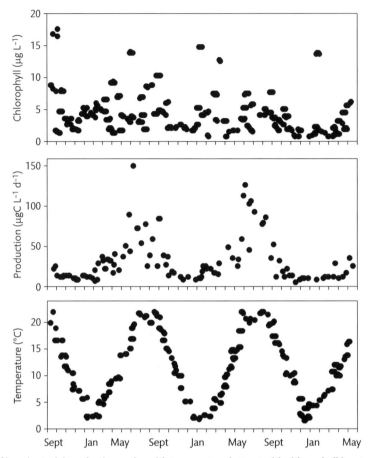

Figure 6.9 Example of how bacterial production varies with temperature but not with chlorophyll in a temperate environment, Narragansett Bay, Rhode Island. Data from Staroscik and Smith (2004).

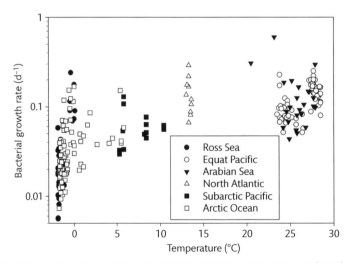

Figure 6.10 Growth rates of bacteria in various oceanic ecosystems. Data from Kirchman et al. (2009).

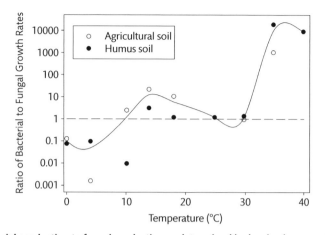

Figure 6.11 Ratio of bacterial production to fungal production as determined by leucine incorporation (bacteria) and acetate-into-ergosterol (fungi). The dashed horizontal line indicates when the ratio of growth rates is one. Data from Pietikainen et al. (2005).

what is predicted from the known response of bacteria to temperature, that is a Q_{10} of about 2. What is more, rates do not get much higher in the warm waters of the equatorial Pacific and Arabian Sea; growth rates remain at about 0.1 d^{-1} (generation time of about seven days), albeit with much variation, for temperatures ranging from 5 to 28 °C. This leveling-off of rates with temperature has also been seen in freshwaters and estuaries.

So, temperature is important, but not necessarily in all ecosystems, and its effect can be overestimated. Even in polar systems, there is evidence that bacteria are in fact adapted to cold temperatures and that growth rates are low for other reasons, most likely low concentrations and supply rates of DOM.

Temperature effects on fungi versus bacteria in soils
Since temperature explains much of the variation in soil respiration, it is likely that it is equally powerful in examining the variation in bacterial and fungal growth rates in soil ecosystems, just as is the case for aquatic habitats. Although soil microbes never experience the extremely hot waters of a hydrothermal vent, temperatures can drop to low levels in soils; there is evidence of bacterial activity even in permafrost colder than −39 °C (Panikov et al., 2006). What may be especially important is the difference in how fungi and bacteria respond to temperature.

We saw in Chapter 3 that bacteria can grow in much hotter water than eukaryotes. This difference between prokaryotes and eukaryotes in temperature tolerance holds true for bacteria and fungi in soils. The optimal temperature for bacterial growth in agricultural and forest soils is about 5 °C warmer than that of fungi (Pietikainen et al., 2005). But fungi do better at the other end of the temperature scale and can grow in soils 4–5 °C colder than can bacteria; one study calculated temperature minima of −12 and −17 °C for bacteria and fungi, respectively (Pietikainen et al., 2005). Consequently, the ratio of bacterial biomass production to fungal biomass production increases with temperature in soils (Fig. 6.11). These results are consistent with the observation that fungi dominate soils in winter but less so in summer when bacterial biomass is higher. Likewise, the ratio of bacterial biomass to fungal biomass is lower in snow-covered soils than in uncovered soils (Schadt et al., 2003).

Limitation by organic carbon
The concentration and supply of organic material are often the most important factors determining the growth of heterotrophic bacteria and fungi in both soils and aquatic systems. As mentioned in Chapter 5, concentrations of organic material and of especially labile components are very low in nature, which explains why growth

rates of heterotrophic microbes are usually far lower in nature than seen in the laboratory. One line of evidence for carbon limitation in aquatic systems comes from studies comparing rates of bacterial biomass production with rates of primary production. As indicated in Figure 6.6, there is an overall correlation between bacterial and primary production in lakes and the oceans. The easiest way to explain this correlation is that primary production determines directly or indirectly the supply of DOM and detritus which in turn drives heterotrophic bacterial activity. Any change in primary production leads to a change in the DOM supply with consequences for heterotrophic bacteria. Few analogous data from soils and for fungal growth are available. There is a correlation between organic matter content and fungal growth (Rousk and Nadkarni, 2009) and between soil respiration and primary production (Sampson et al., 2007), all evidence for organic carbon limitation of soil bacteria and fungi.

Another line of evidence indicating carbon limitation of heterotrophic bacteria is based on addition experiments. In these experiments, organic compounds are added to incubations of water or soil, and microbial production is followed over time. Often bacterial and fungal growth is higher in incubations with the organic compounds than in the no-addition control in experiments with samples from soils and aquatic habitats (Bååth, 2001, Demoling et al., 2007). The addition of organic carbon usually stimulates growth more so than the addition of inorganic nutrients, such as ammonium or phosphate, but there are important exceptions as discussed below.

Both the concentration and the supply rate are important in thinking about limitation by organic carbon and other elements. The relationship between concentrations and growth rates is described by the Monod equation:

$$\mu = \mu_{max} \cdot S/(K_s + S) \qquad (6.10)$$

where μ is the growth rate, μ_{max} the maximum growth rate, S the concentration of the growth-limiting substrate, and K_s the substrate concentration at which the growth rate is half of the maximum (Fig. 6.12). Notice the similarities between the Monod equation and the Michaelis-Menten equation (Equation 4.11). Equation 6.10, however, sometimes does not describe growth in

nature. Growth rates can be low when concentrations are high, such as in early spring in temperate aquatic habitats (Chapter 4), implying a concentration-growth relationship opposite that of the Monod equation. This paradox also applies to limiting substrates other than organic carbon. In some cases, another factor limits growth. Another possibility is that the microbes have not caught up to the high concentrations and have not responded with high growth rates. In this case, growth rates will become high soon after concentrations are high.

In addition to quantity—concentrations and supply rates—of organic material, intuitively, one would think that "quality" of the organic components would have an impact on growth rates of heterotrophic bacteria and fungi. In fact, there is little direct evidence from field studies for this reasonable hypothesis. We do know that degradation rates vary with substrate quality (Chapter 5), implying that heterotrophic microbes grow faster on organic components such as protein and simple polysaccharides rather than on lignin, for example, but the effect of more subtle differences in organic material quality on growth is not clear. Conclusions drawn from studies of degradation in nature may be complicated by differential temperature effects (Craine et al., 2010). In any case, differences in the quality of organic material are likely to lead to variation in growth rates even if concentrations or supply rates of organic material are similar.

Box 6.2 Freedom fighter and microbiologist *par excellence*

The Monod equation is named after Jacob Monod (1910–1976) who won the Nobel Prize (along with his compatriots, François Jacob and André Lwoff) for work on the *lac* operon in *E. coli*. This operon was the one of the first models of gene regulation at the transcription level. Before his work in microbiology, Monod was a member of the French Resistance that fought against the German occupation of France during World War II (1939–1945).

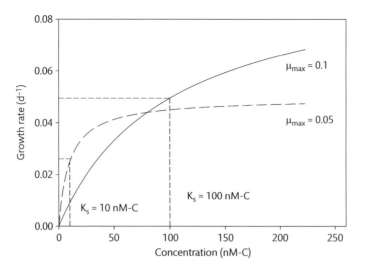

Figure 6.12 The Monod equation describing growth rates by two competing species as a function of limiting substrate concentrations.

Limitation by inorganic nutrients

There are some interesting exceptions to the general rule of organic carbon limitation. The concentration of many inorganic nutrients potentially used by microbes is low in soils, lakes, and the oceans, raising the possibility of these compounds limiting growth of heterotrophic bacteria and fungi. Phosphate seems to limit bacterial growth in the Sargasso Sea and the Mediterranean Sea, based on addition experiments (Fig. 6.13) and on high ratios of carbon to phosphate and nitrogen to phosphate in dissolved compounds, exceeding the ratios in bacteria (Chapter 2). Primary production in both seas is also thought to be limited by phosphate, unlike the general rule of marine waters being limited by nitrogen. The N_2-fixing cyanobacterium *Trichodesmium* is abundant in the Sargasso Sea and may alleviate nitrogen limitation in that system. Parts of the Gulf of Mexico also can be phosphate-limited because nitrogen limitation is alleviated by nitrogen inputs from the Mississippi River.

While a few studies found evidence of heterotrophic bacteria being limited by phosphate, fewer studies have reported that addition of ammonium or other inorganic nitrogen compounds alone stimulates bacterial growth (Church, 2008). This work raises two questions: why is heterotrophic growth generally limited by organic carbon and not by inorganic nutrients? And why is phosphate limitation more common than nitrogen limitation?

One answer is that organic carbon is used by aerobic microbes for both biomass synthesis and respiration, but N and P are used only for biomass synthesis. In Chapter 12, we will see that given typical C:N ratios for the organic material used by microbes and for microbial biomass, both bacteria and fungi should excrete ammonium, not assimilate it, implying that these microbes are not limited by inorganic nitrogen. A similar argument can be built for C:P ratios and phosphate use versus excretion. Another answer involves competition for these inorganic nutrients between the heterotrophic microbes and autotrophic microbes in aquatic systems and higher plants in terrestrial systems. In Chapter 4, we learned that small cells such as heterotrophic bacteria with their high surface area to volume ratios should outcompete large phytoplankton and higher plants for ammonium, phosphate, and other dissolved compounds. However, uptake of inorganic nutrients by heterotrophic microbes eventually would lead to lower growth of autotrophic organisms and lower production of organic material, resulting in organic carbon limitation of the heterotrophs.

The other question is about why phosphate limitation of heterotrophic bacteria is more common than nitrogen limitation. The answer may be that heterotrophic bacteria are exceptionally phosphorus-rich and have very low C:P ratios in waters like the Sargasso Sea, so they need lots of phosphorus for growth. Few data are available to test this

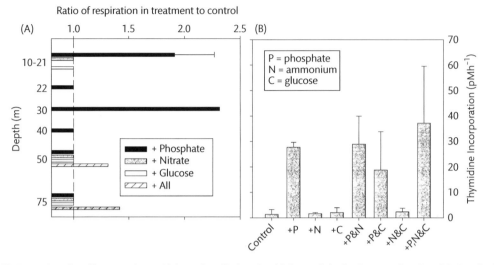

Figure 6.13 Examples of studies to explore which nutrient limits microbial growth in the Sargasso Sea. Panel A: Respiration of the total community was measured in bottles receiving the indicated compounds and in other, untreated bottles over 24 hours. Panel B: Thymidine incorporation was measured with and without the indicated compounds after 24–48 h. Data from Obernosterer et al. (2003) and Cotner et al. (1997).

idea and it may not even be true in some environments (Cotner et al., 2010). Another part of the answer may lie in the biochemicals containing nitrogen and phosphorus in microbes. As mentioned in Chapter 2, nitrogen is mainly in protein, which generally is not degraded and synthesized ("turned over") independent of growth in bacteria, whereas phosphorus is in nucleic acids, lipids, and nucleotides, some of which (e.g. mRNA and ATP) turn over rapidly.

Co-limitation and interactions between controlling factors

Microbes have adapted to live on very low concentrations of many compounds in natural ecosystems, so it can be overly simplistic to focus on a single limiting factor. We see the consequences of these low concentrations in addition experiments where often the addition of both an organic compound and inorganic nutrient stimulates bacterial biomass production more so than the addition of either compound alone. For example, in high nutrient-low chlorophyll oceans (Chapter 4), addition of iron along with an organic carbon source stimulates bacterial production more so than either alone (Church et al., 2000, Kirchman et al., 2000). Some authors call this co-limitation by organic carbon and iron, but it

seems likely that in these experiments, iron just became the next limiting factor, once the addition alleviated organic carbon limitation.

There are several clearer examples of co-limitation for microbes where the limiting factors are physiologically linked (Table 6.4). For example, microbes may be prevented from using nitrate, and thus are limited by nitrogen, because low iron levels interfere with nitrate reductase, an iron-containing enzyme essential for reducing nitrate to ammonium and in so doing making nitrate available for biomass synthesis. Nitrogenase, the critical enzyme for N_2 fixation, is another enzyme that requires iron as a co-factor. Several enzymes require other trace metals, such as cobalt and zinc (Table 6.4), which occur in very low concentrations, especially in the open oceans. These cases are clear examples of co-limitation because one compound or element is required for acquisition of the other.

Two important examples of co-limitation involving temperature should be mentioned. Growth of microbes in polar environments may be co-limited by organic carbon and temperature. One physiological link between the two factors is that low temperature causes stiff membranes and impedes transport of dissolved compounds. According to this hypothesis, higher DOM concentrations are needed

Table 6.4 Some cases of co-limitation of microbial growth by at least two bottom-up factors. Based on Saito et al. (2008).

Microbe	Primary factor	Secondary factor	Comments
Photoautotrophs	Light	Nitrate	Nitrate use requires energy
All microbes	Nitrate	Iron	Nitrate use requires iron-containing nitrate reductase
All microbes	Phosphate	Zinc	Alkaline phosphatase requires zinc
All microbes	Nitrogen (urea)	Nickel	Urease requires nickel
Diazotrophs	Phosphate	Iron	Nitrogenase requires iron
Bacteria	Organic carbon	Temperature	
Soil microbes	Organic carbon	Water	

for a heterotrophic microbe to grow in cold water at the same rate as in warmer waters. In soil microbial ecology, there has been much discussion about whether the sensitivity of organic matter degradation to temperature, as measured by Q_{10}, varies with organic material quality (Knorr et al., 2005, Fang et al., 2005). The other example of co-limitation involving temperature is the interaction between it and water content in controlling microbial activity in soils. We know that addition of water can increase bacterial growth rates in soils (Iovieno and Bååth, 2008), while a glucose addition may not, implying water limitation of growth. There has been more work examining how respiration and decomposition in soils may be affected by both water and temperature (Howard and Howard, 1993). Warmer temperatures alone would stimulate decomposition and presumably microbial growth in soils, but it also leads to more evaporation and less moisture, which potentially limits microbial activity. As mentioned before, the confounding effects of moisture complicate efforts to estimate Q_{10} for soils and to its use in global models to predict the response of terrestrial ecosystems to global warming (Davidson and Janssens, 2006).

Competition and chemical communication between organisms

So far, we have discussed the abiotic factors controlling microbial growth without reference to the abundance of these microbes or of other organisms. In some cases, these factors are referred to as being density-independent, because their effect does not vary with microbial abundance. Temperature is a good example. Predation, on the other hand, is a density-dependent factor because it does vary with predator and prey abundance (Chapter 7). Many abiotic factors are density-independent, but

not all. Physical space or room, for example, may limit microbial growth in a soil micro-environment or in a biofilm. Soil moisture is a product of both density-independent factors, such as the frequency and intensity of rain events, and density-dependent factors, such as the retention of water by microbially produced extracellular polymers within the soil matrix.

Competition is another important density-dependent factor. We have already discussed competition between small and large microbes, such as heterotrophic bacteria competing with eukaryotic phytoplankton for inorganic nutrients. In Chapter 4, competition was examined with the Michaelis-Menten equation describing transport of dissolved nutrients and other compounds, but it can also be viewed in terms of the Monod equation (Fig. 6.12). Depending on the substrate concentration, a microbe with low K_s and high μ_{max} will win over another microbe with high K_s and low μ_{max}. Bacteria and fungi potentially compete for the same organic substrates, and we know that their growth and biomass vary often in opposite directions (Table 6.5), suggesting that the two microbial groups are interacting. However, it is also possible that the two are not competing, but are just responding differently to the same factor. More convincing evidence for competition has come from more direct experiments.

The experiments consist of following bacterial and fungal growth after adding or removing fungi, or by adding inhibitors of bacterial activity (Rousk et al., 2008). In the latter case, stimulation of fungal growth was inversely correlated with inhibition of bacterial growth by the inhibitors (oxytetracycline, tylosin, and bronopol). These experiments show that bacteria affect fungi in a density-dependent fashion, a strong sign of direct competition between the two microbial groups for the same growth-limiting organic substrates. Bacteria and fungi compete with each other in spite of evi-

Table 6.5 Summary of factors affecting bacteria and fungi in soils. The positive effects are indicated by the various number of "+" while "-"and "−"indicate negative and strongly negative impacts.

| Factor | Impact on | | Reference |
	Bacteria	Fungi	
Moisture	+++	++	Bapiri et al. (2010)
Temperature	+++	++	Pietikainen et al. (2005)
Acidity	−	++	Rousk et al. (2009)
Disturbance	++	+	Six et al. (2006)
Metals	−	+	Rajapaksha et al. (2004)
C: N *	-	+	Six et al. (2006)

* Fungi are favored in environments or in experiments with organic material with high C:N ratios. Similarly, addition of nitrogen sometimes leads to lower bacterial growth in soils.

dence that the two microbial groups differ in their capacity to degrade various organic compounds (Chapter 5) and to grow on these compounds (Steinbeiss et al., 2009).

In addition to competition, microbes can interact via chemical cues, which affect growth as well as many other aspects of microbial behavior and metabolism. For example, some types of bacteria can negatively affect fungi by excreting organic compounds, one example being the polyene nystatin. The soil bacterium *Streptomyces* is famous for producing these antifungal compounds as well as other compounds, antibiotics, which work against other bacteria. However, we know little about how these antimicrobial compounds actually work in natural environments, and what happens in the lab or in the human body may not be representative of what happens in nature (Davies, 2009). For example, while nystatin is an effective drug against fungal infections, it also signals some bacteria to form biofilms (López et al., 2009). In addition to chemical warfare, microbes release various organic and inorganic compounds to communicate with themselves and with each other. One form of this communication, quorum sensing, is discussed in Chapter 14.

Summary

1. Many but not all bacteria and fungi in natural environments are actively metabolizing and growing. The state of activity varies from being dead to active cell division and biomass production.

2. Similar to primary production, biomass production of heterotrophic bacteria and fungi can be used to assess their contribution to carbon fluxes. The data are consistent with data from other approaches indicating the high flux of carbon and energy through heterotrophic microbes.

3. Growth rates of bacteria in nature are much slower than rates seen in nutrient-rich laboratory experiments. Bacteria appear to grow faster than fungi in soils and in aquatic habitats, consistent with models of slow and fast carbon pathways in soils.

4. Growth of heterotrophic bacteria and fungi is limited by the supply and quality of organic carbon in most oxic environments, although inorganic nutrients, such as phosphate, can be limiting in some habitats.

5. Temperature also has large but different effects on bacterial and fungal growth. How temperature affects these microbes has many implications for understanding climate change.

6. In addition to competing for limiting organic and inorganic compounds, microbes can directly interact via the secretion of antimicrobial compounds.

CHAPTER 7

Predation and protists

The previous chapter mentioned that growth rates of bacteria, fungi, and algae in natural habitats are generally slow compared to what is possible under optimal conditions in the lab. However, even with slow growth rates, these microbes would quickly fill up the biosphere were it not for some force that kills them off. Some microbes may self-destruct because they lack organic carbon or some limiting nutrient, but many other microbes grow, if only slowly, under the most adverse environmental conditions. Large phytoplankton cells can sink from the upper surface layer of aquatic habitats and eventually die in deep, dark waters, but many phytoplankton, bacteria, and other small microbes do not sink appreciably, nor do microbes of any size in terrestrial systems. The primary mechanism of keeping microbial populations in check is mortality by predation and viral lysis, collectively referred to as "top-down control". How much of mortality is by predators versus viruses is discussed in Chapter 8. Here we focus on predation.

Many organisms are potential predators of microbes, but the most important grazers of bacteria and algae are protists. These single-cell eukaryotes range in size from nanoflagellates nearly as small as bacteria to ciliates over a millimeter in length. The ecological roles of protists are diverse and include primary production, predation, and parasitism (Table 7.1). Small eukaryotic algae are very important in primary production (Chapter 4) while other protists account for much of the grazing on algae (including diatoms), bacteria, and other microbes. Protists have been known for centuries but by other names, such as "animalcules", a term used by Antonie van Leeuwenhoek in the seventeenth century to describe the microbes he saw with a primitive microscope in samples of his stool and of scum from his teeth. Protozoa is another term used by some microbial ecologists, but "protist" is a more appropriate term if the microbe is capable of photosynthesis or if its metabolism is unknown (see Box 7.1). The metabolism and thus ecological roles of many protists are not known mainly because they have not been isolated and grown in the lab, just like many bacteria, although cultivation-independent approaches are now being used to figure out what some protists are doing in nature. The cultivation problem for protists is discussed in

Table 7.1 Ecological roles of protists in nature. Protists other than the one given can take on these roles, and these protists can carry out other ecological functions than those mentioned here.

Ecological Role	Organism name	Comments
Primary production	Phytoplankton and algae	Many autotrophic protists are capable of grazing
Herbivory	Flagellates and ciliates	Protists are the major grazers of phytoplankton
Bacterivory	Nanoflagellates and amoeba	Many protists are capable of grazing on bacteria
Mixotrophy	Several	Mixotrophic organisms obtain energy from both phototrophy and heterotrophy
Carnivory	Ciliates	Large flagellates are also capable of eating small flagellates
Parasitism	Protozoa	Not all protozoa are parasites

Box 7.1 What is a protozoan?

Protozoa, which is the plural of protozoan, comes from the Greek for first (*protohi*) animal (*zoa*). Some microbial ecologists argue against using this term because many of these microbes do not have anything in common with animals. Some carry out photosynthesis and are far from animal-like. These ecologists prefer to use "protists" rather than protozoa. The problem is that protists include a great variety of single-cell eukaryotes, ranging from phytoplankton to heterotrophic grazers. Protozoa is an appropriate and useful term for colorless protists that are not capable of photosynthesis and carry out only heterotrophy.

Chapter 9. Here we concentrate on heterotrophic and photoheterotrophic (mixotrophic) protists.

Bacterivory and herbivory in aquatic habitats

As data came in about bacterial abundance in lakes and marine systems, microbial ecologists discovered that the number was not particularly interesting. About the same bacterial abundance was found wherever and whenever it was measured. Bacterial abundance does vary over time, especially in temperate regions with the seasons (high in summer, low in winter), but it varies less than tenfold, less than that seen in phytoplankton biomass. It is not very interesting to find the same number all the time, but the constancy of bacterial abundance is quite interesting and raises two questions: why is bacterial abundance so constant over time and space? And is there anything special about 10^6 cells per ml? Why this number and not another, radically different? Part of the answer to these questions is bacterivory; that is, the eating of bacteria by protists and other organisms.

To find out who is eating bacteria, new methods had to be developed to examine bacterivory. In contrast to the relative ease in estimating primary production and heterotrophic bacterial production, it is difficult to examine grazing on bacteria and other microbes. Consequently, no single method has emerged as the

one of choice, and all have yielded some information on rates and specific aspects of protistan biology. The methods summarized in Table 7.2 were developed mostly for aquatic habitats and can be applied to soil samples only with difficulty. Some of these methods are not just for examining bacterivory. For example, the dilution method is commonly used to estimate grazing rates on small phytoplankton, including eukaryotes, and the cyanobacteria *Synechococcus* and *Prochlorococcus* (Landry and Hassett, 1982).

Some of the methods give hints about which organisms are the main bacterivores in aquatic habitats, and a simple experiment indicated the size of these grazers (Sherr and Sherr, 1991). The experiment is to remove organisms of various sizes by filtration with filters with various pore sizes, and then bacterial abundance is followed over time. If bacteria increase over time, it implies that the main bacterivore was removed by the filtration step. Application of this approach showed that removal of large organisms such as copepods does not lead to immediate changes in bacterial abundance, implying that the main bacteriovores are still present and are not large zooplankton. However, filtering out organisms less than 5 µm in size does result in an increase in bacterial abundance over time, indicating that the main bacterivores are usually in that size range. Methods using fluorescent bacteria or fluorescent beads, which mimic bacteria, revealed that the small grazers are flagellates, often called heterotrophic nanoflagellates (HNAN or HNF), 3–5 µm in length. Some pictures of flagellates are given in Figure 7.1. These microbes earn their name by having one or more flagella that are used for locomotion through the water column and for feeding. We know that they are heterotrophic because they lack photosynthetic pigments (mainly chlorophyll) and because these microbes only survive when fed bacteria or similar prey in laboratory cultures. Other flagellates are capable of both photosynthesis and of feeding on bacteria and small microbes. Some common flagellates are listed in Table 7.3.

Another potential bacterivore, naked amoeba, are generally thought to be less abundant and less important in the water column of aquatic habitats than in soils. They are less abundant than flagellates in the water column, although they can be as abundant as ciliates (Lesen et al., 2010). The low abundance of amoeba in

Table 7.2 Methods for measuring grazing on bacteria. Class I methods follow the ingestion of a model prey while Class II methods examine the change in bacteria or other prey following a treatment. Based on Strom (2000).

Class	Method	Description	Advantages	Drawbacks
Class I	Fluorescent beads	Appearance of bacteria-sized fluorescent beads in grazers is followed over time	Specific grazers can be examined if identifiable by microscopy	Beads cannot mimic cell surface qualities of bacteria
	Fluorescent bacteria (FLB)	Similar to the bead method, but bacteria labeled with fluorescence are used instead	This method has similar advantages as the bead method	Labeling with fluorescence may change bacterial prey
	Radiolabeled bacteria	Instead of fluorescence, bacteria are labeled by allowing uptake of radiolabeled compounds	Following radioactivity can be easier than fluorescence	Not all bacteria may take up the radiolabeled compound
Class II	Dilution technique	Net growth is followed, after prey are diluted with filtered water	Approach gives both prey growth rate and grazing rate	In addition to being laborious, dilution may change growth rates
	Size fractionation	Bacterial abundance is followed over time after grazers and other large organisms are removed by filtration	This approach is easy and depends only on measuring bacterial abundance	Filtration may not remove all grazers and will not remove any viruses
	Metabolic inhibitors	Bacterial abundance is followed over time after inhibitors of bacterial growth or of grazers are added	This approach is easy and depends only on measuring bacterial abundance	The inhibitors may affect non-targeted cells (specificity problem) or may not be effective inhibiting the targeted cells

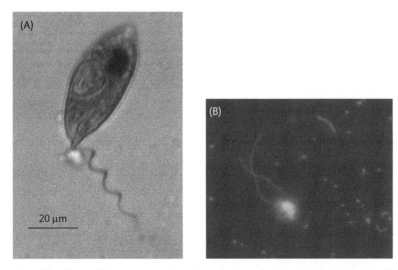

Figure 7.1 Some examples of flagellates able to graze on bacteria and other microbes. Panel A: Heterotrophic euglenoid flagellate from the Bering Sea, viewed by light microscopy. Courtesy of Evelyn Sherr. Panel B: Unidentified flagellate from the Sargasso Sea, stained with a DNA stain and viewed by epifluorescence microscopy. The small cells are bacteria about 0.5 μm. Courtesy of Craig Carlson.

Table 7.3 Some flagellates in nature. Flagellates come from many phyla spread across the Tree of Life. Dinoflagellates are discussed separately. From Sherr and Sherr (2000) and Howe et al. (2009).

Group	Example Genus	Habitat	Characteristics
Chrysomonads	*Paraphysomonas*	Freshwater and marine	Commonly isolated heterotroph
Chrysomonads	*Ochromonas*	Freshwater and marine	Mixotrophic
Euglenozoa	*Euglena*	Freshwater	Mixotrophic
Bicosoecids	*Cafeteria*	Freshwater and marine	Heterotroph with two unequal flagella
Pedinellids	*Ciliophys*	Freshwater and marine	Heterotroph with one flagellum
Choanoflagellates	*Monosiga*	Freshwater and marine	Heterotroph with one flagellum and collar
Cercozoans	*Heteromita*	Soils	Dominant heterotrophic flagellate
Cercozoans	*Bodomorpha*	Soils	Dominant heterotrophic flagellate
Parabasalia	*Trichomitopsis*	Termite gut	Hydrolyzes cellulose
Zoomastigophora	*Giardia*	Mammalian intestines	Parasite without mitochondria

the water column may reflect adaptation for growth on particulate detritus and other surfaces, which are much less common in the water column than in soils and sediments of course. At times, however, amoeba can contribute significantly to grazing on bacteria and other small microbes. More work is needed on these fragile microbes.

In addition to flagellates and amoeba, several other organisms potentially graze on bacteria and similar-sized microbes (Strom, 2000). These other organisms are not protists, but can be important in top-down control of bacteria and other microbes in some aquatic habitats. In freshwaters, non-protist bacteriovores include some zooplankton belonging to the order Cladocera, such as the genus *Daphnia*. These zooplankton feed by filtering out prey with a mesh of hair-like structures (setae) which are spaced closely enough to capture micron-sized particles, including bacteria. In addition to feeding on bacteria, *Daphnia* are important herbivores and grazers of phytoplankton in lakes and other freshwaters. In marine waters, other potential bacteriovores include gelatinous zooplankton, such as larvaceans, salps, and doliolids, all belonging to the phylum Chordata (which includes *Homo sapiens*), quite different from "true jellyfish" in the phylum Cnidaria. Larvaceans, for example, live in gelatinous houses and feed on bacteria and other small microbes by catching them in a fine-meshed, sticky, filtering structure. When the filtering mesh becomes clogged, larvaceans throw it away and build a new one, often several times a day in productive waters. Notably, larvaceans and other bacterivorous gelatinous zooplankton are much larger (millimeters to centimeters) than

flagellates (microns). This discrepancy in the size of the prey (bacteria) and predator (gelatinous zooplankton) has several implications for thinking about how material and energy move through food webs, a topic addressed below. As with much of microbial ecology, size matters.

Grazers of bacteria and fungi in soils and sediments

The main grazers of microbes in soils include protozoa, nematodes, and arthropods (Fig. 7.2). The main soil protozoa were once classified as being in two phyla, Sarcomastigophora and Ciliophora (Coleman and Wall, 2007), but Sarcomastigophora is now considered to be archaic. The taxonomy of these microbes continues to be debated (Adl et al., 2005). They still can be put into one of four ecological groups: the flagellates, naked amoebae, testate amoebae, and ciliates. Flagellates are functionally similar to those seen in aquatic habitats, eating bacteria as their main prey. They can reach densities of 10^5 cells g^{-1} in forest soils. Unlike aquatic systems, naked amoebae are very abundant and active in many types of soils, eating not only bacteria, but also fungi, algae, and even small detrital particles. Because the lack of rigid cell walls makes them very flexible, naked amoebae are able to explore small crevices and pores in soils where other grazers cannot go. In contrast to the naked amoebae, the testate amoebae have a rigid external "house" and usually are not as abundant as the naked variety, except in some forest soils. Microbes in the fourth protozoan class, the ciliates, also eat bacteria, but they are likely to be less important than flagellates as

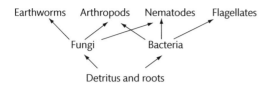

Figure 7.2 Some of the predators of bacteria and fungi in soils. The same arthropod or nematode species may not eat both fungi and bacteria. The arthropods here include mites and collembolans. Based on Chapin et al. (2002).

bacterivores because ciliate abundance is only 10 to 500 cells g^{-1} in soils, much lower than that of flagellates.

While flagellates are important grazers of bacteria, they are not effective at eating the other major soil heterotrophic microbe, the fungi (Ekelund and Rønn, 1994). Those eukaryotic microbes are grazed on by nematodes, one of the most abundant and diverse groups of multicellular organisms in the biosphere (Coleman and Wall, 2007). Along with protozoa and rotifers, nematodes live in aqueous films and water-filled pores in soils where they feed on a variety of prey, including fungi. Nematodes differ from each other by the food they eat. Some specialize on bacteria, while others graze on specific fungal taxa, including saprophytic, mycorrhizal, and pathogenic fungi (Wardle, 2006). Some nematodes are omnivores. Various nematodes have a hollow stylet, a dagger-like structure for piercing fungal hyphae or roots and root hairs. Feeding on fungi has been examined by following the appearance of fungus-specific fatty acids in nematodes (Ruess and Chamberlain, 2010). Other grazers of fungi include arthropods, such as mites and collembolans, many of which can graze on prey in air-filled pores.

In sediments of aquatic habitats, flagellates are major grazers of bacteria (Kemp, 1990), but as in soils, some meiofauna and macrofauna (Chapter 5) can also eat bacteria (Pascal et al., 2009). In a mudflat, the dominant meiofaunal grazers were found to be a foraminifer, nematodes, and harpacticoid copepods, while one macrofaunal type, a mudsnail, also ingested bacteria and probably detritus as well. In both soils and sediments, it is difficult to determine if grazing is enough to balance bacterial growth (First and Hollibaugh, 2008). Often, grazing seems low compared to bacterial growth, implying that another form of mortality accounts for top-down control in these ecosystems. The most likely other form is viral lysis. However, the many methodological difficulties in working

---

> **Box 7.2 Role reversal**
>
> This chapter on predation is the place to mention that some fungi can turn the tables on nematodes and microarthropods. There has been much research on nematophagous fungi that trap and digest nematodes for food, and there is at least one case of a fungus apparently eating collembolans (Klironomos and Hart, 2001), although usually these microarthropods feed on fungi. Fungi are suspected to attack nematodes and microarthropods for their nitrogen as well as for carbon and energy.

with samples with heavy particle loads may lead to underestimating grazing in soils and sediments.

Grazing mechanism for protists

The previous sections mentioned some of the organisms that are important grazers of bacteria and fungi. To gain more insight into these grazers and into grazers of other microbes in aquatic habitats and soils, let us consider the grazing mechanisms used by protists. Many protists feed by phagocytosis (Fig. 7.3), a process by which microbes engulf particles and digest them in a food vacuole. Understanding phagocytosis helps to explain several aspects of protistan biology and ecology. While there are many parallels between predation by protists and that by metazoans, phagocytosis is fundamentally different from how a macroscopic carnivore eats a herbivore.

The first problem faced by a protist grazing on a prey is finding and encountering it. To understand the first step, it is crucial to remember that protists and their prey live in a low Reynolds number world where viscous forces dominate, quite unlike the world of their macroscopic counterparts. As mentioned in Chapter 3, to imagine life in this world, think of swimming in molasses or hot tar. To feed in this low Reynolds number world, protists have to get water flowing past them in one direction. They achieve this unidirectional flow by moving asymmetrically (Strom, 2000). Flagellates do so by swimming in a corkscrew pattern or by moving their flagella asymmetrically. Ciliates beat their cilia like rowing a boat.

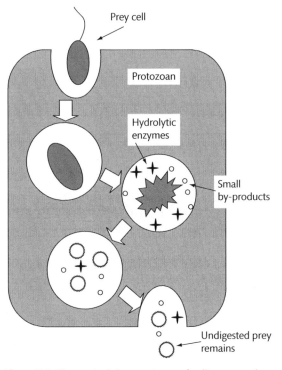

Figure 7.3 Phagocytosis by a protozoan feeding on another microbe. Adapted from Nagata (2000).

Some protists have specialized feeding apparatuses, such as the collar of choanoflagellates.

Heterotrophic protists can be classified by how they obtain their prey (Montagnes et al., 2008). Filter feeders, such as some ciliates and flagellates, produce feeding currents, while diffusion feeders, such as heliozoans, stick out stiff arm-like structures (axopods) into which prey collide. Raptorial feeders, which include some ciliates, flagellates, and naked amoebae, actively hunt and capture prey. Once captured by any of these mechanisms, the prey particle is packaged into a food vacuole, formed by the protist's outer membrane stretching around the prey particle. This process of phagocytosis is similar to what happens when a mammalian lymphocyte encounters a foreign particle. In protists, the entire process must be very efficient and fast to account for observed feeding rates at prey concentrations typically seen in nature.

Once inside the food vacuole, digestion can begin (Fenchel, 1987). This consists of the release of various extracellular enzymes, such as proteases and lysozyme (for bacterial prey), into the food vacuole to attack and break down the entrapped prey. The acidity of the food

vacuole also helps to disable the prey and to assist in digestion, analogous to the mammalian digestive system. The products from the digestion process are carried into the cytoplasm by pinocytotic vesicles, analogous to food vacuoles, except that the pinocytotic vesicles are much smaller. During the entire digestion process, the food vacuole moves around the protistan cell until digestion is completed, at which time it fuses with the protist outer membrane; in ciliates, this fusion occurs at a miniature anus, the cytoproct. The undigested contents of the food vacuole are then expelled to the outside environment. The time required by a protist to digest a prey item varies with food abundance and quality and protist growth, but is on the order of minutes to hours.

Factors affecting grazing

When confronted with changes in their prey, protistan grazers respond in two general ways. Most of this section will focus on functional responses, which are how the grazing rate responds to changes in prey abundance over a short timescale. The second way is the numerical response, which is how grazer growth rate changes in response to prey abundance. This response occurs over longer timescales (the generation time of the protist which is roughly a day) than the functional response occurring within minutes to hours. Both types of responses are important in thinking about the ecological roles of protists and indeed of all grazers in soils and aquatic habitats. The factors affecting protistan grazing are prey numbers, prey size, and chemical composition of the prey. After discussing these factors, we will consider how grazing is affected by prey defenses.

Prey number and predator-prey cycles

One of the simplest but most important factors affecting grazing is the number of prey. From first principles, we would expect grazing in the microbial world to increase as prey abundance increases because larger numbers of both prey and predator increase the chance of encounters between them. However, the rate cannot increase indefinitely due to the limit set for protists by the rate of phagocytosis. So, after increasing with prey concentrations, the grazing rate reaches a maximum at some point (Fig. 7.4). In the example given in Figure 7.4, grazing is expressed as the ingestion rate, which is the number or amount of prey taken in by each predator per unit time. Another commonly used metric is the clearance rate, which is the volume of water cleared of all particles per unit time. It is calculated by dividing the ingestion rate (prey per unit time) by prey density (prey per unit volume). The clearance rate has units of per time, such as per hour. Rather than numbers of prey, the biomass of prey, expressed as grams of carbon per unit volume, is often used to examine predation (Fig. 7.4).

The general shape of the ingestion versus prey curve is very similar to what we have seen before, such as the uptake of a dissolved compound as a function of its concentration. An equation similar to the Michaelis-Menten equation (Equation 4.11) can be written to describe the ingestion rate as a function of prey abundance. Unlike uptake, however, protists can stop feeding at low prey densities. The mathematical result is the curve crossing the x-axis in Figure 7.4 at low but positive prey numbers because the ingestion rate becomes zero even though prey are present at some threshold level. This response is in effect the end result of a cost-benefit analysis by the protist. It may cease feeding when energetic costs outweigh the benefits of grazing on scarce prey.

The existence of grazing thresholds is one answer to the question of why bacterial abundance is about 10^6 cells ml^{-1} in many aquatic environments and 10^9 cells g^{-1} in soils and sediments. These abundances may reflect grazing thresholds. Growth brings up bacterial abundance to the threshold levels, but grazing prevents these microbes from exceeding the threshold for long. Assuming they exist, the thresholds may be set at abundances of 10^6 cells ml^{-1} and 10^9 cells g^{-1} because of fundamental limitations in feeding behavior and energetics of bacterivores. Grazing thresholds can also account for why protists and other microbes can exist in nature in spite of grazing by hungry carnivores searching for food. These carnivores may go after other prey when prey numbers drop below the threshold level.

The threshold idea seems reasonable and would explain some facets of life in the microbial world, but the data supporting it are weak. As illustrated in Figure 7.4,

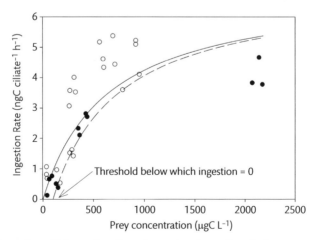

Figure 7.4 Ingestion of algal prey by a ciliate as a function of prey abundance. Two algal prey were used: *Nannochloropsis* (filled in circles) and *Isochrysis* (open circles). The solid line was determined by regression analysis of the actual data: I = 8.96•P/ (641 + P) where I is the ingestion rate and P the prey concentration. The dotted line illustrates the effect of a threshold on ingestion, though there is no evidence of it in the actual data. Data from Chen et al. (2010).

it is difficult to demonstrate thresholds experimentally because errors are large when rates are very low.

Figure 7.4 illustrates a functional response to prey abundance. The graph of the numerical response of protists to the prey would look very similar to Figure 7.4; that is, protist growth rates also increase with prey concentrations before reaching a maximum, analogous to the response of heterotrophic bacterial growth to organic carbon concentrations or algal growth to inorganic nutrient concentrations (see Figure 4.12). The equation for describing growth rates as a function of prey abundance is exactly the same as that for ingestion rate as function of prey abundance. In contrast to ingestion rates, however, it is often easier to demonstrate experimentally a threshold in prey abundance below which protist growth ceases. But a graph of protist growth versus prey abundance is a static picture of how protists respond to initial prey level. In nature, both predator and prey abundance vary continuously because of one population impacting the other.

One of the first models to describe this interaction is the Lotka-Volterra model, developed independently by the American biophysicist Alfred Lotka in 1925 and the Italian mathematical biologist Vito Volterra in 1926. This model has been used to examine all sorts of predator-prey relationships with the classic one being snow lynxes and hares in Canada. Here we will apply the model to the microbial world. The Lotka-Volterra model consists of two differential equations, with the first describing how the prey changes as a function of its growth rate (r) and a grazing rate constant (a) multiplied by the prey abundance (H) and the predator abundance (P):

$$dH/dt = r \cdot H - a \cdot H \cdot P \qquad (7.1).$$

In words, the change in prey abundance over time is equal to prey growth in prey minus predation on the prey. The second equation is for the predator:

$$dP/dt = b \cdot H \cdot P - m \cdot P \qquad (7.2)$$

where b is the growth rate of the predator and m the specific mortality rate for the predator. In words, the change in predator abundance over time is equal to predator growth minus mortality of the predator. Among many assumptions, this model assumes that the prey grows exponentially, that rates are proportional to population sizes, that the rate constants do not vary with population

size (in spite of the known numerical response), and that predation is the only ecological process at work; there is no competition, for example. The mathematics and implications of the Lotka-Volterra model have been examined in great depth. Here we concentrate on only a couple of predictions from this model.

The first is that predator and prey abundances oscillate over time (Fig. 7.5). Some values for the model parameters lead to unstable solutions, meaning the populations either go extinct or increase to infinitely large levels. The stable solutions imply that the predator population lags behind the prey population and that both vary around each other forever. The model can also be used to predict the period (how long it takes for a population to return to a starting value after increasing and decreasing) and the amplitude (the difference between minimum and maximum abundance) of the oscillation. The period is the same for both predator and prey, while the amplitude differs.

Oscillations seen in the classic predator-prey relationship are rarely observed with microbes in nature, although these oscillations sometime are seen in controlled experiments in the lab. More precisely, it would be difficult to see in nature the oscillations illustrated in Figure 7.5, even if they exist. The amplitude in oscillation of prey abundance is only on the order of 20% in this example, which is small and would be difficult to detect in nature. More problematic is the timescale of these oscillations.

The Lotka-Volterra model predicts that the period of the oscillation should be about 20 days, given rates and population levels typical of natural habitats. To be really convincing, data from two, or better, three cycles, equivalent to 40–60 days of observations, are needed to test the model. That would be difficult even for a study of soils or of a small lake, nearly impossible for an ocean where a long scientific cruise is 30 days. So, grazing could account for the apparent constancy of bacterial abundance in many natural ecosystems even if the Lotka-Volterra model applies because the predicted oscillations are relatively small and hard to see. A more fundamental problem, however, is that the Lotka-Volterra model is not realistic. It does not include many ecological processes, such as switching by the predator to other prey, competition among predators and prey, and bottom-up controls. Many of these other processes would tend to dampen

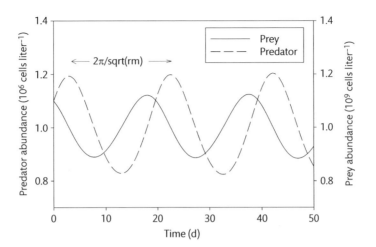

Figure 7.5 Variation in the abundance of predator and prey over time according to the Lotka-Volterra model. The period of the oscillation is $2\pi/\sqrt{(rm)}$ (2π/sqrt(rm) in the figure) while the amplitude for the prey population is proportional to m/b and that for the predator population is r√m/(a√r); the parameters r, m, and b are defined in the text. Parameters r and m were assumed to be 0.2 and 0.5 d^{-1}, which are typical microbial growth rates (Chapter 6). Parameters m and b were then calculated with equations describing the critical values for the predator and prey populations (found by setting Equations 7.1 and 7.2 to zero), assuming that the predator and prey abundances were those of a typical bacterivorous protozoan (10^6 cells per liter) and of bacteria (10^9 cells per liter).

the oscillations and disrupt the timing between predator and prey variations.

Still, the Lotka-Volterra model is a useful starting point for more realistic and sophisticated models of predator-prey relationships.

Size relationships of predator and prey

In addition to prey and predator numbers, size matters. We saw that the dominant grazers of bacteria, which are about 0.5 μm in nature, are 1–5 μm flagellates. This observation raises a general question about the relationship between sizes of other prey and predators and whether there is a general rule that can predict who is eating whom in the microbial world. The Danish microbial ecologist Tom Fenchel (1940–) suggested an elegant answer to this question (Fenchel, 1987). He considered a hypothetical spherical protozoan with a radius R eating an equally hypothetical spherical prey with a radius r, and argued that the clearance rate of a predator should vary as a function of r/R. He went on to provide empirical evidence that the ratio is on the order of 0.1 (Fig. 7.6). That is, the predator has to be about tenfold bigger than its prey.

How has the tenfold bigger rule fared over the 20 years since Fenchel first proposed it? Overall, it has held up rather well, and the exceptions are more interesting to consider now than the rule itself. Some exceptions include microbial predators that are much larger than expected, such as gelatinous bacterivorous zooplankton like larvaceans. Also, some species of mussels and large detritivores in soils and sediments are more important grazers of bacteria and equally small microbes than expected from the tenfold rule. These are understandable exceptions. Perhaps more problematic are cases in which the predator is substantially less than tenfold larger than its prey. Examples include 1–3 μm protists that eat 0.5 μm sized bacteria or 50 μm ciliates and heterotrophic dinoflagellates that prey on diatom chains extending for >100 μm (Sherr and Sherr, 2009). Even so, the tenfold bigger rule is still a useful order of magnitude guideline in thinking about predator-prey relationships in the microbial world.

A related issue is to consider how grazing by one predator of a particular size varies as a function of prey size. Fenchel's 10:1 rule leads to the prediction that grazing is low on very small prey and also low on very large prey, relative to an optimal prey size for a particular

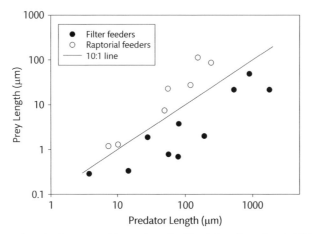

Figure 7.6 Relationship between length of a prey and its predator (a ciliate). Data from Fenchel (1987).

<table>
<tr><td>

Box 7.4 Consider a spherical protist

In thinking about how size would affect grazing, Fenchel assumed that all prey and predators in the microbial world are spheres. Of course, this is quite wrong, as he knew. But the assumption simplified calculations and led to some powerful predictions about grazing by protozoa and other protists. More so than biologists, physical scientists and modelers make simplifying assumptions to address complicated questions, often ending up with useful results and insights. The book *Consider a Spherical Cow* by John Harte (1985) discusses this way of thinking about complex problems. The title comes from an approach for estimating how many shoes could be made from one cow. The answer comes from a "back of the envelope" or an "order of magnitude" calculation. Even if the calculation needs a computer, the key is to avoid unnecessary detail and to concentrate on the important aspects of the problem.

</td></tr>
</table>

by feeding one predator prey of different sizes, as was done to obtain the data in Figure 7.7. Another experiment is to observe the change in the size distribution of a bacterial assemblage or a bacterial culture over time with and without a protozoan. The presence of a grazer can result in the bacterium forming long chains or aggregates that are too big for protozoa to eat.

So, size is a key factor in determining grazing rates and who is eating whom in the microbial world; it explains many observations. Size is why plastic beads are ingested by some protozoa at nearly the same rate as microbes of the same size. Because of this, fluorescent plastic beads can be used as a surrogate food to estimate grazing rates. The size effect also implies that prey of similar size should be eaten by the same predator. For example, heterotrophic bacteria and the cyanobacteria *Prochlorococcus* and *Synechococcus* are probably eaten by the same suite of protists, given that cells in the three microbial groups are similar in size. Likewise, carnivores that eat heterotrophic nanoflagellates should also eat similar-sized phytoplankton.

predator. Big prey are beyond the capacity of a predator to ingest while small prey are captured too inefficiently. Somewhere in between the two extremes in prey size is the optimum that results in the highest grazing rate. As a predator increases in size, the optimal prey size also increases. Experimentally, this size effect is demonstrated

Chemical recognition and composition
While size is a powerful predictor of grazing behavior, microbial ecologists knew early on that it was not the only factor. The chemical composition of the prey also has a role (Jürgens and Massana, 2008). The early evidence for this conclusion came from the experience

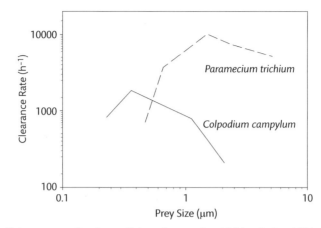

Figure 7.7 Grazing by two ciliates on prey of various cell sizes. *Paramecium trichium* is about 90 by 55 µm while *Colpodium campylum* is about 60 by 25 µm. Data from Fenchel (1980).

of growing protozoa on different bacterial species in the lab. Some bacteria were better food than others, even though they seemed to share the same size and appearance. Also, protozoa sometimes grow more slowly on heat-killed bacteria than on live food. One explanation is that heating changes prey chemistry, similar to what cooking does to our food. Think of the difference between fresh and hard-boiled eggs. Finally, there is evidence of discrimination against plastic beads and fluorescent-labeled bacteria; grazing rates on these particles can be lower than on natural, unaltered bacteria. These findings indicate that chemical properties of prey, especially the composition of the prey cell surface, affects grazing and growth of protists. It seems that somehow protozoa and other protists can "taste" their food.

More support for this idea comes from experiments that examine feeding on plastic beads coated with various organic compounds (Wootton et al., 2007). These experiments also suggest a mechanism (Fig. 7.8). Wootton and colleagues found that the marine dinoflagellate *Oxyrrhis marina* ingested more plastic beads coated with mannose than with other sugars. Furthermore, feeding by the dinoflagellate on its regular phytoplankton prey was inhibited when the dinoflagellate was exposed to mannose, but again not by other sugars. These experiments are classic ones to demonstrate that a cell-cell interaction is mediated by a type of cell surface receptor called a lectin. In general, lectins are sugar-binding proteins which are involved in a great variety of cell-cell

interactions in organisms ranging from plants to humans. In this case, the dinoflagellate uses a mannose-binding lectin as a receptor to recognize mannose on the prey cell surface. Mannose itself is not why the grazer selects the prey, but rather its presence indicates a desirable food item for the grazer.

Once inside the food vacuole, cell surface chemistry of the prey can also determine digestion by the protozoan grazer (Jürgens and Massana, 2008). The prey cell surface can affect the efficiency of digestion; that is, the amount of prey carbon incorporated into protozoan cytoplasm. For example, some bacteria are resistant to grazing because of unique cell walls (Tarao et al., 2009). Some prey cells are rejected by the protist and are ejected ("egested") from the food vacuole back to the outside environment. Some bacteria such as *Legionella* survive chlorination or other disinfectant treatments while inside protozoan food vacuoles, resulting in human health problems. Some prey components may not be digested and are also egested back to the outside (Fig. 7.3).

Among several reasons why grazers may select for some prey over others, one is that the preferred microbial prey may be more nutritious for the protist. "More nutritious" here means that the chemical composition of the preferred prey leads to faster growth of the protist. There is some evidence of protists favoring prey with a low C:N ratio (Montagnes et al., 2008), probably due to higher protein content of these prey. Lipid content of the prey also appears to have a big impact. Diets of prey with

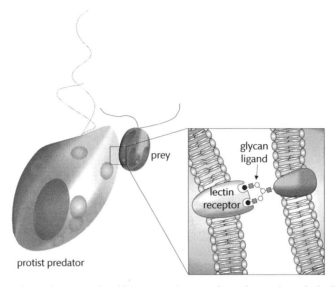

Figure 7.8 Model of how protist predators may be able to recognize prey through protein-carbohydrate interactions. Lectin receptors on the cell surface of the predator bind to specific carbohydrate conformations on the surface of the prey. Lectins are carbohydrate-binding proteins and glycans are the carbohydrate portions of glycoproteins and glycolipids. Figure provided by Emily Roberts. See Wootton et al. (2007) for more details.

certain polyunsaturated fatty acids, for example, promote faster growth of protists and metazoans (Strom, 2000), and the lack of these fatty acids in bacteria may limit fluxes through food webs based on bacteria (von Elert et al., 2003). Bacteria, both heterotrophic and cyanobacteria, also do not have the sterols that are needed by protozoa and other eukaryotes (Martin-Creuzburg and Elert, 2009). Oddly, lipids unique to bacteria can be retained in the lipids of eukaryotic predators, thus providing evidence that the predator grazes on bacteria.

Defenses against grazing

Bacteria and other microbial prey have some defenses to ward off predation by protists. One strategy relies on size. At one extreme, being small may help some microbes escape predation by large grazers. The best-studied example is that of heterotrophic bacteria. The lower grazing pressure on small heterotrophic bacteria may be part of the explanation for their high but relatively constant abundance in nature. At the other extreme, prokaryotes and many other microbes may increase their cell size to avoid predators, but this strat-

egy is constrained by other factors. Prokaryotic cells especially cannot get too large because diffusion within the cell becomes limiting, and low concentrations of organic carbon or inorganic nutrients select against both large prokaryotic and eukaryotic cells. Also, a prey may be too big for one predator to eat but is just the right size for another predator. There are other ways, however, to exceed the size limits of grazers. As mentioned before, bacteria in chains or aggregates are too big to be eaten by nanoflagellates. Aggregate formation may be triggered by chemical cues released during grazing (Blom et al., 2010). Exopolymers may also hinder grazing by a protistan predator because of how the polymers affect prey size or chemical recognition by the predator. In nature, however, these features of microbes may be in response to factors other than grazing. Bacteria may be small in many natural environments, for example, because of low organic carbon concentrations, not as a direct result of grazing pressure. Still, since predation exerts such strong selection pressure, prey have evolved strategies to avoid or at least to minimize predation.

Another line of defense is chemical warfare. The freshwater bacteria *Janthinobacterium lividum* and *Chromobacterium violaceum* produce the purple pigment violacein which kills off nanoflagellates, rotifers, and *Daphnia* predators (Deines et al., 2009), and some soil fungi appear to secrete anti-predator compounds into the soil and to build crystalline structures in their cell walls to deter grazing by soil microarthropods (Bollmann et al., 2010). Another example of chemical defensive strategy comes from experiments with *Oxyrrhis marina* (Martel, 2009). In these experiments, the protist ate some strains of the alga *Emiliania huxleyi* but not other strains. Those that were not eaten produced dimethylsulphoni-opropionate (DMSP), implying that this sulfur compound is an anti-herbivore defense mechanism, adding to its other proposed roles in algal physiology, as discussed in Chapter 4. On the other hand, DMSP attracts some microbes, including possible bacterivores and herbivores (Seymour et al., 2010). Since DMSP can be converted to the climate-active gas dimethyl sulfide (DMS), as mentioned in Chapter 4, there may be a loose connection between grazing and atmospheric processes.

Effect of grazing on prey growth

Predation by protists has other effects on microbial prey, in addition to killing off some members of the prey community and evoking defensive countermeasures. One positive effect on growth is the release by predators of inorganic and organic compounds that are used by those prey escaping predation. If the compound is limiting, the fortunate prey cells not being eaten can take up the released compounds and grow faster in the presence of grazing. This effect may be especially beneficial to the prey population when grazers feed on dormant or very slowly growing cells, leading to more prey cells being active (Berman et al., 2001, del Giorgio et al., 1996). This culling of inactive cells helps to explain why grazing increases decomposition rates of detritus in soils and in aquatic habitats. Grazing can have even more direct impacts on microbial growth.

One mechanism for a very direct impact is based on the relationship between the cell size of prey and predation rates. Remember that grazing increases with prey size within limits and is one reason why so many microbes are small in nature. But in order to reproduce, many microbes have to get bigger first, by roughly twofold, before dividing into two cells. These big, about-to-divide cells are especially vulnerable to being eaten, more so

than after division when each cell is once again small. Experiments with flagellates feeding on bacteria in coastal marine waters have provided some support for the idea that protists feed preferentially on about-to-divide cells (Sherr et al., 1992). This mechanism means that grazers could select for fast-growing cells over slow-growing cells. The fast-growing cells have to increase in cell size more frequently than slow-growing cells. There is also evidence that fast-growing cells are bigger than slow-growing cells (Gasol et al., 1995), although there are many exceptions to this rule. But if it holds, the rule would result in another relationship between growth and grazing. Regardless, the negative tie between grazing and growth means that there are advantages to growing slowly, and could explain how slow-growing microbes can survive and coexist with fast-growing competitors.

Grazing by ciliates and dinoflagellates

If flagellates 1–5 μm in size are the main grazers of heterotrophic bacteria, coccoid cyanobacteria, and micron-sized eukaryotic algae, who eats the flagellates? The "tenfold bigger" rule would predict a carnivore about 10–50 μm. In soils, these carnivores include nematodes and amoebae. In the water column of aquatic habitats, a complex suite of microzooplankton protists in the 20–200 μm size range are potential predators of flagellates, including big flagellates grazing on small flagellates (Fig. 7.9). Some of the first microzooplankton protists studied by microbial ecologists were tintinnids, which are choreotrich ciliates, characterized by their elaborate houses (loricae). The sturdy houses enabled microbial ecologists to collect these otherwise fragile ciliates with fine-mesh plankton nets and to identify them. The houses vary among species of tintinnids. Other collection and fixation methods led to the discovery that ciliates other than tintinnids are generally much more abundant than tintinnids. These other ciliates are sometimes referred to being aloricate or "naked" because of their houseless lifestyle.

Ciliates as herbivores in aquatic ecosystems
Ciliates and other protistan grazers are now known to have roles once thought to be those of crustacean zooplankton such as copepods in the oceans and

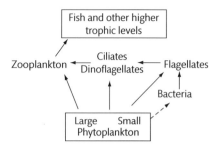

Figure 7.9 Model of a microbial food web to illustrate some of the roles of ciliates and dinoflagellates in aquatic habitats. "Bacteria" here refers to heterotrophic bacteria; coccoid cyanobacteria are included with the small phytoplankton. The dotted line linking phytoplankton and bacteria indicates that the connection between these organisms is indirect and involves dissolved organic material and detritus. Zooplankton refers to crustaceans and other metazoan grazers.

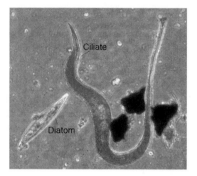

Figure 7.10 Protostomatid ciliate of the genus *Tracheloraphis* found in marine sediments, along with a pennate diatom. The ciliate, which is contracted, is about 500 μm long. If stretched out, it would be over 1 mm. The large dark blobs next to the ciliate are sand grains. Image by David J. Patterson, used courtesy of micro*scope (microscope.mbl.edu). See Fig. 1.1D for another example of a ciliate.

Box 7.5 Whirling whips

The "flagellate" (from the Latin for whip) part of "dinoflagellates" makes sense, knowing that these microbes have two of them. But what about "dino"? That comes from a Greek word for "whirling", an appropriate name given that dinoflagellates are strong, fast swimmers, with some big species capable of speeds up to 1 meter per hour.

---

cladocerans in freshwaters (Strom et al., 2007). In addition to grazing on bacterivorous nanoflagellates, ciliates and other microzooplankton are important herbivores and graze on many types of phytoplankton, including long chains of diatoms. This top-down control of phytoplankton was previously thought to be by crustacean zooplankton larger than the microzooplankton. Ecologists now think, however, that zooplankton such as copepods may be more important as carnivores eating microzooplankton than as herbivores grazing on phytoplankton. In addition to herbivory, small enough ciliates may graze directly on bacteria and similar-sized phytoplankton, such as the coccoid cyanobacteria *Synechococcus* and *Prochlorococcus*.

When classified by morphology the most common ciliates in aquatic habitats are in the subclass Choreotricha (Sherr and Sherr, 2000), which is in the class Spirotrichea, the phylum Ciliophora and the superphylum Alveolata. These microbes are round or oval with a crown of cilia at the oral cavity. In addition to the tintinnids, common aloricate genera include *Strombidinopsis* and *Strobilidium*. Not all ciliates are strict heterotrophs. Some are mixotrophic and use both phototrophy and heterotrophy (see below) while still others are mostly autotrophic. Examples of ciliates are given in Figure 7.10 and Figure 1.1D.

Ciliates in soils and sediments

Ciliates are not as ecologically important in soils as they are in aquatic habitats, one reason being that they are restricted to very moist soils (Foissner, 1987). In contrast to the water column of aquatic habitats, ciliates share with nematodes and other organisms the ecological role of feeding on flagellates. The abundance of ciliates is lower than that of flagellates in soils, which is the case in aquatic habitats as well. Some soil ciliates feed on bacteria trapped in soil cavities and pores, while larger species can eat flagellates. Soil ciliates are able to form cysts, a type of resting stage that is resistant to desiccation, sometimes for decades. Aquatic ciliates don't have a problem with desiccation, but some still form cysts when growth conditions turn adverse. Soil ciliates are also adapted to life on surfaces of soil particles (thigmotactic). About half of all soil species are from the class Colpodea and nearly the rest of them are surface-associated Stichotrichia, a subclass of the Spirotrichea.

In the benthic habitat of the ocean bottom, ciliates are important in sandy sediments with large interstitial spaces (Fenchel, 1987). Sediments dominated by silt and clay have small interstitial spaces and lower ciliate numbers than sandy sediments. As in soils, sediment ciliates have several adaptations to life on surfaces and in interstitial spaces. Some are long (millimeters in length) with cilia only on one side of the cell. These microbes feed on benthic algae, flagellates, other ciliates, and bacteria. Depending on time of year and location, ciliates can be the dominant consumers of algae and bacteria in sediments of aquatic habitats.

Heterotrophic dinoflagellates

Other prominent members of the microzooplankton protistan community include heterotrophic dinoflagellates. We first encountered dinoflagellates in Chapter 4 in the discussion of phytoplankton and primary production. In addition to their contribution to primary production, many dinoflagellates are mixotrophic while others are strictly heterotrophic. If there is a "typical" dinoflagellate, it is pear-shaped with one groove, the girdle, around the middle of the cell and another groove, the sulcus, going down from the girdle (Fig. 7.11). The microbe moves thanks to two flagella, one wrapped around the girdle, the other beating in the sulcus. Some species are armored with thecal plates composed of cellulose while others are "naked" or "unarmored". Another unusual feature of all dinoflagellates is that the chromosomes in their nucleus (the dinokaryon) never completely unwind when not dividing, in contrast to other eukaryotes.

Dinoflagellates vary greatly in size and in metabolism (Table 7.4). Species in the genera *Amphidinium* are >20 µm, have thecal plates, and are capable of feeding on nanoflagellates with their peduncles, tube-like structures that are stuck into prey. Other species feed by more conventional phagocytosis on diatoms and ciliates (Hansen, 1991). Still others are referred to as "veil feeders" because they can exude a pseudopodial cytoplasmic sheet that can envelop even large diatom chains, digesting them extracellularly and absorbing nutrients before being hauled back into the cell. Some heterotrophic dinoflagellates are bioluminescent, with *Noctiluca scintillans* being the most famous example. This microbe is about 500 µm or bigger and produces light by the luciferin-luciferase system. In response to unfavorable growth conditions, dinoflagellates can form cysts which sink to and survive in bottom sediments.

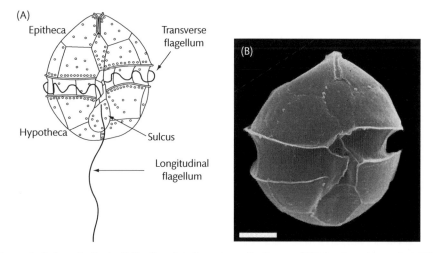

Figure 7.11 The ventral view of a "typical" dinoflagellate in a schematic diagram (A) taken from Jeong et al. (2005) and in a scanning electron micrograph (B). This particular species (*Stoeckeria algicida*) is about 17 µm by 13 µm. Used with permission from Hae Jin Jeong.

Table 7.4 Some common genera of dinoflagellates. For heterotrophic species, the main feeding mechanism is also given. Data from Hansen (1991).

Genus	Feeding mechanism	Prey	Comments
Gonyaulax	Photoautotrophic	None	Some toxic species
Peridinium	Photoautotrophic	None	Armored, mostly freshwater species
Ceratium	Photoautotrophic	None	Armored
Dinophysis	Predation by a peduncle	Ciliates	Armored, some toxic species
Amphidinium	Predation by a peduncle	Nanoflagellates	Armored
Gymnodinium	Predation by a peduncle	Diatoms	Unarmored
Noctiluca	Predation by engulfment	Nearly all particles of the right size	Some bioluminescent
Oxyrrhis	Predation by engulfment	Nanoflagellates, diatoms	Easily grown in the lab, but poor model for other taxa in nature
Protoperidinium	Predation by a feeding veil	Diatom chains, other large phytoplankton	First attaches to prey with a "tow filament"

Fluxes from microbial food webs to higher trophic levels

The previous sections identified, at least at a crude level, many of the major predators in the microbial world. In soils, grazers of microbes are eaten by earthworms, insects, and nematodes (Fig. 7.2), which in turn are eaten by still larger terrestrial organisms. In aquatic habitats, zooplankton are the link between the microbial world and larger organisms at higher trophic levels, including fish of commercial importance. Given these links, how much material and energy can be channeled from microbial food webs to large organisms and higher trophic levels? The answer depends on the number of trophic transfers, which is one organism eating another, and the efficiency of this transfer. This efficiency, the

trophic transfer efficiency, is the amount of carbon, other material, and energy potentially passed on to the next trophic level. It is similar to the growth efficiency defined for bacteria and fungi as the ratio of biomass production to the total use of the organic carbon (Chapter 5). Figure 7.12 summarizes growth efficiencies for various parts of the aquatic food chain other than bacteria and fungi. The trophic transfer efficiency may be lower than the growth efficiency if processes like viral lysis reduce the amount of food available to the next trophic level. For the following argument, for simplicity we assume that the trophic transfer efficiency is 30%. With these assumptions, it is possible to derive a simple equation to describe the amount of carbon available to a particular trophic level (the "ith" one), given a starting level of primary production (P).

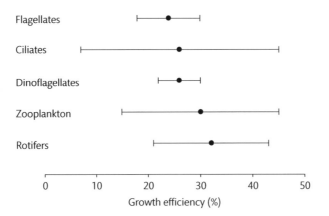

Figure 7.12 Growth efficiencies of protist grazers and crustacean zooplankton. The circle is the mean and ends of the bar indicate the 25% and 75% quartiles of the data. Data from Straile (1997).

PREDATION AND PROTISTS **133**

Assuming that all biomass production at each trophic level is used, then the amount of carbon available to a carnivore (the second trophic level) eating a herbivore is E·P where E is the trophic transfer efficiency and P is the rate of primary production. The amount of carbon available to the third trophic level is E·(P·E) or P·E². In short, the amount of primary production available to the ith trophic level (H_i) is:

$$H_i = P \cdot E^{i-1} \qquad (7.3).$$

Assuming 30% of what is eaten ends up as new biomass of the next trophic level (E = 0.3), we can see that the grazer food chain in habitats is much more efficient than microbial food chains, simply because of the large number of steps in the latter (Fig. 7.13). According to Equation 7.3, the two-step grazer food chain allows 9% ($0.09 = 0.3^2$) of primary production to reach a top carnivore (fish, for example), much more than the 2.7% ($0.027 = 0.3^3$) of primary production routed through a microbial food chain. The additional steps are needed to link microbial food chains with higher trophic levels because of the effect of prey size on the size of predators capable of eating that prey.

Seeing the various steps lined up in a food chain raises another question: can changes in one level—for example, in carnivorous zooplankton or an earthworm— affect organisms two or more trophic levels below it, such as the primary producers or heterotrophic bacteria? The short answer is, yes. The effect of one trophic level affecting another several steps removed is called a "trophic cascade" (Pace et al., 1999). It is part of one explanation, the "green world hypothesis", for why the terrestrial world

is green. It is due to carnivores keeping the herbivores in check and allowing plants to flourish. In aquatic habitats, a trophic cascade explains why removing large carnivorous zooplankton can affect nanoplankton, bacteria, and other microbes several trophic levels below (Zöllner et al., 2009). On land, trophic cascade mechanisms can connect organisms in the above-ground community with microbes and other organisms in the soil below, in the below-ground community (Wardle et al., 2005). Even cattle can affect soil microbial biomass and growth by altering the input of organic material used by bacteria and fungi, which in turn affect bacterivorous and fungivorous nematodes (Wang et al., 2006).

Mixotrophic protists and endosymbiosis

Chapter 4 concentrated on cyanobacteria and protists (phytoplankton) that are photoautotrophic while most of this chapter has focused on heterotrophic protists which feed on other microbes and small particulate organic detritus. Protists devoted 100% either to photoautotrophy or to heterotrophy are at the extremes of a continuum (Fig. 7.14). Diatoms, for example, cannot graze on other microbes because their thick, siliceous cell walls make phagocytosis impossible. The colorless protists, the protozoa, on the other hand, lack chlorophyll and cannot carry out photosynthesis, making them strict heterotrophs. But in between the two extremes are mixotrophic protists. These microbes can carry out photosynthesis while also being able to capture prey by phagocytosis. Mixotrophic microbes are among the things frequently left out of food web diagrams such as in Figure 7.9.

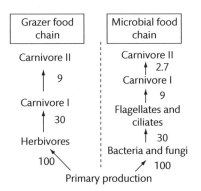

Figure 7.13 Net efficiency of transfer by a grazer food chain versus a microbial food chain.

Figure 7.14 Protist metabolisms, ranging from strict photoautotrophic organisms incapable of growing without light to strict heterotrophs incapable of growing without prey. The various types of mixotrophic protists are in between these two extremes. Modified from Caron (2000).

Near the strict phototrophic end of the continuum are the phagotrophic algae. These protists are basically phototrophic but are thought to carry out some phagotrophy to obtain essential vitamins, specific lipids, or organic material rich in nitrogen and phosphorus. Prey fed on by phagotrophic algae are rich nuggets of nitrogen and other elements and may supplement the elements supplied by dissolved compounds such as ammonium and phosphate, which are often in very low concentrations. One of the first hints of mixotrophy was the inability of microbiologists to grow some algae in the absence of bacteria and other microbes. Many species of phagotrophic algae are now known (Table 7.5). This type of "phytoplankton" can be quite common and account for up to 50% of the entire community in freshwaters and are equally important in marine systems as well. Their grazing impact can also be substantial, with rates comparable to those of strict heterotrophic bacteriovores (Zubkov and Tarran, 2008). Mixotrophic protists can occur in the top layer of soils receiving sufficient sunlight. An example of a mixotrophic protist is given in Figure 7.15.

Another type of mixotrophic protist feeds on phytoplankton and digests everything, except the chloroplasts, at least for a short time. These undigested chloroplasts retained by the protist are referred to as kleptochloroplasts or cleptochloroplasts (Caron, 2000), with "klepto" coming from the Greek for an irresistible urge to steal. Some examples of protists with kleptochloroplasts include the ciliates *Strombidium* and *Mesodinium*, and the dinoflagellates *Gymnodinium* and *Amphidinium* (Table 7.5). Protists with kleptochloroplasts can obtain some organic carbon from these chloroplasts, with the extreme example being *M. rubrum* (now called

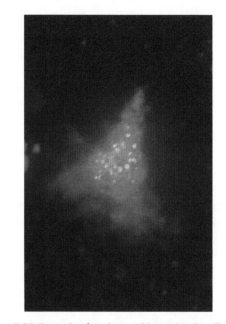

Figure 7.15 Example of a mixotrophic protist, the ciliate *Laboea strobila* in the Gulf of Alaska. The dark spots at the top of the triangle (red cells in the original) are chloroplasts while the white spots (originally orange) in the middle are ingested algal prey. The ciliate is about 100 × 40 μm. Picture used with permission from Brady Olson.

Myrionecta rubra) which obtains nearly all of its carbon from kleptochloroplasts. These chloroplasts differ in many fundamental ways from those found in a strict phototrophic protist. One difference is that kleptochloroplasts are usually lost over a few days if the protist does not feed again. An exception is the ciliate *M. rubra*, which retains the nucleus of its prey along with the chloroplasts, so its kleptochloroplasts can reproduce within the ciliate

Table 7.5 Some examples of phototrophic protists capable of phagocytosis. The main habitat of each organism is given, but many of these have species found in other aquatic habitats. Taken from Caron (2000).

Class	Genus	Habitat	Comment
Chrysophyceae	*Dinobryon*	Freshwaters	Usually occurs as colonies
Chrysophyceae	*Poterioochromonas*	Freshwaters	Some species mainly heterotrophic
Chrysophyceae	*Ochromonas*	Freshwaters and marine	Some species found in soils
Dinophyceae	*Gymnodinium*	Marine	Can have kleptochloroplasts
Dinophyceae	*Amphidinium*	Marine	Can have kleptochloroplasts
Dinophyceae	*Gonyaulax*	Freshwaters and marine	Some red tide species
Prymnesiophyceae	*Prymnesium*	Marine	Some toxic species
Raphidophyceae	*Heterosigma*	Marine	Some red tide species
Spirotrichea	*Strombidium*	Marine	Can have kleptochloroplasts
Litostomatea	*Mesodinium*	Marine	Can have kleptochloroplasts

(Johnson et al., 2007). So, these protists are basically heterotrophic and opportunistically take advantage of photosynthesis by chloroplasts from their prey. Chloroplast-retention is best known among the ciliates, and species with this form of metabolism can make up nearly half of the total ciliate community in estuaries and oceanic environments. It is also common among dinoflagellates. In fact, some species were erroneously thought to be photoautotrophs before it was discovered that these dinoflagellates depended on other microbes for their chloroplasts. Now it is unclear if any dinoflagellate is a strict autotroph incapable of phagotrophy.

Phagotrophy, endosymbiosis, and algal evolution

The relationship between a chloroplast-retaining phagotrophic protist and its prey is one-sided, but that is not the case with protists bearing endosymbiotic algae. Instead of digesting the prey, these phagotrophic protists have evolved mechanisms to retain and nurture the alga, playing host to only a single algal species, in contrast to chloroplast-retaining protists which may steal chloroplasts from several types of algae. The symbiotic algae vary greatly in taxonomy and include chlorophytes, prymnesiophytes, prasinophytes, diatoms, and many dinoflagellates (Caron, 2000). But some algal lineages seem to be especially common in endosymbiotic relationships. Among the dinoflagellates, for example,

Gymnodinium beii is found in four species of planktonic foraminiferans, and several species of radiolarians harbor *Scrippsiella nutricula*. The algal symbiont of corals is a dinoflagellate in the genus *Symbiodinium*.

The phagotrophic protist and its endosymbiotic algae probably each enjoy several benefits from the relationship. The otherwise heterotrophic protist gains another source of carbon and energy from organic material synthesized and exuded by the phototrophic algae. The host may gain additional material and energy by digesting some of the algae from time to time, which may be necessary to keep the endosymbiont population to a manageable level. The algal symbiont may also absorb ultraviolet light and protect the protist (Sonntag et al., 2007). On the other side of the relationship, the alga benefits by being protected from predation (ignoring the occasional digestion by the protistan host) and perhaps from viral lysis. The endosymbiotic alga is also physically close to a ready source of inorganic nutrients, such as ammonium and phosphate, released as wastes by its host. The large number of algae, reaching several thousand cells in some cases, inside of one protistan host is an indication that the relationship is benefitting the alga.

In addition to their ecological importance, phagotrophic protists with endosymbiotic algae are important examples in support of the endosymbiotic theory for the evolution of algae (Fig. 7.16). One of the first endosymbiosis events over the course of evolution was

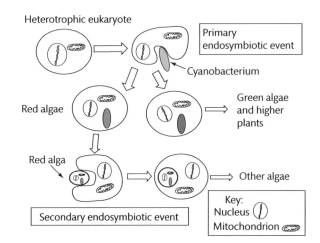

Figure 7.16 The endosymbiotic events leading to algae and higher plants. One of the primary endosymbiotic events not shown is the one leading to mitochondria. The scientific names for red and green algae are Rhodophyta and Chlorophyta. Based on Delwiche (1999) and Worden and Not (2008).

between a heterotrophic protist and a cyanobacterium, which eventually became the chloroplast, a type of plastid in plants. Additional endosymbiosis events are needed to explain other features of some algae, such as the presence of three to four membranes around chloroplasts in dinoflagellates, haptophytes, and cryptophytes. The plastids in these algae are thought to have arisen by secondary or even tertiary endosymbiotic events; that is, the phagocytosis of an alga other than a cyanobacterium. In the case of the Chromalveolata, the plastids are thought to have come from a red algal ancestor (Reyes-Prieto et al., 2007). Today, these chloroplasts are fully integrated into the algal cell, a relationship cemented by the transfer of genes from the former symbiont to the host nucleus. But the partnership started off as a symbiosis. The existence of endosymbiotic protists in today's world is strong evidence that these endosymbiotic events occurred during evolution of algae now common in aquatic habitats and soils. The act of predation and retention of plastids, still going on in contemporary protists, underlies the evolution of microbes in our world today.

Summary

1. Many organisms are capable of grazing on bacteria (bacterivory), but the main grazers of bacteria in most aquatic and soil habitats are flagellates 1–5 µm in length. Larger organisms, such as nematodes, eat fungi in soils and sediments.

2. Protozoa and other protists usually feed by phagocytosis. The entire process consists of three phases: encounter and cell-cell recognition of the prey by the protist; engulfment (phagocytosis) of the prey particle; and digestion of the prey in the food vacuole.

3. Grazing rates are affected by prey size, prey numbers, and chemical composition of the prey.

4. Among the many impacts on prey populations, grazing affects cell size and growth by selecting for large, about-to divide prey cells. It may also evoke anti-grazing chemical defense mechanisms.

5. Many protists are mixotrophic, capable of both photoautotrophy and predation on other microbes. Some mixotrophic protists have standard chloroplasts while others have chloroplasts (kleptochloroplasts) taken from partially digested photoautotrophic prey.

CHAPTER 8

Ecology of viruses

One of the most abundant biological entities in the biosphere is viruses. There are about 10^{31} of them on the planet, enough to extend past the nearest 60 galaxies in the universe, if lined up and strung end to end (Suttle, 2005). In spite of being inert particles outside their hosts, incapable of catalyzing a chemical reaction, viruses in fact directly and indirectly affect many biogeochemical processes, including the carbon cycle. That ecological view of viruses, however, is not shared by all biologists, evident from remarks by two Nobel laureates. Peter Medawar described a virus as "a piece of bad news wrapped up in a protein", while David Baltimore thought that "if they weren't here, we wouldn't miss them" (Ingraham, 2010). We may not miss the diseases caused by viruses, but eventually we would see very different ecosystems if viruses were magically blotted out. This chapter will discuss how viruses play irreplaceable roles in the ecology and evolution of microbes and of all organisms. Arguably life could not exist without viruses.

A defining characteristic of a virus is that it has to infect a host in order to replicate, although a virus of a hyper-thermophilic archaeon seems an exception to this rule (Haring et al., 2005). Since infection by a virus can be fatal to the host cell, viruses are a form of top-down control of microbial populations, as already mentioned in Chapter 7. In this chapter, we learn more about viruses to understand this top-down control and to explore other ecological roles of viruses in soils and aquatic habitats. There is probably a virus or several for every organism on the planet, but the most common viruses in nature are thought to be those that infect bacteria because bacteria are the most abundant form of cellular life. These viruses are called bacteriophages or simply phages.

What are viruses?

In some ways, viruses are very simple, consisting of only nucleic acids (Medawar's "bad news") surrounded by a protein coat, the capsid, and by membranes for some viruses. The protein coat is needed to protect the viral nucleic acids from degradation by microbes, host defenses, and physical forces. Viruses come in four basic morphologies, ranging from simple geometric shapes to more complicated structures that resemble a lunar landing craft (Fig. 8.1). Some of these shapes are determined by how capsid protein subunits are arranged to house the viral genome, accounting for two of the four viral morphologies. One arrangement leads to helical viruses in which the capsid subunits spiral around the genome, while another results in polyhedral viruses with the proteins forming geometric shapes with flat surfaces. One such shape is an icosahedron, which has 20 equilateral triangle faces and 12 corners. Viruses in the third morphological category take on the lipid membrane from their hosts and are called enveloped viruses. The human immunodeficiency virus (HIV) is an example of this type of virus. The fourth morphology, found in complex viruses, includes tails and other structures in addition to the capsid. Many viruses of archaea have weird shapes, including those of spindles and bottles (Prangishvili et al., 2006). These structures and regular geometric shapes become important when using electron microscopy to distinguish viruses from detrital particles in samples from natural environments.

More dramatic and fundamental than differences in morphologies are the differences in the viral genomic material. In stark contrast to prokaryotes and eukaryotes,

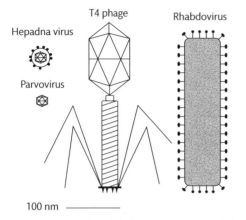

100 nm

Figure 8.1 Some examples of viral shapes and sizes. T-4 phage and hepadna virus have dsDNA while parvovirus has ssDNA. Note the icosahedral coat for all three. Rhabdovirus has (-) ssRNA and is encased in a lipid envelope with surface glycoproteins. Adapted from Wagner et al. (2008).

viral genomes are not just double-stranded DNA, but occur in every possible variation of nucleic acid: double-stranded DNA, single-stranded DNA, double-stranded RNA, or single-stranded RNA (Table 8.1). Some of the RNA viruses have positive-sense RNA, which is essentially mRNA, and can use it to synthesize proteins immediately after entry of the viral genome into the host cell. The negative-sense RNA viruses must first convert their RNA to the positive sense using an RNA polymerase. Still other RNA

viruses, retroviruses, first make DNA, commonly called cDNA, using the enzyme reverse transcriptase, and the resulting cDNA is incorporated into the host chromosome. Most of the viruses examined in the laboratory are RNA viruses while in contrast we know the most about double-stranded DNA viruses in nature. There have been few studies of RNA viruses in natural ecosystems (see below).

The size of viruses varies greatly, in part due to the size of the viral genome (Table 8.1). The smallest virus, circovirus, has only two genes (<2000 nucleotides) and is only 20 nm in diameter. Viruses need only a few genes because they rely on host genes for reproduction. Other viruses are very large. Some approach the size of a bacterial cell with a genome bigger than that of some bacteria. Mimiviruses are nearly one micron in diameter with a 1.2 Mb genome (Claverie and Abergel, 2009). But the viruses studied in the laboratory are bigger than those found in nature (Weinbauer, 2004) and most viruses in nature are small. As a general rule, viruses, even those in laboratory cultures, are tenfold smaller than bacteria in size and have many fewer genes in less genetic material.

Viral replication

Viruses are obligate parasites that must invade a host cell and take over the host cell biochemical machinery with the ultimate goal of synthesizing more viruses. Viruses accomplish this goal by two general strategies. One

Table 8.1 Some types of viruses, classified by nucleic acid (the Baltimore system). There are seven classes in the Baltimore system. The genetic material can be double-stranded (ds) or single-stranded (ss). Single-stranded nucleic acids can be present in viruses as the positive strand (+) or the negative strand (−). Some of this genomic material is first transcribed to DNA by reverse transcriptase (RT). Genome size is measured in thousands of base pairs (kbp) or thousands of bases (kb) in the case of single-stranded nucleic acid viruses. The examples given here are biased towards those causing diseases in humans or economically important organisms. Data from Wagner et al. (2008). See also the Universal Virus Database of the International Committee on Taxonomy of Viruses (www.ictvdb.org).

Type	Genetic material	Example family	Example virus	Genome Size (kbp or kb)	Host
I	dsDNA	Myoviridae	T4	39–169	Bacteria, Archaea, algae
II	(+) ssDNA	Parvoviridae	Aleutian mink disease (AMDV)	4–6	Vertebrates, invertebrates
III	dsRNA	Reoviridae	Rotavirus A	19–32	Vertebrates, invertebrates, plants
IV	(+) ssRNA	Picornaviridae	Hepatitis C virus	7–8	Vertebrates
VI	ssRNA-RT	Retroviridae	HIV	7–12	Vertebrates

strategy is taken by virulent viruses which go through only the lytic phase. During this phase, the virus immediately begins the process of viral replication inside the host cell after infection. Although the entire virus, genome and capsid, of some viral types may enter the host cell, other viruses inject only their genetic material into the host. For some amount of time, the viral particle is not visible and the host will not appear to be infected. This period is known as the latent phase. The virus then directs the host cell to make more viral genomic material. The virus genome may encode for some of the enzymes necessary for viral replication, but the virus also takes over key genes of the host and coerces it to make more viral genomic material and the protein building blocks of the viral capsid. When sufficient genomic material and protein coats are ready, the viral genomic material is packaged into the coats, resulting in many fully formed viruses inside the host cell. Now the host cell will appear to be infected if viewed by electron microscopy. The viruses then break out of and thus break up the host cell (lysis), expelling many viruses into the environment. The number of viruses manufactured by a single host cell, which is called the burst size, varies from a few to >100, but often is around 50 (Fuhrman, 2000).

The temperate viruses use a different strategy and have another phase, lysogeny, before the lytic one. During the lysogenic phase, the viral genomic material is integrated into the host genome and is replicated along with it for an indeterminate time after the virus infects the host cell (Fig. 8.2). In a sense, the virus goes into hiding within the host. When the host is a bacterium, the integrated viral genome is called a prophage, but there are also prophage analogs (proviruses) in eukaryotic hosts. The viral genome is replicated along with the host genome and is transmitted to daughter cells as the host reproduces, peacefully coexisting with the host. At some point, however, given the right environmental cues, the prophage wakes up and starts to make new virus particles, thus beginning the lytic phase. The switch from the lysogenic phase to the lytic phase has been examined in great detail in the laboratory because it was an early model for the regulation of gene expression. The end result is the same as for virulent phages: the host is lysed, releasing new viruses into the environment.

The lytic and lysogenic phases are extremes among several versions of how viruses replicate. Pseudolysogeny

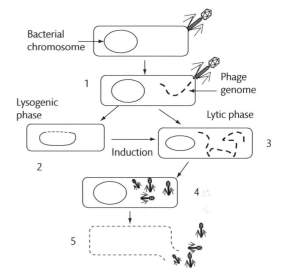

Figure 8.2 Lysogenic and lytic phases of a temperate phage. Step 1 consists of the injection of phage nucleic acid into the bacterium. In the lysogenic phase (Step 2), the phage nucleic acid is incorporated into the bacterial chromosome. The incorporated phage DNA, the prophage, is indicated by a dashed line. During induction, the prophage is excised from the bacterial chromosome and replicates itself (Step 3). During Step 4, the phage nucleic acid is packaged into phage heads. Once packaging is complete, the bacterial cell is lysed (Step 5).

is similar to lysogeny in that the pseudolysogenic virus replicates inside and along with the host for some time before starting the lytic phase. Unlike true lysogeny, however, the genome of the pseudolysogenic virus is not integrated into the host genome. Some viruses do not lyse or otherwise kill their host when ready to re-enter the environment. Rather, the host cell releases these viruses by extruding them through the membrane, or viruses are encapsulated in host membranes and are budded off away from the host cell into the environment. This process may occur over several generations, with the host in the state of chronic infection. In these cases, the host cell is being parasitized by the virus but is not killed by it.

Temperate viruses in nature

In addition to being a fundamental aspect of viral ecology, the prevalence of temperate viruses in nature has implications for relationships between viruses and their hosts. Virulent viruses kill the host soon after infection while that is not necessarily the case for temperate viruses. Consequently, temperate viruses may have a smaller immediate impact than virulent viruses, but a larger effect in shaping host populations over the long term. The number of temperate viruses in a sample from natural environments is estimated by counting viruses before and after the addition of mitomycin C or with exposure to UV light or both, which induce the switch from the lysogenic to the lytic phase. An increase in viral counts after the induction treatment indicates the presence of temperate viruses in that habitat.

The prevalence of temperate viruses, or more precisely, the ability to detect them by the induction assay, seems to be low in the aquatic habitats examined so far. For example, lysogeny was observed in less than half the samples taken during a two year study in the Mediterranean Sea (Boras et al., 2009) and in only 20% of the samples from a year-long study in the Gulf of Mexico (Williamson et al., 2002). In the latter study, the samples positive for lysogeny came from waters with low microbial growth, consistent with the principle that the switch from lysogeny to the lytic phase occurs when growth by the host is compromised. However, addition of phosphorus can induce the lytic phase in the phosphorus-stressed waters of the Gulf of Mexico. The phosphorus addition apparently enhanced metabolic activity of the host and thus enabled a higher replication rate of the viruses. Taken together, the data indicate that a virus goes into the lytic phase when its host is either growing poorly or very well. When the host cell is growing poorly, it is advantageous to leave the host when host metabolism slows and thus prophage replication lessens, but the virus has to leave before the host becomes too inactive and cannot support the final steps of viral replication. At the other extreme, when the host cell is growing well (the intracellular ATP concentration is one clue), it is advantageous to leave the host because the new viruses are likely to encounter other hosts which are also growing quickly.

As is the case for other aspects of viral ecology, little is known about the extent of lysogeny in soils, but there is some evidence that it is very prevalent. One study extracted bacteria from soils using beads and found a very high fraction (85%) of lysogenized bacteria and temperate viruses (Fig. 8.3), much higher than seen in the aquatic habitats examined to-date. Temperate viruses

may be selected for by the heterogeneous environment of soils and by the high diversity of host microbial communities. It is also possible that high detection of lysogeny in soils is an artifact of how the soil bacteria were isolated in this study.

Contact between host and virus at the molecular scale

How a virus recognizes and invades a host cell has several ramifications for thinking about host-virus interactions and viral ecology. With no means of motility, a virus depends on random motion to bring it and a host together. The two are separated by at least 30 μm in aquatic habitats, less for soils and sediments with higher numbers of viruses and microbes (Chapter 3). And the virus cannot just hit the host cell anywhere. It must bump into and recognize a specific component, a receptor, in the host outer membrane. Hitting this receptor initiates attachment by the virus to the host, followed eventually by invasion of the virus or just the viral genome into the host. The receptors are not made by the host to encourage viral attack. Rather these membrane components, hijacked by the virus to gain entry into the host, have some other function of importance to the host. A classic example is a protein encoded by the gene *lamB* for maltose transport by *E. coli*. It is at this membrane protein that the lambda phage attaches to *E. coli*. These receptors on the host surface are often proteins, but they can be the carbohydrate part of glycoproteins or glycolipids. However, proteins make for more specific receptors than carbohydrates.

Because of the specific molecular interactions between viruses and hosts, generally a virus attacks only one type of host while a host is potentially attacked by several types of viruses, each targeting a different receptor molecule in the host membrane. It is because of this specificity that we can swim in a lake or weed a garden without any worries of aquatic or soil viruses attacking us. Hosts attacked by the same virus generally belong to the same species or are even more closely related. This specialization is crucial in thinking about how viruses potentially control microbial communities (Chapter 9). One implication is that the distance between a virus and its specific host would be much more than the 30 μm calculated with total viral and bacterial abundances. Still, viruses

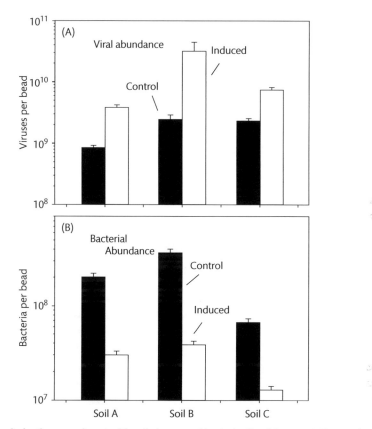

Figure 8.3 An example of an induction experiment with soil viruses and bacteria. Panel A presents the number of viruses in the control (no treatment) and in incubations with mitomycin C, which induces temperate viruses in lysogenized bacteria to switch to the lytic phase, leading to an increase in total viral abundance. The number of bacteria (panel B) decreased at the same time because of viral lysis. The unusual "per bead" units reflect how viruses and bacteria were isolated from these soil samples. Data from Ghosh et al. (2008).

affect the success of specific microbes and shape the overall composition of microbial communities. However, some viruses can attack several hosts that are not closely related. An example is some phages that attack cyanobacteria (Weinbauer, 2004). Most viruses in nature are thought to be specialized for one host, but in fact the prevalence of host specialization in nature is not really known (Winter et al., 2010).

The overall picture of virus-host interactions just sketched out applies to microbial and animal cells, but not to higher plants with thick external coverings. Viruses that attack higher plants do not use specific receptors because plants cells closest to the external environment are protected by waxes and pectin. The cellulose wall of plant cells is another barrier against viral attack. Unlike viruses of bacteria and animals, plant viruses rely on insects or mechanical breakage to get past the outer barriers of higher plants.

The number of viruses in natural environments

The methods for estimating viral abundance are analogous to those for counting bacteria and other microbes. The difficulties in isolating microbes have a direct effect on how viruses in nature are studied. Problems with these methods explain in part why we do not know more about viruses in nature.

Counting viruses by the plaque assay

This classic assay is usually used for counting phages in laboratory experiments, but it could be applied to viruses that attack any microbe capable of growth on a solid media like agar (Fig. 8.4). For the plaque assay, the host microbe is grown on an agar plate so that it forms a dense, continuous lawn of host cells. The sample containing the viruses is then poured over the top of this lawn and is incubated overnight. As the viruses replicate and lyse the host cells, holes or plaques in the lawn become visible. It is assumed that one plaque started off as one virus, which gives rise eventually to the many viruses needed to lyse enough host cells to make the plaque visible to the naked eye. So, the number of plaques on the bacterial lawn equals the number of viruses in the original sample. In addition to counting viruses, the plaque assay can be used to isolate different viruses from complex mixtures. Each plaque contains many clones of the original virus first landing on that section of the lawn. The plaque can be subsampled and the viruses from it grown in liquid culture with the host cells.

The plaque assay greatly underestimates the total number of viruses actually in a natural sample. In fact, viruses must first be concentrated before it is even possible to detect viruses capable of infecting the single bac-

terial strain used in the plaque assay. In one study of the Chesapeake Bay, for example, only 10 of 36 samples yielded detectable viruses and the overall estimate of viral abundance was seven plaque forming units (pfu) per liter (Wommack and Colwell, 2000). In fact, there were about 10^{10} viruses per liter in those samples. The reason for the severe underestimation by the plaque assay is the inability to grow the right host cell on solid media, another consequence of the culturability problem encountered in Chapter 1. As mentioned several times already in this book, nearly all of the bacteria and most of the other microbes known to be present in nature cannot be grown on solid media. Non-traditional cultivation approaches that are successful in growing these microbes in the laboratory are not easily modified to include the plaque assay or anything similar to it. Consequently, approaches based on counting plaques miss the many viruses that infect uncultivated microbes.

The inadequacy of the plaque assay approach has a huge consequence for examining viruses in nature. It means that most viruses in nature cannot be isolated, identified, and studied in the laboratory by traditional methods, as is the case for nearly all microbes found in nature. If the host cannot be grown in the laboratory, and many microbes cannot, then the virus cannot be isolated and identified, at least by traditional methods. Cultivation-independent approaches are starting to reveal much about viruses (Chapter 10), but we still know little about the types of viruses and their ecological roles in nature. We know little because there is no analog to the rRNA-based approaches used to examine uncultivated prokaryotes and eukaryotes. Whereas every microbe has at least one rRNA gene, no gene is common to all viruses. Their genomes vary too much, even in the type of nucleic acid. The variation in viral genomes reflects the diversity of strategies used by viruses to infect their hosts. The diversity of viruses in nature has been explored using a few genes, such as those for DNA polymerases and capsid proteins (Rowe et al., 2011). But there was never any illusion that these phylogenetic markers covered all or even most viruses in a habitat. One solution to this problem is given in Chapter 10.

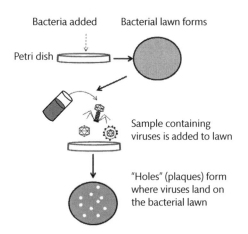

Figure 8.4 The plaque assay for counting and isolating viruses from a sample. Each of the holes or plaques in the lawn corresponds to where a virus landed on the bacterial lawn. The virus can be purified and studied in more detail by subsampling the plaque. Here the host is a bacterium, but it could be any microbe capable of growing on an agar plate and forming a continuous lawn of cells.

Bacteria added   Bacterial lawn forms
Petri dish
Sample containing viruses is added to lawn
"Holes" (plaques) form where viruses land on the bacterial lawn

Counting viruses by microscopy

The plaque assay indicated that viruses were present in nature, but the assay indicated very low numbers. It was not until transmission electron microscopy (TEM) was used to examine samples from coastal marine habitats that the high abundance of viruses was discovered (Torrella and Morita, 1979, Bergh et al., 1989). The first step in the TEM method is to spin down viruses by centrifugation onto a small grid placed at the bottom of the centrifuge tube. When the centrifuge run is done, the small grid is taken out, prepared, and viewed by TEM. The TEM pictures revealed many amorphous particles of unknown origin, but also particles with shapes and sizes identical to known viruses (Fig. 8.5A), including classic ones like the T4 virus. If it is assumed that all viruses in a known volume of water are collected onto the TEM grid, the number of viruses in the original sam-

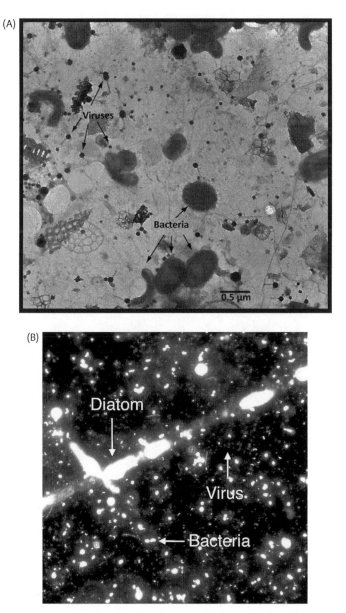

Figure 8.5 Examples of viruses in nature, as revealed by transmission electron microscopy (A), and epifluorescence microscopy (B). Panel A is used with permission from K.E. Wommack and Panel B is used with permission from M. T. Cottrell.

ple can be estimated by counting the viruses in a known area of the grid. Microbes infected by viruses can be seen with or without thin sectioning of the microbes and examination by TEM. While quite powerful, the TEM approach has its problems. It requires expensive instruments which are difficult to operate. Another problem is that viruses obscured by particles or cells would not be counted.

Enumerating viruses by epifluorescence microscopy gets around both problems. The approach is nearly the same as that used to count bacteria and other microbes, with one critical difference being that the sample is stained with a very bright nucleic acid stain such as SYBR Green I. When viewed by epifluorescence microscopy viruses stained with SYBR Green I look like small pin points of green light while bacteria and other microbes appear huge in comparison (Fig. 8.5B). The smallest particles are viruses, a conclusion that has been confirmed by other tests. Remarkably, estimates of viruses by this approach are similar (only about 30% higher) to those by the TEM method (Fuhrman, 2000). It is likely that the TEM method underestimates viral abundance.

Still another method for enumerating viruses in aquatic habitats is by flow cytometry. As with other non-pigmented organisms counted by this method, the viruses stained with a bright fluorescence stain, such as SYBR Green I, are distinguished from other microbes and particles by fluorescence and side scatter in the flow cytometer. The instrument must be very clean and free of all other particles in order to lower the background noise and to detect viruses. When used properly, flow cytometry yields similar estimates of viral abundance as epifluorescence microscopy but with higher precision and greater ease and speed, all important when processing many samples.

Variation in viral abundance in nature

Because of studies using the TEM and microscopic direct count approaches, we can now make intergalactic analogies about viral abundance in nature. These studies revealed very high numbers of viruses in virtually all habitats of the biosphere, ranging from about 10^7 per milliliter in aquatic habitats to 10^{10} per gram of sediments (Table 8.2), in stark contrast to the plaque assay results. Viruses are found everywhere microbes live, including

extreme environments. Some viruses with weird shapes attack hyperthermophilic archaea in hot springs (Prangishvili et al., 2006). The discovery of high abundance prompted many studies of viruses and their role in nature. An informative way to express viral abundance is relative to bacterial abundance; that is, the virus to bacteria ratio (VBR). As mentioned before, most viruses are thought to use bacteria as hosts because bacteria are usually much more abundant than other microbes and organisms in nature. In many environments, VBR is about 10, quite commonly so in aquatic systems, which is not entirely understood. But there is much variation in this ratio over time and among environments. The ratio varies up to a thousandfold for aquatic systems, but nearly 10 000 for soils (Srinivasiah et al., 2008).

Soils, especially agricultural soils, have some of the highest levels of viruses found so far, with VBR values exceeding 2500 (Table 8.2). Perhaps the high and highly variable VBR values in soils are because some of the viruses are targeting fungi. As mentioned in Chapter 5, fungi are often quite abundant in soils and the ratio of fungi to bacteria varies, which could lead to higher and

---

Box 8.1 Detection limits of light and electron microscopy

In theory, the smallest particle that can be resolved clearly by light and other electromagnetic radiation is very roughly one half its wavelength. This means that the detection limit of electron microscopy (the wave-like nature of electrons being important here) is about 0.2 nm, depending on the voltage of the microscope while the smallest particle that could be seen by visible light microscopy is about 200 nm (0.2 μm). Given these limits set by physics, it would seem physically impossible to see most viruses by light microscopy. However, viruses, even those as small as 50 nm, are visible under epifluorescence microscopy when stained with bright nucleic acid stains because the light from the fluorescing virus flares out, forming an image larger than the actual size. Still, very small viruses and those with single-strand nucleic acids are likely to be missed by epifluorescence microscopy.

Table 8.2 Number of viruses and bacteria in some natural environments, per milliliter for the aquatic habitats or per gram for soils and sediments. "Viruses:Bacteria" refers to the ratio of viral abundance to bacterial abundance. The aquatic data are from Wommack and Colwell (2000), the soil data are from Williamson et al. (2005) and the sediment data are from Danovaro et al. (2008). See also Srinivasiah et al. (2008).

	Habitat		Number of viruses (10^6 per ml or per g)	Virus:Bacteria
Freshwater	Lake Plußsee	Spring	254	41
	Quebec lakes	Summer	110	23
	Danube River	Entire year	12–61	2–17
Marine	Chesapeake Bay	Spring	10	3.2
	South California	Spring	18	14.2
	North Pacific	Spring-fall	1.4–40	2.3–18
Sediments	Lake Gilbert, Quebec	2–14 meters	720–20 300	0.8–25.7
	Chesapeake Bay	1–17 m	340–810	57
	Sagami Bay, Japan	1450 m	290–2560	8.0–35.0
Soils	Silt loam	Corn field	1100	2750
	Loamy sand	Corn field	870	3346
	Silt loam	Forest	2940	11
	Piedmont wetland	Forest	4170	12

more variable VBR values in soils. On the other hand, many fungal viruses ("mycoviruses") may not be counted by the standard epifluorescence microscopic technique because many of the known mycoviruses have genomes composed of RNA (Yu et al., 2010b) which would be stained poorly if at all by common DNA stains. Also, many mycoviruses are transmitted intracellularly (Ghabrial, 1998) and would not even be countable by standard techniques regardless of stain. Another reason for the high variability of VBR in soils is that viruses are less sensitive than bacteria to desiccation and other environmental factors. So, total bacterial abundance may vary independently of total viral abundance, leading to high variability in VBR.

As a general rule, conditions that promote the growth of bacteria and other microbes lead indirectly to more viruses, and time periods and locations with high microbial abundances are also the times and locations with many viruses. Over a day, however, viral abundance can lag behind bacterial abundance in aquatic habitats. In the example given in Figure 8.6, the maxima in bacterial abundance at 1300 h, and 2400 h were followed by maxima in virus abundance a couple of hours later. This lag is reminiscent of predator-prey cycles in which prey (bacteria) abundance goes up and down, followed by changes in predator (protist grazers) abundance, sepa-

rated in time. Bacteria-virus interactions are equally dynamic in soils (Srinivasiah et al., 2008). In one experiment, yeast extract added to soil caused an increase in bacterial abundance within a day, followed by a corresponding increase in virus abundance, similar to what is illustrated in Figure 8.6.

The lag between changes in abundances over time leads to variation in VBR. This ratio tends to be low in times and places with high bacterial abundance (Wommack and Colwell, 2000), which can be simply a consequence of the predator-prey nature of virus-bacteria interactions. However, these interactions are more complicated than implied by a single predator eating a single type of prey. Because specific viruses attack specific hosts, the diversity of the bacterial community has an impact on viral abundance. Low diversity and fewer bacterial phylotypes mean fewer viruses, all else being equal. So, VBR would be low if high bacterial abundance is not accompanied by high diversity. On the other hand, VBR can be high for some nutrient-rich, productive environments in which hosts are capable of high growth rates. In these environments, more bacteria and other microbes lead to higher infection rates (more encounters between hosts and viruses), and a fast-growing bacterium can produce more viruses (larger burst size).

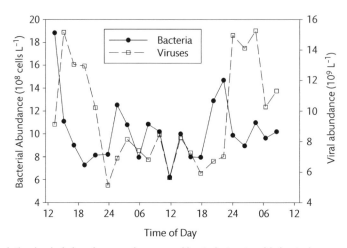

Figure 8.6 Example of variation in viral abundance and presumed hosts, heterotrophic bacteria, over nearly two days. Data from Weinbauer et al. (1995).

Mortality of bacteria due to viruses

One of the main roles of viruses in nature is similar to that of grazers. Like heterotrophic protists, viruses help to control biomass levels of their hosts. Any increase in the number of a particular bacterial strain attracts more viruses and viral lysis, leading to the reduction in that particular strain. This form of control has been called "killing the winner" (Chapter 9). But how important is this control mechanism? How much of total microbial mortality is due to viral lysis versus grazing?

The high number of viruses suggests that viral lysis accounts for a large fraction of bacterial mortality, but that is not necessarily the case. A free virus may be non-infectious and incapable of attacking a host cell. Even if all free viruses are infectious, a rate cannot be estimated from a standing stock measurement (here, the number of viruses) without many assumptions. So, we need a more direct method. There are at least six of these methods (Weinbauer, 2004), one indication of the difficulty in measuring viral processes. Two methods are discussed here as a way to learn more about viral ecology.

Percentage of infected cells

Perhaps the most direct method to estimate viral lysis is to count the number of bacterial cells that are infected by viruses. Since an infected cell is doomed for eventual death by lysis, the fraction of the bacterial community that is infected is the fraction of bacterial growth being controlled by viruses. The fraction of infected cells, visible by TEM, is low, about 1–5% in nature (Proctor and Fuhrman, 1990, Weinbauer and Peduzzi, 1994), which would seem to imply an equally low effect on bacteria. However, the observed fraction does not take into account infected cells without visible viral particles. Cells could have viruses that have not yet reached the stage of having viral capsids and are not yet recognizable as viruses. The final stage of viral infection when viral particles are visible inside of host cells takes up only about 10–20% of the entire viral life cycle. So, the fraction of infected cells has to be increased, basically by the 10–20% factor, to estimate the full impact of viruses on their host. When thus corrected, the estimate of 1–5% infected cells implies that 5–50% of bacterial mortality is due to viruses.

The viral reduction method

As implied by the name, in this approach the number of free viruses is reduced (ideally, to zero) by filtering water through filters with pore sizes small enough to remove viruses while retaining bacteria and other microbes. (As with many methods in microbial ecology, this one does not work with soils or sediments). Virus-free water is then added to a concentrated bacterial fraction, with or without mitomycin C, to induce the lytic phase of temperate phages in lysogenized bacteria. Total viral

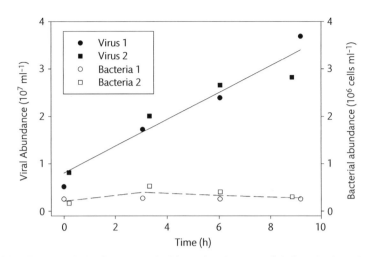

Figure 8.7 Example of data from the viral reduction method for estimating rates of viral production. Virus and bacterial abundance are reduced by filtration and dilution with virus-free water. In this case, the abundances of both viruses and bacteria were reduced by 60–90%. The increase in viral abundance is due to their release from lysogenized bacteria and bacteria lysed by virulent viruses. Bacterial abundance does not change on this timescale. Results from two separate incubations are given. Data from Wilhelm et al. (2002).

abundance should increase over time, as both virulent and temperate phages (in the presence of mitomycin C) are released from hosts infected before the experiment began. The very low number of viruses left in the sample ensures that no viruses are lost due to adsorption and no new viruses are produced by new attacks on bacteria. Also, it is easier to measure an increase in viral particles when initial levels are very low. An example of this method is given in Figure 8.7. Estimating viral mortality with the viral reduction method requires estimates of the burst size. Although far from perfect, this method is one of the best for estimating rates of viral mortality (Boras et al., 2009). This method confirmed earlier reports that viruses were responsible for a large fraction (10–50%) of bacterial mortality.

Contribution of viruses versus grazers to bacterial mortality

The results from the two methods just discussed, plus from all of the other methods, indicate that viruses can account for a large fraction of bacterial mortality. Roughly about half of all bacterial mortality can be attributed to viruses, the other half to grazing by various protists, but these percentages vary greatly. The relatively few habi-

tats examined so far are nearly all marine, and there are no data from soils. Even among these environments, the fractions attributed to viruses and grazers vary greatly.

Some of the environments in which viruses are especially important tend to be eutrophic. Grazing accounted for all of the measured bacterial production in oligotrophic marine environments, but not in eutrophic ones, according to one analysis of several studies published at the time (Strom, 2000). Since that analysis was done, other studies found that viruses were responsible for about half of all bacterial mortality in the oligotrophic Mediterranean Sea and North Atlantic Ocean (Boras et al., 2009, Boras et al., 2010). Still, it makes sense that the impact of viruses would be higher in environments where high nutrients promote more cell production and higher biomass, leading to more contact between viruses and hosts and larger burst sizes.

Viruses also contribute more to bacterial mortality in habitats where protists do not grow well. Potentially, these habitats include Arctic sea ice where viral abundance is very high (Maranger et al., 1994) and environments with low pH, high salt concentrations, or high temperatures, all factors that select against protists and other eukaryotes. Viruses are less abundant in extreme environments, but probably because host cells are less abundant. Adsorption of viruses to host cells and other particles is affected by pH and salt. Specifically, divalent cations, such as Mg^{2+}, can promote adsorption of viruses to particles (Weinbauer, 2004). Perhaps most importantly, viruses are likely to be the major, if not the only form of mortality of bacteria in anoxic habitats where the lack of oxygen excludes nearly all eukaryotes. Some of the highest levels of viruses and rates of viral lysis were found in an anoxic hypolimnion of a lake (Weinbauer and Höfle, 1998).

Viral production and turnover

Viral production is analogous to cell production by bacteria and other microbes. Data on this parameter is one way to explore the impact of viruses on bacteria. Analogous to bacterial production from rates of DNA synthesis, viral production is estimated from the incorporation of radiolabeled thymidine (for the DNA viruses) or phosphate (for viruses with either DNA or RNA) into the viral size fraction (Steward et al., 1992). The method confirmed the initial reports that viruses were responsible for 10–50% of bacterial mortality. The data also can be used to get a sense of how fast the pool of viruses turns over. What is the half-life of a free virus in nature? Let us answer that question relative to the turnover of the host of many viruses, the heterotrophic bacteria.

The rate of viral production (P_V) depends on rate of bacterial production (P_B), the burst size (S), and the fraction of bacterial mortality due to viral lysis (F):

$$P_V = P_B \cdot S \cdot F \qquad (8.1).$$

This equation means that every bacterial cell that is produced but doomed to be lysed will produce S number of viruses. We know that

$$P_B = \mu \cdot B \qquad (8.2)$$

where μ is the bacterial growth rate and B is bacterial abundance. Let us define a turnover time for viruses (V) as

$$T_V = V \cdot P_V^{-1} \qquad (8.3)$$

and the analogous one for bacteria, which is nearly equal to the generation time:

$$T_B = B \cdot P_B^{-1} \qquad (8.4).$$

Substituting Equations 8.2–8.4 into 8.1, using the ratio of virus to bacterial abundance (VBR) and rearranging, we end up with:

$$T_V = T_B \cdot VBR \cdot S^{-1} \cdot F^{-1} \qquad (8.5).$$

Equation 8.5 implies that the turnover time of viruses varies with the bacterial turnover time, but as a function also of VBR and the inverse of the burst size and the fraction of bacterial production killed off by viruses. If we use typical values, such as VBR = 10, S = 50, and F = 0.5, then the turnover time of viruses would be 25-fold faster than that of bacteria. For a bacterial population growing on the order of a day, viruses would be turning over about once every hour. Except for low values of F, most estimates of VBR and S indicate that viruses turn over much faster than bacteria. These calculations yield estimates similar to actual data indicating turnover of over one to three times per day in productive estuarine waters (Winget and Wommack, 2009).

Viral decay and loss

The fast turnover of viruses just calculated seems consistent with the high number of viruses in nature. But without any mechanism to get rid of viruses, they would be even more abundant, filling up every available cubic micron in the biosphere. Fortunately, there are ways in which viruses are neutralized and eliminated. Decay refers to a decrease in the capacity for viruses to infect their hosts (infectivity) while loss is the reduction in the number of viral particles. Measuring viral decay and loss is one way in which to explore the impact of viruses on their bacterial hosts (Heldal and Bratbak, 1991). If the viral abundance is roughly constant over time, then viral decay and loss must be equal to the rate of viral production, which in turn is a measure of mortality due to viruses. The viral decay approach, however, has yielded suspiciously high rates of viral mortality (Fuhrman, 2000). Regardless of methodology, decay and loss are important processes in thinking about the ecology of viruses.

The most important factor causing viral decay in aquatic systems is sunlight, specifically ultraviolet (UV) light, that damages viral genomic material beyond repair. Similar to its effect on microbes, light can damage viral nucleic acids directly or indirectly via the formation of reactive compounds such as superoxides produced by

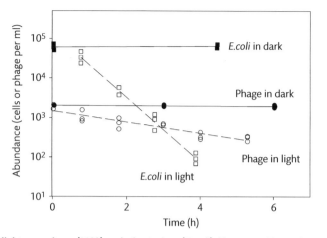

Figure 8.8 The effect of sunlight on a phage (MS2) and a bacterium (*E. coli*). Phages and bacteria were enumerated by methods that depend on infectivity (plaque assay for the phage) or capacity to grow (plate count assay for bacteria). Data from Kapuscinski and Mitchell (1983).

sunlight. Inactivation and decay have been well studied for viruses (coliphages) infecting coliform bacteria, such as *E. coli*, because coliphages have been used as models for understanding the fate of human pathogenic viruses in the environment. In the example shown in Figure 8.8, light kills off both the phage and *E. coli*. But sunlight alone would not necessarily result in loss of viral particles, and it cannot explain viral decay in soils, sediments, and subsurface habitats where sunlight never reaches.

In soils and sediments, adsorption to colloids and other particles inactivates viruses and is likely a major loss mechanism. Other physical-chemical factors such as pH and salt concentrations affect viral infectivity because of how they affect adsorption, in addition to direct effects on viruses and their host (Kimura et al., 2008). Drying of soils also leads to inactivation of viruses, as assessed by cultivation-dependent assays, but viral particles still could remain and be counted, leading to high estimates of viral abundance and VBR in some dry soils (Srinivasiah et al., 2008). Temperature is often the property that explains most of the variation in cultivation-dependent assays of virus infectivity in soils (Kimura et al., 2008); infectivity decreases with higher temperature, even though burst size increases with temperature. Some of the other mechanisms accounting for viral loss involve microbes. Heterotrophic bacteria may be able to degrade viruses, treating them as just another nutrient-rich parti-

cle, and protists may graze on large viruses. But there is not much evidence for these processes except for experiments indicating a role for biotic processes of some sort (Weinbauer, 2004).

Viruses of phytoplankton

The previous sections focused on viruses of bacteria, especially heterotrophic bacteria, but viruses attack every organism in the biosphere with potentially large impacts on the biology and ecology of those organisms. Of course, *Homo sapiens* is no exception. More people died during the influenza pandemic of 1918–1920 (50 million worldwide) than were killed in the trenches of World War I (16 million). Today, AIDS caused by HIV remains a deadly foe in developed countries and a devastating one in several developing countries, especially in sub-Saharan Africa. Viruses, such as the one responsible for foot and mouth disease, cause billions of dollars of damage in livestock each year. Viruses also kill off wildlife, but these are not as well studied. What are known are the many connections between viruses of wild and domesticated animals, and humans. Many influenza viruses start off in pigs and birds before evolving to infect humans, and other viruses now capable of infecting humans apparently first used wild primates or other mammals as hosts. These viruses include the Ebola and

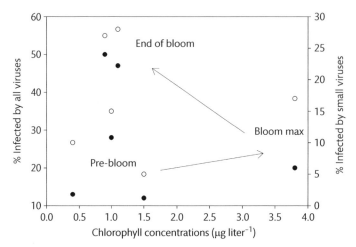

Figure 8.9 Viruses in the coccolithophorid *Emiliania huxleyi* during a bloom in the North Sea. The presence of small viruses (filled-in circles) and total viruses (open circles) in the alga was examined by transmission electron microscopy. The stage of the bloom was deduced from reflectance measured by satellites and nitrate and chlorophyll concentrations. Nitrate concentrations were low at the end of the bloom but high in pre-bloom waters. Data from Brussaard et al. (1996).

Marburg viruses, as well as HIV (Daszak et al., 2000). A disease caused by viruses originating from these other animals is called a zoonosis.

Viruses also affect cyanobacteria and eukaryotic algae. One motivation for examining viruses of photoautotrophic microbes is that viral lysis could be a mechanism of controlling algal growth and stopping phytoplankton blooms, including harmful ones, such as the brown tide alga. Reports appeared in the early 1960s of viruses capable of infecting what were called at the time, "blue-green algae", and there was already discussion of how these viruses (cyanophages) might affect blooms of cyanobacteria (Suttle, 2000). We now classify these cyanophages in the Myoviridae, Siphoviridae, and Podoviridae families. Viruses of the green alga *Chorella* were isolated soon after the cyanobacterial work, but evidence of viruses attacking ecologically more important eukaryotic algae did not appear for several years. Now we know of many viruses attacking algal species important in soils, lakes, and the oceans. Viruses are known to infect even diatoms and coccolithophores, somehow getting past the silicate and calcium carbonate coats of these algae (Tomaru et al., 2009). Many viruses of algae are large with double-stranded DNA in the Phycodnaviridae family, but others have single-stranded DNA or single-stranded RNA.

Viruses attacking the coccolithophore *Emiliania huxleyi* are a good example of virus-algal interactions. This phytoplankton is common in the oceans and often is the dominant alga in spring blooms in the North Atlantic Ocean. Figure 8.9 illustrates how as many as 50% of all *E. huxleyi* cells were visibly infected by viruses at the end of a phytoplankton bloom in the North Sea, implying very high lysis rates. Later work demonstrated that an *E. huxleyi* virus has genes for glycosphingolipid synthesis that are turned on when the virus infects the alga (Vardi et al., 2009). The isolated glycosphingolipid alone is sufficient to kill the alga by setting off a series of biochemical events similar to programmed cell death. A survey in the North Atlantic Ocean found concentrations of the glycosphingolipid to be high where a coccolithophore pigment (19′-hexanoyloxyfucoxanthin) was low (Vardi et al., 2009), implying that viral lysis caused variation in the abundance of coccolithophores in this oceanic region.

Viruses are not grazers

One ecological role of viruses, that of killing off hosts and effecting a form of top-down control, is similar to that of grazers. We saw that roughly half of all bacterial mortality is due to viruses, the other half being taken care of by grazing. But this similarity between viruses and grazers is rather superficial; the ecological roles of viruses differ from those of grazers in many ways. The rest of this chapter discusses these differences and the other ecological roles of viruses.

Viral shunt and DOM production

Grazing and viral lysis both kill prey and host cells, but what physically remains after the two processes are completed differs greatly. A grazer can completely consume its prey, oxidizes organic carbon to carbon dioxide, and mineralizes the organic nitrogen and phosphorus to ammonium and phosphate and any other inorganic nutrients not needed by the grazer. In contrast, a virus needs the host cells to remain viable up to the end, throughout the synthesis of viral components and assembly of new viral particles. The biochemical machinery of the host cell is hijacked by the virus and transformed to suit the purpose of the virus, but it is not totally destroyed. Consequently, lysis by the virus releases the entire cellular contents of the host cell into the environment with little oxidation or mineralization. In soils, the released cellular contents may adsorb onto surfaces whereas in aquatic habitats they become part of the dissolved organic material (DOM) pool.

This production of DOM by viral lysis and its subsequent use by microbes is called the viral shunt (Fig. 8.10). Most of the organic compounds released by viral lysis are thought to be labile and readily used by microbes. Lysis of algae or higher plant cells would make organic material available for bacteria and fungi that otherwise may have gone to an herbivore. Viral lysis of heterotrophic bacteria and fungi, which use DOM and detritus anyway, does not make "new" organic material available

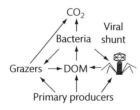

Figure 8.10 The viral shunt and the rest of the microbial food web. This diagram implies that only bacteria and phytoplankton are attacked by viruses, but in fact viruses infect all organisms, potentially releasing dissolved organic material (DOM) and other detrital organic carbon.

to the surviving heterotrophic microbes, but it still could have a positive impact on their growth. Viral lysis may release material containing potentially limiting elements, like phosphorus and iron (Riemann et al., 2009, Poorvin et al., 2004). Some evidence suggests that average bacterial growth is faster with viruses than without them because those bacteria not being lysed feed off the DOM produced by viral lysis. DOM from viral lysis potentially is a large part of total DOM production (Evans et al., 2009). The relative contribution of DOM from viral lysis is roughly equal to the fraction of primary production and bacterial production that is lysed by viruses, since nearly all of the algal and bacterial biomass in cells lysed by viruses would enter the DOM pool.

Population dynamics of a virus and its host

Phage-bacteria systems have served as models for exploring theoretical questions about predator-prey interactions (Kerr et al., 2008). The same mathematical model for predator-prey interactions can be applied to exploring the population dynamics of a virus and its host. In both cases, the basic feature of these dynamics is that one population oscillates out of phase with the other. However, there are crucial differences between virus-host interactions and predator-prey interactions.

One difference is that the host can evolve defenses against the virus more easily than prey can fend off predators. A single mutation can lead to the host being impervious to viral attack. In laboratory experiments with a single bacterium and a phage, a spontaneous mutant of the bacterium often arises that is resistant against the phage. Among several mechanisms for resistance, one is simply not to synthesize the protein or other membrane components used by the virus to recognize the host. For example, *E. coli* without the maltose-transport protein cannot be attacked by phage lambda. Even if the virus gets past the first line of defense by the host, the viral genome may be degraded by enzymes ("restrictionases") and inactivated. So, a microbe cannot be driven to extinction by a virus, as has been demonstrated for a virulent virus infecting a phytoplankton species (Thyrhaug et al., 2003). In contrast, under the right conditions grazing can wipe out the prey, reducing its numbers beyond recovery. Some microbes can evolve mechanisms to avoid grazing, such as forming large

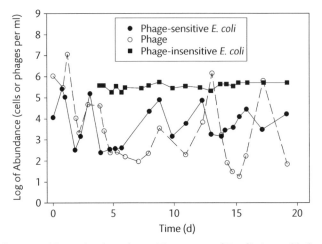

Figure 8.11 Coexistence of phage-sensitive and a phage-insensitive mutant of *E. coli*, along with the phage in an organic carbon-limited continuous culture. The phage-insensitive mutant does not take over because it is a weak competitor for organic carbon. Note the predator-prey oscillations between the phage and the phage-sensitive *E. coli* strain. Data from Bohannan and Lenski (1999).

chains or clumps that are too big for the grazer to eat. But there are fewer of these mechanisms than in the antiviral repertoire of microbes.

Why then are not all microbes resistant against viruses? How can viruses exist? The answer has two parts.

First, there is a cost associated with resisting viruses. Because of these costs, virus-resistant microbes grow more slowly than the original, virus-sensitive strain. Consequently, virus-resistant microbes do not necessarily become the dominant members of the community. Simple laboratory experiments have shown that virus-resistant and virus-sensitive microbes can coexist (Fig. 8.11), with the former being regulated by bottom-up control (organic carbon) and the latter by top-down control (virus lysis). Elimination or even just modification of the virus cell-surface receptor could affect how the host interacts with its environment. For example, an *E. coli* strain without the maltose-transport protein cannot be attacked by phage, but it also would no longer be able to take up maltose. The cost of the antiviral defense may not be so obvious; the phage-resistant *E. coli* strain may still grow more slowly than the parent strain even if the main carbon source is not maltose. The cost depends on the type of mutation, the metabolic capacity of the host, and the environment.

The other part of the answer is that viruses evolve in response to the microbial host. The complete loss of a host receptor would be difficult to overcome, but more subtle changes in the host receptor would select for mutant viruses able to recognize the mutated receptor. The mutant virus now is able to attack the once resistant mutant host. The cost for the mutant virus is that it would not be able to recognize the original receptor or attack that host. In addition to selection due to virus-host receptor interactions, there may be selection for mutations in the latency period or the burst size. Evolution doesn't just work on viruses, of course; mutations in a virus may select for mutations in its microbial host, and so on and so on. The end result is an evolutionary arms race between viruses and their microbial hosts.

Genetic exchange mediated by viruses

So far, it seems that viruses have only negative impacts on host populations, and it is certainly true that microbes have evolved many mechanisms to avoid viral lysis, as implied by the arms race metaphor. However, viruses also have a potential positive impact with many far-reaching implications for the ecology and evolution of microbes and higher organisms. Viruses are a form of sex for microbes. In more scientific terms, viruses mediate the exchange of genetic material among microbes. We will focus on bacteria for now.

Bacteria normally reproduce by asexual cell division, meaning that the same genetic material is passed from mother to daughter without the mixing of genes from another cell. However, bacteria can take on genes from other cells via three mechanisms. The first, transformation, is the uptake of free DNA from the surrounding environment. The second, conjugation, involves exchange of DNA from one cell to another via a protein-aceous tube (pilus) that connects the two cells. The third mechanism, transduction, involves viruses. It is not clear which mechanism is most common or important in nature, but transformation is probably the rarest. Microbes seem more likely to degrade DNA and use it as a source of carbon, nitrogen, and phosphorus than as a source of new genetic material. In contrast, viral infection of microbes is quite common, and viral genes are abundant in the genomes of microbes and larger organisms. The exchange rate among microbes mediated by viruses is thought to be very high in nature (Jiang and Paul, 1998). Some of this exchange is by genetic transfer agents (GTAs), which are virus-like particles that contain only host DNA (McDaniel et al., 2010).

Viruses can mediate the exchange of microbial genes because of "mistakes" during the packaging of the viral genetic material into viral particles (Fig. 8.12). While the viral genetic material is being placed into the capsid, genes from the microbial host genome can be included as well. The host genes are either randomly selected from the entire genome (general transduction) or the genes may be specific ones because the virus inserts itself into specific sites within the host chromosome (specialized transduction). In either case, the newly formed virus will now carry those host genes into a new host after infection. The newly infected host could express the virus-borne host genes and potentially gain a new metabolic capacity, a new version of metabolism it already had, or simply more of the same metabolism.

The best-studied cases of bacteria taking on new metabolic pathways involve the conversion of a nonpathogenic strain of a bacterium to a pathogenic form caused by virulence genes introduced into the bacterium by a phage. One example is the conversion of *Vibrio cholerae*, which is usually an innocuous estuarine bacterium, to a cholera-causing pathogen due to infection by the filamentous phage CTXphi. The phage carries a "pathogenicity island" consisting of the cholera toxin and other genes including those for pili that facilitate attachment. Infection by the viruses promotes survival of this bacterium by enhancing attachment to chitin and gut cells and by increasing the capacity to embed itself in biofilms (Pruzzo et al., 2008, Faruque et al., 2006).

Because of transduction, bacterial genes have been discovered in viruses isolated from various natural habitats. One example is genes for photosynthesis found in cyanophages that infect *Prochlorococcus* and *Synechococcus* (Lindell et al., 2005), the most abundant photoautotrophs in the biosphere (Chapter 4). The genes include *psbA* encoding the photosystem II core reaction center protein D1 and high light-inducible (*hli*) genes. These genes, carried in by the cyanophage, are co-transcribed along with phage genes and are expressed in the cyanobacteria during infection. This expression leads to higher rates of photosynthesis and of phage reproduction, benefitting both the host (if only temporarily) and the phage.

It seems inevitable that a viral infection leads to the death of the host, yet that is often not the case. For many virus-host interactions, cell death may be the exception rather than the rule. Temperate viruses may lose the capacity to excise themselves out of the host chromosome, resulting in the lytic phase never getting started. In addition to sparing the host cell from being lysed, another consequence is that the viral genetic material, along with genes from prior hosts, becomes a permanent fixture in the host genetic material. In fact, the genomes of bacteria, other microbes, and indeed all organisms, including humans, are littered with the remains of viruses. As many as 70% of all sequenced bacterial genomes have prophages (Paul, 2008). Being replicated along with host genetic material is one mechanism for a viral gene to reach its selfish goal: to make more copies of itself. That goal can be attained by the gene inhabiting a free viral particle or by being tucked away among the many genes of a host genome. The microbe also wins. It gains potentially new genetic material—genes that may prove advan-

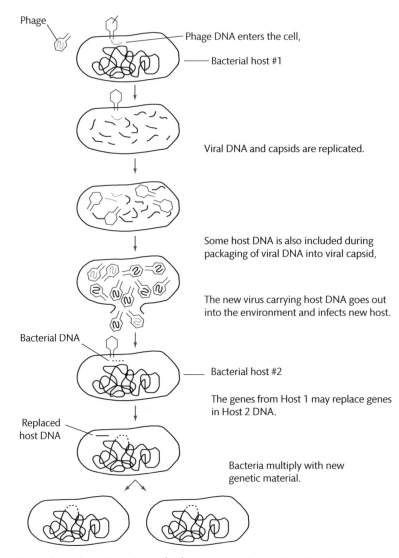

Figure 8.12 Transduction, a form of genetic exchange (sex) among microbes mediated by viruses.

tageous as the microbe is challenged by changing environmental conditions.

The acquisition of new genetic material is the stuff of evolution and one reason why Baltimore is misguided in saying we wouldn't miss viruses if they weren't here. Of course we should work to eradicate viruses that cause

diseases in humans in order to minimize the "bad news" disliked by Medawar. But the vast majority of viruses do not cause disease in humans or in agriculture, and many have positive impacts on other organisms (Roossinck, 2011). Viruses play essential roles in nature and are essential to life as we know it.

Summary

1. Viruses consist of nucleic acids surrounded by a protein coat and for some, membrane material from a previous host. The nucleic acids occur in every possible variety of DNA and RNA and vary among different types of viruses.

2. Viruses reproduce by taking over the biochemical machinery of the infected hosts. Virulent viruses have only a lytic phase whereas temperate viruses have a lysogenic phase as well as the lytic phase. Lysogeny may be favored in oligotrophic environments.

3. Microscopic methods indicate that viruses are very abundant, about tenfold more abundant than bacteria, their most probable hosts. In contrast, the plaque assay recovers very few viruses from natural environments, which means that few viruses can be isolated and grown in the lab.

4. Viral lysis accounts for about 50% of bacterial mortality with the other 50% due to grazing, although these percentages vary greatly among environments. Viruses can also be important in stopping phytoplankton blooms. They affect the biology and ecology of all organisms in the biosphere.

5. Hosts can evolve defenses against viruses, such as changing the receptors used by viruses to recognize the host. These defenses come with costs. Viruses also can evolve in response to the host.

6. Viruses have many positive impacts on microbes, including the release of DOM during lysis. They also mediate the exchange of genetic material among hosts, affecting the evolution of microbes and other organisms.

CHAPTER 9

Community structure of microbes in natural environments

The previous chapters discussed the ecological and biogeochemical roles of microbes in natural environments while mentioning only a few names of genera and species. The early studies of eukaryotic phytoplankton, protozoa, and other protists did identify some of these organisms, evident from the number of taxonomic names given in the previous chapters, but even for these microbes, it is possible to discuss many processes without referring to specific taxa and without knowing which species are present. As mentioned in Chapter 1, this approach is sometimes called "black box microbial ecology".

In this chapter, we will start to open up the black box and learn about the types of microbes that dominate natural environments. Microbial ecologists sometimes use the term "community structure" when referring to the list of organism names, their phylogenetic relationships, and abundances in an environment. One motivation for exploring community structure is to gain insights into biogeochemical processes and other "functions" mediated by microbes. The connection between community structure and function is a great unsolved problem in microbial ecology today. Even if there wasn't a problem, there are other reasons to learn about microbial community structure. We should want to know the names of the most abundant organisms on the planet, the microbes. Naming organisms is part of putting them in order, the goal of taxonomy, and it is also part of understanding the evolutionary relationships among organisms, their phylogeny. Finding out about the diversity of microbes in nature is an intriguing puzzle and an essential part of understanding the diversity of life on earth.

Taxonomy and phylogeny via genes

The traditional way of classifying an organism is simply by looking at it, taking note of various characteristics, such as the number of legs, and the presence of hair, scales, seeds, or flowers. Internal features, such as backbones or cell walls, also are important, but even these are found by visually inspecting the organism. This general approach works for just a few microbes. The problem is not only their small size, but the lack of distinguishing features. While some protists and fungi have distinctive features and can be identified by their appearance, many cannot. Even similar appearing eukaryotic microbes, "morphospecies", may actually be different species, as we will soon see.

The problem is especially acute for bacteria and archaea. Even with an electron microscope, these microbes look like spheres (coccoids) or rods (bacilli), and rarely is there much to learn from their shape. To identify these microbes, the traditional approach was to do a battery of biochemical tests: Gram staining, the capacity to degrade key compounds, enzymatic activity, and so on. The tests rely on the phenotype of the microbes, the analogues of the hair and seeds of animals and plants. However, nearly all of the biochemical characteristics used for traditional identification are observable only for microbes that have been cultivated and grown in the lab in a pure culture. They cannot be

Box 9.1 Bible of bacterial taxonomy

One important repository for bacterial taxonomic information is *Bergey's Manual*. First produced by a committee chaired by David H. Bergey, the manual was published in 1923 under the auspices of the Society of American Bacteriologists, now called the American Society for Microbiology. Nine editions of *Bergey's Manual of Determinative Bacteriology* followed with the last published in 1994. It is now a multivolume book called *Bergey's Manual of Systematic Bacteriology*. The first volume published in 2001 is devoted to archaea and some phototrophic bacteria, and the fourth volume on the Bacteroidetes and other phlya was published in 2010 (http://www.bergeys.org/).

examined in uncultivated microbes. The problem then is how to identify the vast majority of microbes that cannot be cultivated.

The solution is to use a gene. Simply put, organisms with similar sequences of this gene are more closely related than organisms with different sequences. In addition to its use in microbiology, the basic idea is also used for exploring the taxonomy and phylogeny of larger organisms. The Barcode of Life Project uses sequences of the cytochrome c oxidase subunit 1 (COI) gene and other genes found in mitochondria to identify and classify invertebrates and vertebrates (Bucklin et al., 2011). As mentioned in Chapter 1, the gene most commonly used for prokaryotes is that for 16S rRNA, and for eukaryotic microbes it is the 18S rRNA gene.

The gene approach does not really replace having a microbe in pure culture. We can now identify microbes in nature without cultivation, but it is still very difficult to deduce a microbe's physiology and its ecological role without being able to grow it and do experiments with it in the lab. As methods for getting around the culturability problem were being developed, microbiologists continued to tackle the cultivation problem head-on by attempting to reproduce the natural environment of the microbe in the lab. Many microbiologists believe that all living microbes found in nature can be cultivated, if the right conditions are found, and that there is no such thing

as an unculturable microbe (Fig. 9.1), a microbe that cannot be grown by itself in the lab under any circumstances. However, some symbiotic bacteria and those living in complex consortia are candidates for being unculturable. More debatable are seemingly independently growing, free-living microbes which have not yet been cultivated. Perhaps some of these can be cultivated if we knew the right approach, but others may remain uncultivated regardless of the approach. In the public health field, these microbes are sometimes referred to as being viable but not culturable (VBNC) (Oliver, 2010).

Introduction to 16S rRNA-based methods

Chapter 1 pointed out the work of Carl Woese who first used 16S rRNA sequences for exploring the taxonomy and phylogeny of cultivated bacteria and archaea. Why the 16S rRNA gene? There are several reasons:

- It is found in all bacteria and archaea. Even eukaryotes have it in mitochondria and chloroplasts.
- Different regions of the gene have different levels of variability, ranging from highly conserved regions that are very similar in all organisms to other regions that are highly variable and differ greatly between distantly related organisms. Both types of regions are needed. The highly conserved regions are very useful for finding all 16S rRNA genes in complex samples while variable regions are essential for distinguishing one microbial group from another.

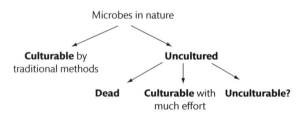

Figure 9.1 Distinctions between microbes that are culturable, uncultured, or unculturable. Among the microbes called uncultured, some are unculturable by traditional methods, but can be grown as pure cultures with extraordinary efforts and innovative approaches. The microbes resisting even these efforts may be really unculturable, but the possibility always remains that apparently unculturable bacteria will prove to be culturable in the future by a method not known today. Inspired by a figure in Madsen (2008).

- Phylogenies derived from the 16S rRNA gene agree well with phylogenies derived from many other genes, and are therefore likely to represent the evolutionary history of the organism as a whole.

Figure 9.2 outlines the main approaches for examining 16S rRNA genes in natural microbial communities. As mentioned in Chapter 1, approaches that do not rely on cultivation are collectively called cultivation-independent methods. Some authors use "to culture" and its grammatical relatives instead of "to cultivate", but the meaning is the same. "Cultivation-independent methods" is more inclusive and informative than another commonly used phrase, "molecular methods". Here we focus on bacteria and archaea, but these methods are also used to examine protists and other eukaryotes.

The problem with analyzing a natural sample is to retrieve 16S rRNA genes from complex communities consisting of many organisms with many other genes. An important step in many cultivation-independent methods is to use the polymerase chain reaction (PCR) to retrieve the genes being targeted; in this case, 16S rRNA genes. In addition to sample DNA, PCR requires two oligonucleotides (primers), each being about 20 base pairs (bp) in length, designed to match specific regions of the gene. Here is where the conserved regions of the 16S rRNA molecule pay off, as they serve as targets for the PCR primers. Several primer sets are available for

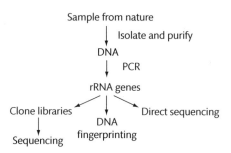

Figure 9.2 Approaches for examining microbial community structure by PCR-based methods of rRNA genes. A "clone library" is a collection of *E. coli* colonies containing the cloned DNA, in this case the PCR-generated fragment of the rRNA gene from the microbial community. "DNA fingerprinting" refers to a collection of methods, such as denaturing gradient gel electrophoresis (DGGE) and terminal Restriction Fragment Length Polymorphism (t-RFLP) that explore variation among communities by examining the pattern of physically separated rRNA gene fragments.

> **Box 9.2 Xeroxing a gene**
>
> Invented by Kary Mullis in 1987 (Nobel Prize, 1993), PCR is one of the most commonly used techniques in molecular microbial ecology. A key ingredient in PCR is a heat-resistant DNA polymerase. The one used by Mullis was from *Thermophilus aquaticus* (Taq), which was first isolated from a Yellowstone Park hot spring by Thomas Brock, one of the founders of microbial ecology.

these regions; the right one to use depends on the application and how the PCR products or amplicons (the DNA produced by PCR) are analyzed. Many other textbooks and web sites describe how PCR works. What is critical to know is that the end result of PCR is lots and lots of the desired gene (the amplicons) from all organisms in the sample; more precisely, what is synthesized is the DNA fragment between the two primer sites. Once the PCR is completed, the problem is then how to separate out and analyze the PCR products. Figure 9.2 outlines the main approaches.

The species problem

Before discussing what has been learned from using cultivation-independent approaches, we need to confront the species problem. The problem is that there is no clear definition of what "species" means for microbes. Without a clear definition, it can be difficult to discuss pathogenicity, the specificity of symbiotic relationships, and the role of microbes in the environment: any topic where it is important to distinguish between "same" and "different" microbes. In classic ecology, the definition of a niche, which is the multidimensional space of resources available to and used by a species, is tied to the species concept.

The classical definition, often called the biological species concept, is that a species is a collection of individuals that can mate and produce offspring themselves capable of reproduction. This definition is meaningless for prokaryotes, many other microbes, and even for large eukaryotes that reproduce asexually. Other definitions have been suggested but none is universally accepted.

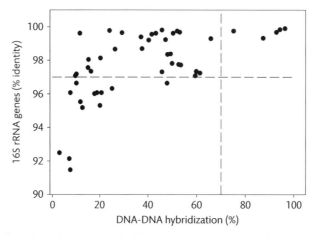

Figure 9.3 16S rRNA genes of one bacterium compared with another versus the level of DNA-DNA hybridization between the same two bacteria. The horizontal line at 97% is the level of identity often used to define phylotypes and is probably closest to a "species" based on this gene. Points to the left of the vertical line at 70% are different species as defined by DNA-DNA hybridization. Data from Stackebrandt and Goebel (1994).

Microbiologists once thought that two bacteria or two archaea belong to the same species if their 16S rRNA genes are ≥97% identical. This ≥97% cut off is based on data comparing 16S rRNA similarity versus DNA-DNA hybridization for bacteria in pure cultures; DNA-DNA hybridization is expressed as a percentage of how much of the genome from one organism hybridizes or binds to the genome of another organism. Some of the data suggested that two organisms belonged to the same species if their DNA-DNA hybridization was at least 70%. This 70% level matches with the 97% threshold in the 16S rRNA data (Fig. 9.3). However, Figure 9.3 shows much scatter around 97%, and other data clearly indicate the problems with this threshold. It is true that two organisms with 16S rRNA genes that are not >97% identical do not belong to the same species. The problem is that two organisms sharing ≥97% identical 16S rRNA genes may or may not belong to the same species. One example is the ecotypes of the cyanobacterium *Prochlorococcus*, as mentioned in Chapter 4. The 16S rRNA genes of three *Bacillus* species (*B. anthracis*, *B. thuringiensis*, and *B. subtilis*) are >99% identical, yet key features of their physiology differ greatly, reasons why they are treated as separate species.

Many microbial ecologists avoid using "species" and instead use other terms that can be defined operationally by the investigator. The terms include ribotype, phylotype, and operational taxonomic unit (OTU). Usually these terms are used to describe organisms with 16S rRNA genes that are ≥97% similar. In this book, "phylotype" will be used. "Clade", which is also used here, is a closely related group of organisms descended from a common ancestor. Table 9.1 summarizes cut off values in 16S rRNA gene identity for various taxonomic levels.

Diversity of bacterial communities

Sequences of 16S rRNA and other genes are used to explore the diversity of microbial communities. There are two facets of diversity within a community, also referred to as "alpha diversity": species richness and evenness. Species richness is simply the number of phylotypes in a community. One way to characterize phylotype richness is with rarefaction curves, also called collection curves. An example is given below. These curves are constructed by calculating the number of phylotypes that would be found for a particular sample size. The other facet of diversity is evenness, which takes into account the number of individuals per phylotype. A highly even community would have the same number of individuals for each phylotype. By contrast, a highly uneven community would be dominated by a few phylotypes represented by many individuals while the other phylotypes would have few individuals. These two aspects of diversity are captured by several diversity

Table 9.1 Taxonomic levels and corresponding 16S rRNA identity for bacteria. Because other genes are needed to distinguish between prokaryotes and eukaryotes, the degree of 16S rRNA identity is not applicable (NA) at the domain level. Data from Konstantinidis and Tiedje (2007) and Brenner and Farmer (2005).

Taxonomic level	Example	% identity	Number per level*
Domain	Bacteria	NA	3
Phylum	Proteobacteria	75	90
Class	Gammaproteobacteria	78	7
Order	Enterobacteriales	84	18
Family	Enterobacteriaceae	90	1
Genus	*Escherichia*	95	38
Species	*Escherichia coli*	97–99	5
Strain	*E. coli* O157	>97	?

* The number of elements at each level. For example, there are three domains of life, of which Bacteria is one. There are about 90 phyla of bacteria (the exact number varies among bacterial taxonomists and how candidate phyla are counted), of which Proteobacteria is one.

Table 9.2 Some measures of diversity. Some of these indices are designed to measure species richness while others are measures of the evenness of the community. Still others try to do both. Chao1 is named after Anne Chao whereas ACE is an acronym for Abundance Coverage Estimator. Magurran (2004) discusses several other indices.

Measure	Symbol	Purpose	Equation
Shannon	H	Both	$-\sum p_i \log(p_i)$
Simpson	D_1	Mostly evenness	$1 - \sum p_i^2$
Inverse Simpson	D_2	Mostly evenness	$1/\sum p_i^2$
Shannon Evenness	J'	Evenness	$H/\ln S$
Simpson Evenness	E_D	Evenness	D_1/S
Chao1	S_{choa1}	Richness	$S_{obs} + f^2(1)/2f(2)$
ACE	S_{ACE}	Richness	$S_{abund} + S_{rare}/C_{ace} + f(1)/C_{ace}\,\gamma^2_{ace}$

Definition of symbols:

S = Observed number of species in the community

$f(1)$ and $f(2)$ = the number of species that are represented once or twice, respectively

p_i = the proportion of the ith species making up the community. $p_i = n_i/N$ where n_i is the number of individuals in the ith species and N is the total number of all individuals in entire community

$C_{ace} = 1 - f(1)/N_{rare}$ where N_{rare} is the number of rare species, defined arbitrarily as having fewer than 10 individuals

S_{abund} = the number of species represented by more than 10 individuals

S_{rare} = the number of species represented by fewer than 10 individuals

S_{obs} = the number of observed species

$$\gamma^2_{Ace} = max\left[\frac{S_{rare}\sum_{i=1}^{10}i(i-1)F_i}{C_{ACE}(N_{rare})(N_{rare}-1)} - 1, 0 \right]$$ where F_i is the number of species with i individuals

indices used in ecology and by microbial ecologists (Table 9.2). These indices do not take into account the phylogenetic relationships among the phylotypes. Two phylotypes that are closely related are counted the same as two other phylotypes that are distantly related. Phylogenetic differences are captured by other indices, such as phylogenetic diversity (PD) and phylogenetic species variability (PSV) (Cadotte et al., 2010).

Often diversity indices are calculated with data from sequencing the 16S rRNA genes and other phylogenetic markers. Once laborious and expensive, sequencing is now easy and cheap, thanks to the development of "next-generation" sequencing approaches. These new approaches generate millions of sequences per sample in hours, thousands of times more sequences than the traditional method. One next-generation approach is tag

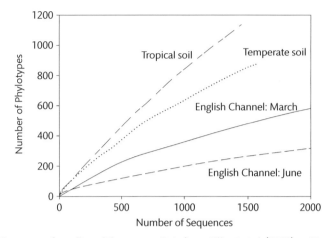

Figure 9.4 Typical rarefaction curves for soils and the oceans. Data from Gilbert et al. (2009) and Lauber et al. (2009).

pyrosequencing. This approach consists of pyrosequencing fragments (tags) of the 16S rRNA gene retrieved by PCR. Pyrosequencing, also referred to as 454 sequencing after the company that first commercialized the method, yields millions of sequences quickly and cheaply. Two observations became apparent when the pyrosequencing data first rolled in.

The first is the high diversity of microbial communities. In fact, most bacterial communities have not been completely sampled to-date, even by approaches using high throughput sequencing methods. Most rarefaction curves do not level off unless phylotypes are grouped together at a very low percent identity (Fig. 9.4). The precise estimate for the number of phylotypes in bacterial communities varies with the environment, but usually it is in the thousands, if not tens of thousands for a particular environment and time point. The other observation is that bacterial communities are dominated by a few very abundant phylotypes and many rare ones; the 100 most abundant phylotypes may account for 80% or more of the entire community while the remaining 20% is spread out over thousands of phylotypes. This collection of rare phylotypes has been called the "rare biosphere" (Sogin et al., 2006). The distribution of the abundances of each phylotype is illustrated in rank abundance curves which plot the relative abundance versus the rank of the phylotype in the community, starting with the most abundant and ending with the least abundant.

The shape of these rank abundance curves varies with the environment, reflecting the diversity of com-

munities in these environments (Fig. 9.5). In the English Channel, for example, there are a few very abundant types of bacteria, each making up 10% or more of the total community whereas the other phylotypes are less abundant, making up 1% or much less of the total. In contrast, soil communities are more even with fewer highly abundant phylotypes (Lauber et al., 2009), so soil rank abundance curves are below those for the English Channel and other marine systems. In addition to being more even, soil communities are richer in phylotypes. Tag pyrosequencing data indicate that soils generally have twice or more phylotypes grouped at ≥97% identity than do the oceans (Fig. 9.4), although these numbers vary greatly among environments and within each environment over time due to many factors. The high diversity in soils is possible because of its many micro-environments, which are less common in the much more homogeneous world of water. Chapter 3 discusses some of the patchiness occurring on the micron scale in aquatic habitats, but this patchiness still pales in comparison to the spatially complex environment of soils.

The paradox of the plankton

While aquatic microbial communities are less diverse than soils, they still are quite diverse, so diverse that it has troubled aquatic ecologists. The limnologist G.E. Hutchinson (1903–1991) called this the "paradox of the plankton" (Hutchinson, 1961). He pointed out that

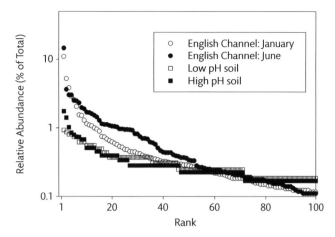

Figure 9.5 Relative abundance of phylotypes in the English Channel and two soils. The soils were from a forest with low pH and from scrubland with high pH. The relative abundance is the number of individuals in a particular phylotype divided by the total individuals in the community. These data are plotted versus their rank in abundance, starting with the most abundant type in each community. Only the 100 most abundant phylotypes are given; each community has thousands of rare bacteria. Data from Gilbert et al. (2009) and Lauber et al. (2009).

aquatic microbes seem to be competing for the same limited number of resources in a physically simple, unstructured environment. These microbial communities should not be very diverse. Competition should result in only a few successful species that out-compete and exclude all others, according to the competitive exclusion principle. In fact, aquatic communities are much more diverse than expected, resulting in the paradox: how can so many species coexist in such a simple environment? Hutchinson had a few suggestions for resolving the paradox, one being that aquatic environments vary substantially over time, thus allowing for coexistence. He was thinking of phytoplankton, but the paradox and Hutchinson's answer also apply to heterotrophic bacteria and other microbes as well.

There are other ways out of the paradox for bacteria. One is that the environment of microbes even in water may be quite complex at the micron scale (Chapter 3). Also, unlike phytoplankton which mostly use the same carbon source (CO_2) and a few inorganic nutrients, heterotrophic bacteria can use a myriad of organic compounds, which may select for specialization and allow the coexistence of many different types of bacteria. Another resolution to the paradox is that many bacteria are not active and thus are not in direct competition with each other. Finally, top-down control of grazers and

viruses may allow the coexistence of bacteria that otherwise could not persist together.

There is a bit of the same diversity paradox in soils and sediments. When discussing larger organisms, the problem has been called the "the enigma of soil animal species diversity" (Coleman, 2008). Whether a paradox or an enigma, micro-environments go only so far in explaining the large number of microbial types in soils and sediments. The diversity of these systems is higher than can be explained by micro-environments. The resolution of the paradox/enigma is similar to that for aquatic environments: temporal variation, diversity in organic material, and top-down processes. In addition, perhaps more so than in aquatic habitats, in soils and sediments many microbes avoid the competitive exclusion principle by being dormant.

Differences between cultivated and uncultivated microbes

One important question addressed with cultivation-independent methodology is whether or not cultivated prokaryotes are the same as the uncultivated ones. While this question has been examined most intensively for aerobic heterotrophic bacteria, it applies also to other microbes. The culturability problem means that we

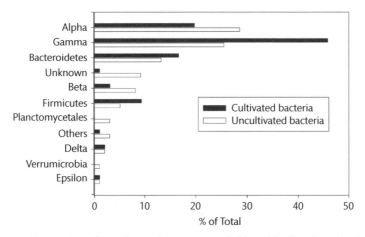

Figure 9.6 Bacterial community structure of a marine environment sampled by cultivation-dependent and cultivation-independent approaches of 16S rRNA genes. "Unknown" are 16S rRNA genes that could not be classified. "Others" refers to bacteria that could be classified, but individually were <2% of the total community. Alpha, Beta, Delta, Epsilon, and Gamma refer to classes of Proteobacteria. Data from Hagström et al. (2002).

cannot use agar plates to enumerate bacteria, but it leaves open the question of whether cultivated bacteria are representative of the uncultivated bacteria found in nature. Perhaps the bacteria that can be cultivated are closely related to those bacteria that cannot be cultivated.

To see if that is the case, microbial ecologists have compared the 16S rRNA genes of cultivated bacteria with those of uncultivated bacteria obtained by cultivation-independent methodology. The answer is, bacteria cultivated by traditional methods are not very similar to those assayed by cultivation-independent methodology. In soils, for example, the bacteria most commonly culti-vated on agar plates are those in the genera *Streptomyces* and *Bacillus*, whereas these bacteria are not abundant in natural communities, according to clone libraries of 16S rRNA genes or other cultivation-independent app-roaches (Janssen, 2006). Likewise, in coastal oceanic waters, bacteria in the genera *Pseudomonas* and *Vibrio* are often isolated and grown on agar plates, but their 16S rRNA genes are not common, if they are found at all, by cultivation-independent approaches (Fig. 9.6).

In both soils and aquatic habitats, the differences between cultivated and uncultivated bacteria are evident at high phylogenetic levels (phylum and subphylum), not just the presence or absence of a few species. In soils, Gram-positive bacteria are much more commonly

sampled by cultivation-dependent methods than by cul-tivation-independent methods. In aquatic systems, microbes within the Gammaproteobacteria commonly grow on agar plates. These can be abundant according to cultivation-independent methods, but other phyla, such as Acidobacteria in soils, and other proteobacterial classes, such as Alphaproteobacteria and Betaproteo-bacteria in aquatic habitats, are often more abundant. So, the types of bacteria cultivated by standard approaches are quite different, at all phylogenetic levels, from the uncultivated microbes found in natural communities.

Like bacteria, few archaea in cultivation are represent-ative of the archaea found in natural environments. That is also the case for protists, although this is debated, as discussed below. The situation for soil fungi has not been examined as extensively as for other microbes, but it would be surprising if there isn't a cultivation problem with soil fungi as well.

Types of bacteria in soils, freshwaters, and the oceans

Microbial ecologists are still trying to learn about the dis-tribution of various types of microbes among the major habitats of the planet. This is an important question in biogeography, the study of the geographic distribution of organisms. Although much work remains to be done,

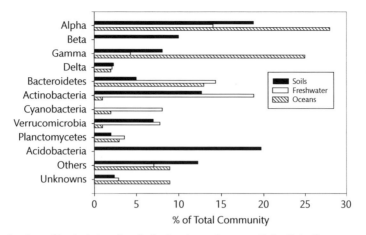

Figure 9.7 Community structure of bacteria in oxic soils, freshwater and oceans. Alpha, Beta, Gamma, and Delta refer to classes of the Proteobacteria. Data from Hagström et al. (2002), Zwart et al. (2002), and Janssen (2006). See also Tamames et al. (2010).

we do know of differences between soils, lakes, and the oceans even at the level of phylum and other high phylogenetic groups.

Of the 50–100 known phyla of bacteria in nature, only a few are abundant in any particular environment. A couple of phyla are found in many environments, while the abundances of the others vary among the main habitats of the biosphere (Fig. 9.7). The Proteobacteria phylum is found virtually everywhere, but different proteobacterial classes dominate freshwaters, the oceans, and soils. In freshwaters, Betaproteobacteria are most abundant, followed by Gammaproteobacteria and Alphaproteobacteria. In marine waters, however, Alphaproteobacteria are usually most abundant (especially the SAR11 clade), with Gammaproteobacteria in second place among the proteobacterial classes; Betaproteobacteria are much less abundant. These classes of Proteobacteria are also abundant in soils. Deltaproteobacteria are prevalent in anoxic sediments (Chapter 11). Several cultivated members of the Proteobacteria have been examined intensively in laboratory experiments for decades. These include the alphaproteobacterium *Rhizobium* (Chapter 14), the betaproteobacterium *Burkholderia*, and many gammaproteobacterial genera such as *Alteromonas*, *Escherichia*, *Pseudomonas*, *Salmonella*, and *Vibrio*.

In addition to Proteobacteria, the phylum Bacteroidetes is abundant in freshwaters and some oceanic systems. Among the Bacteriodetes, *Flavobacteria* are abundant in the oceans while *Sphingobacteria* are

found in lakes (Barberan and Casamayor, 2010). Figure 9.7 implies that Bacteroidetes is not very abundant in soils, but this phylum can be abundant in some soils (Lauber et al., 2009). The phylum is quite complex with both anaerobic and aerobic members. The anaerobic members include those in the genus *Bacteroides*, which dominate the human intestines (Karlsson et al., 2011), and *Porphyromonas*, well known because of its role in dental problems. Microbial ecologists often use a hyphenated name, *Cytophaga-Flavobacteria*, to refer to the aerobic members of the Bacteroidetes.

Another phylum, Actinobacteria, is abundant in freshwaters and in soils, but less so in the oceans. A few groups of Actinobacteria, such as those in the acI and acIV clades, can be especially abundant in freshwaters (Newton et al., 2011). The actinomycetes, which belong to the order Actinomycetales, are abundant in soils and are responsible for the distinctive odor of freshly wetted soils. These organisms have branching filaments, analogous to the hyphae of fungi. Organisms in the Actinobacteria phylum used to be referred to as being high G-C Gram-positive bacteria; G-C is the proportion of guanine and cytosine in the DNA of these organisms. Low G-C Gram-positive bacteria, now recognized as being the phylum Firmicutes, are often retrieved from soils by cultivation-dependent approaches, but are not particularly abundant in soils or aquatic habitats when assayed by cultivation-independent methods. Some cultivated representatives of the genus *Bacillus* have been extensively studied. These

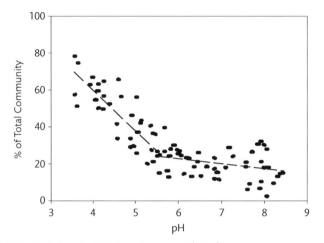

Figure 9.8 Abundance of Acidobacteria in soils. Data from Jones et al. (2009).

include *B. anthracis* (causes anthrax), *B. thuringiensis* (the biological insecticide "b.t."), and *B. subtilis* (model organism for cellular development).

The phylum Acidobacteria is quite abundant in soils (Jones et al., 2009b), but it is rarely seen in freshwaters and the oceans. This group of bacteria was recognized as a separate phylum only in 1995, because few can be easily cultivated and grown in the lab. Of the over 60 000 characterized isolates of bacteria, only about 70 have been classified as being in the Acidobacteria phylum. As implied by the name, these bacteria grow best in acidic media, and cultivation-independent approaches have demonstrated that Acidobacteria make up a very large fraction, well over 50%, of the total community in low pH soils. As pH increases, their relative abundance decreases, but even for soils with pH values above about 6, Acidobacteria are still quite abundant, making up 20% of the community (Fig. 9.8). Low pH values are less common in lakes than in soils and very rare in marine systems, explaining why this phylum is not abundant in most aquatic systems. The exceptions are acidic bogs and lakes polluted by the acids from mining operations (Kleinsteuber et al., 2008).

So, different phyla and subphyla of bacteria are found in different environments. This is one piece of evidence that these high phylogenetic levels share some ecological traits (Philippot et al., 2010), in spite of the fact that there are many examples of bacteria with very similar 16S rRNA genes that differ in some key metabolic function. Microbes that are phylogenetically related should share some similarities in their ecology. This issue is part of the broader question about links between community structure and function.

Archaea in non-extreme environments

DNA sequence data from a few isolates in pure cultures were the basis for putting archaea into its own kingdom of life, quite distinct from bacteria and eukaryotes. Based on these isolates, it was once thought that archaea thrive in only extreme environments and were analogs to early life on the planet, as mentioned in Chapter 1. However, studies using cultivation-independent methods have demonstrated that archaea live in nearly all natural environments, not just extreme ones. Still, usually archaeal abundance is low relative to bacteria in most natural environments. In soils and surface waters of the oceans, for example, archaea make up <5% of total microbial abundance (Ochsenreiter et al., 2003, Karner et al., 2001).

The big exception is the deep ocean (Fig. 9.9). In waters below about 500 m, archaea can account for as much as 50% of all microbes. Of the two main archaeal phyla, Crenarchaeota is usually more abundant, but Euryarchaeota abundance is high in some oceanic basins (Stoica and Herndl, 2007, Arístegui et al., 2009). Since the deep ocean makes up 75% of the total biosphere (Chapter 1), total abundance and biomass of

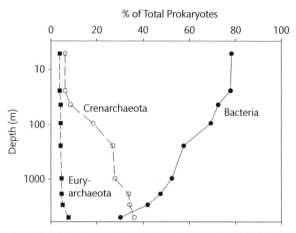

Figure 9.9 Abundance of bacteria and two archaeal phyla in the North Pacific Ocean. Data from Karner et al. (2001).

archaea in the entire biosphere are substantial. Archaea are also abundant in surface waters in winter near Antarctica and in the Arctic Ocean, whereas abundance is low in the summer in both polar systems (Church et al., 2003). Like the deep ocean, polar waters in winter are dark, with low concentrations of organic material and phytoplankton biomass.

Key to the distribution of archaea in nature is their physiology. As will be discussed in Chapter 12 in more detail, many archaea appear to be chemoautotrophs that oxidize ammonia to obtain energy. This helps to explain the apparent negative relationship between archaea and phytoplankton and light in the oceans. As a fraction of the total prokaryotic community, chemoautotrophic archaea are not abundant in many environments because they are out-competed by algae and heterotrophic bacteria for ammonium. Light also inhibits ammonia oxidizers (Chapter 12). Likewise, in soils, chemoautotrophic archaea are unable to compete effectively for ammonium with not only heterotrophic bacteria, but also fungi and plants. There is some evidence indicating that archaeal ammonium oxidizers do better when ammonium concentrations are low (Erguder et al., 2009), explaining the success of chemoautotrophic archaea in the deep ocean with its extremely low ammonium concentrations. Other hypotheses are needed to explain why heterotrophic archaea are not more abundant in the surface layer of aquatic habitats and in soils.

Everything, everywhere?

It is not surprising that bacteria and archaea vary among different environments even at the level of phylum and other high phylogenetic levels because these environments differ greatly in many physical and biological properties. A harder question in microbial biogeography is whether two microbial communities differ when environmental conditions are the same for two locations that are separated by some geographic distance. With many large organisms, it is clear that habitats with the same environmental conditions but on different continents often are not home to the same communities. Gazelles are found in the savannas of Africa and pronghorn antelope live in North America, and not the other way around. These animals evolved independently on distant continents, separated by an ocean and by the 250–500 million years since Africa and the Americas split apart by continental drift. In soils or water with the same environment, would the same microbes be found?

One answer is an aphorism attributed to Lourens G.M. Bass Becking (1895–1963); "everything is everywhere, but the environment selects". That is, environmental conditions determine whether a phylotype is abundant or not in a particular habitat; geography and evolutionary history play no role (Fig. 9.10). There are good reasons to believe that the Bass Becking hypothesis is correct. Microbes are incredibly abundant, and dispersion seems unstoppable by distance or by mountains, oceans, or other geographical features. Once a microbe arrives, in theory only one is necessary to reproduce and to populate a new habitat. High abundance, easy

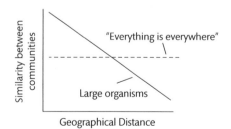

Figure 9.10 Similarity between communities in similar environments separated by geographical distance. "Everything is everywhere" applies to many microbial communities, while communities of large organisms often differ when separated by large distances.

dispersion, and asexual replication all argue for the "everything is everywhere" hypothesis.

The weight of the data to-date supports the hypothesis, at least for bacteria and archaea (Martiny et al., 2006). Bacterial communities from similar soil environments, for example, are similar even if from different latitudes (Fierer and Jackson, 2006). Also, bacteria in the Arctic Ocean and arctic lakes appear similar to those in corresponding Antarctic waters (Pearce et al., 2007), and likewise for archaea in soda lakes of Mongolia and Argentina (Pagaling et al., 2009). A final example is the bacterium living symbiotically on the giant marine ciliate, *Zoothamnium niveum*. The 16S rRNA genes and genes for carbon and sulfur metabolism appear to be the same for this bacterium in both the Mediterranean Sea and in the Caribbean Sea (Rinke et al., 2009). In all of these cases, the local environment, not geography, explains the high similarity of bacteria in these distant habitats. But there are counterexamples and arguments against the Bass Becking hypothesis (Pagaling et al., 2009, Martiny et al., 2006).

Similar arguments for and against the Bass Becking hypothesis are going on for protists. What complicates the discussion is that unlike prokaryotes, some protists can be distinguished based on morphology. Some protists with very similar appearances seem to be cosmopolitan (Fenchel, 2005) and have been used to argue for the Bass Becking hypothesis. On the other side of the debate are studies using gene sequences for rRNA and other phylogenetic markers (McManus and Katz, 2009). Some of these studies also support the Bass Becking hypothesis, but others argue against it; rRNA genes from protist communities differ with geography, suggesting that everything, at least every protist, is not everywhere. Likewise, fungi in soils, when identified by morphology and other traditional methods, seem to have restricted distributions and are not cosmopolitan (Foissner, 1999).

As with many questions in microbial ecology, the scale of the comparison matters. The answer to the "everything is everywhere" question depends on how phylotypes are defined by the 16S rRNA gene level of identity (97%, 99%, or 100%?) or by other genes with even greater phylogenetic resolution. It also depends on the scale of the sampling. Large samples encompassing many microhabitats may obscure differences between communities that would be observed if smaller samples were taken.

What controls diversity levels and bacterial community structure?

Since different environments often have different microbial communities, there must be something about the physical, chemical, and biological features of these environments that lead to variation in the make-up of microbial communities. We do know that the same top-down and bottom-up factors that affect total biomass production and standing stocks also affect the success of individual members of the community, the sum of which makes up community structure. The challenge is figuring out which factors are most important in which environments. The difference between communities is sometimes referred to as beta diversity.

Temperature, salinity, and pH

These physical factors often explain a large fraction of the variation in the properties of the whole bacterial community (Chapter 3). They directly affect microbes and indirectly affect them via how they control other properties, such as the effect of pH and salinity on adsorption. These three physical factors also have the potential to shape bacterial community structure, although not necessarily the same for soils and aquatic habitats.

Temperature, for example, explains a large fraction of the variation in bacterial community structure in the oceans (Fuhrman et al., 2008) and in hot springs (Miller et al., 2009), but there is no apparent relationship between diversity and temperature in soils (Fierer and Jackson, 2006). Likewise, bacterial diversity tends to be higher in tropical waters than in polar seas, but again this is not the case for soils (Fig. 9.11). Microbial communities may be more diverse in warmer waters because faster metabolic rates lead to higher rates of speciation, but the support for this hypothesis remains unclear. Even if speciation is slower in cold waters, bacterial communities still have had thousands to millions of years to develop high diversity. And the temperature hypothesis should be applicable in soils, but it does not seem to be the case; diversity of soil communities varies primarily with pH, not temperature (Fierer and Jackson, 2006, Lehtovirta et al., 2009).

Salinity is another factor likely to affect community structure. It may explain the high abundance of

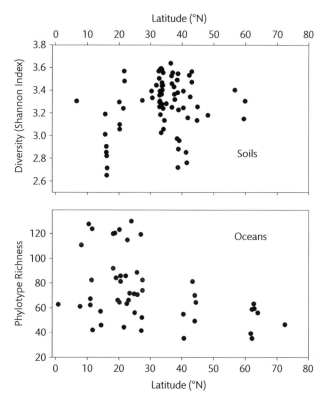

Figure 9.11 Diversity of bacterial communities in soils and the oceans from the tropics to the arctic. There is a significant decline in diversity (phylotype richness) with increasing latitude in the oceans, but not in soil communities. Data from Fierer and Jackson (2006); Fuhrman et al. (2008).

Betaproteobacteria in lakes and low abundance in the oceans, and it accounts for large-scale differences in community structure of all bacteria and archaea (Lozupone and Knight, 2007, Tamames et al., 2010, Auguet et al., 2010). There is no evidence, however, that salt affects phylotype richness. Oceans and freshwater lakes are equally diverse, and the number of phylotypes does not vary systematically within estuaries where salinity changes from near freshwater to oceanic levels.

Extremes in salinity and in other physical factors do lead to low diversity. That is the case for salt pans with very high salt concentrations—near saturation for NaCl. These systems have only a couple of types of bacteria and archaea (Anton et al., 2000). Waters with very low pH are also not very diverse. Ponds polluted by run off from mines can have pH near 1 or lower and harbor only a couple of prokaryotic taxa, although biomass may be quite high (Bond et al., 2000). High temperature, especially above about 65 °C, also leads to low diversity in hot springs (Miller et al., 2009).

Moisture and soil microbial communities
Water content has a large impact on microbial activity and diversity in soils. Results from cultivation-independent approaches indicate that bacterial communities are more even with higher phylotype richness in unsaturated surface soils than in saturated (waterlogged) soils which are dominated by a few bacterial types (Zhou et al., 2004). Community structure also varies with soil texture due to silt and clay content (Carson et al., 2010), an effect related to water content. Soils with high silt and clay content have low water potential and pore connectivity. In these soils, competition is reduced because pores with water and thus active bacteria are separated by dry stretches, creating micro-environments where some bacteria can

flourish. The end result is higher bacterial diversity in drier soils. In contrast, soil fungal diversity was lower during a drought, according to a study using a DNA fingerprinting approach (Toberman et al., 2008). The two studies indicate differences in how bacterial and fungal communities respond to soil moisture, but a study examining bacteria and fungi simultaneously in the same soil and moisture conditions is needed to say for certain.

Organic material and primary production

One explanation for the high diversity of tropical rain forests and coral reefs is their high rates of primary production. High productivity may support a large number of diverse herbivores which in turn feed many carnivores and so on up the food chain, but only up to a point. At very high levels of productivity, diversity decreases, resulting in a hump-shaped relationship between productivity and diversity. This pattern is also seen for some microbes in some places. Figure 9.12a shows the hump-shaped relationship for fungal diversity, but it is not the norm for the communities examined so far. There is a significant positive relationship between bacterial diversity and phytoplankton biomass in the oceans along a latitudinal gradient without any hint of a hump. Overall, however, negative relationships between diversity and productivity (Fig. 9.12b) are actually more common in freshwater and marine habitats (Smith, 2007). In contrast, diversity of soil bacterial communities does not seem to vary with soil organic content or with plant diversity (Fierer and Jackson, 2006), although organic carbon additions do affect the make-up of soil communities (Fierer et al., 2007a).

In addition to amounts, the type of organic compounds in an environment may affect bacterial community structure. We expect selection for bacteria capable of growing quickly on those types of organic material that require specific enzymes in order to be used. We do know that addition of one or two organic compounds often leads to dominance of the community by only a few types of bacteria; this is the principle behind enrichment cultures in which growth conditions are set to select for, to "enrich" for, the growth of only a few bacterial taxa. The presence of, for example, structural polysaccharides from higher plants, such as cellulose and hemi-cellulose, undoubtedly selects for certain bacteria, and some data suggest that different proteobacterial classes and aerobic Bacteroidetes differ in use of biopolymers and other organic material in aquatic systems (Kirchman, 2002a). It is unclear, however, if diversity in natural organic material leads to higher bacterial diversity.

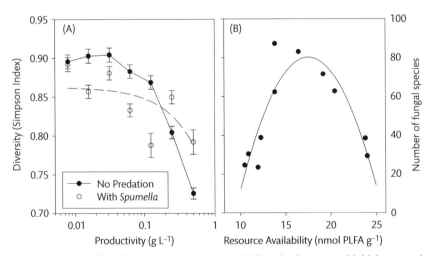

Figure 9.12 Diversity as a function of productivity. In Panel A, bacterial diversity decreases with higher organic material, although the effect depends on grazing by protists, such as *Spumella*. In Panel B, the typical "hump" relationship was seen between the number of fungal species and a measure of resource availability, total microbial biomass as measured by phospholipid fatty acids (PLFA). Data from Bell et al. (2010) and Waldrop et al. (2006).

Predation and viral lysis

Top-down control by grazers and viruses limits standing stocks of microbes and can affect their growth rates in natural environments. Both grazing and viral lysis have the potential for determining the success of specific microbes and thus in shaping diversity levels and overall community structure of microbes. Although more data are needed to say for certain, most arguments point to viruses having a larger impact than grazers on community structure.

How grazing affects community structure involves the same factors known to affect overall grazing rates. Cell size is one such factor. Because grazing varies strongly with the size of the prey (Chapter 7), the abundance and thus evolutionary success of a microbial group will also depend on cell size. The chemical composition of cellular surfaces is another property of prey that affects grazing and is likely to account for why a particular grazer preys more heavily on one microbial group than another. For these two reasons and others, some bacteria are known to be relatively resistant to grazing, helping to explain their abundance in freshwater environments (Jürgens and Massana, 2008). Resistance to grazing helps to explain the success of the betaproteobacterial genus *Polynucleobacter* and of the phylum Actinobacteria in lakes. Similar mechanisms are undoubtedly operating in the oceans and in soils.

The impact of viruses is probably stronger than that of grazers, however, if for no other reason than that viruses substantially outnumber grazers. Viruses are more abundant than prospective hosts by about tenfold (Chapter 8), while in contrast, grazers are much less abundant than their prey. Put together, there are about 10^4 more viruses than grazers on average in nature (Fig. 9.13). Not all of those viruses can attack a particular host because of specific virus-host specialization, but the high abundances ensure that each microbe is potentially attacked by many viruses. The number of viruses and the specificity in virus-host interactions select for bacteria to develop strategies to minimize viral attack, which in turn forces viruses to evolve. This evolutionary arms race is thought to lead to bacterial diversity.

Another important point is that viral lysis increases as the abundance of the host microbe goes up because encounter rates between viruses and hosts depend on the abundance of each. As a result, a highly abundant type of microbe is likely to encounter and select for more

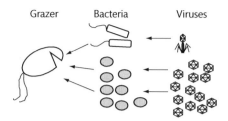

Figure 9.13 The "kill the winner" hypothesis. In this version of the model, the grazer does not discriminate between the two types of bacteria while the viruses do. The viruses are bigger and less numerous in this diagram than in reality; the virus to bacteria ratio is about 10:1 in most natural environments. The most abundant bacterial type, the "winner", attracts the most viruses.

viruses. This concept has been called the "kill the winner" hypothesis (Thingstad, 2000). The microbe that wins the competition for limiting dissolved compounds becomes abundant, but that success results in more viruses and higher viral lysis, which in turn depresses standing stocks of the winner. Consequently, the superior competitor does not crowd out inferior competitors, allowing more species to coexist. Unfortunately, there is little experimental data to support this hypothesis.

Problems with 16S rRNA as a taxonomic and phylogenetic tool

All of the conclusions about microbial community structure just discussed were based nearly entirely on data about 16S rRNA genes. However, there are a few well-known problems with the 16S rRNA gene tool for examining community structure. One is that many bacteria have several copies of this gene. Among cultivated bacteria, the 16S rRNA gene copy number varies from one, as is the case for *Pelagibacter ubique* (a SAR11 representative) to as many as 15, a record now held by *Clostridium paradoxum* and *Photobacterium profundum* (Lee et al., 2009). The average is about four versions of the gene per cell among cultivated bacteria. A study using metagenomic approaches (Chapter 10) found that on average each oceanic bacterium has 1.8 16S rRNA genes per genome (Biers et al., 2009). Uncertainty about the number of 16S rRNA genes per bacterium puts limits on using the abundance of a 16S rRNA gene to estimate the abundance of that bacterium in nature.

Another problem with the 16S rRNA gene is that two microbes that differ substantially in physiology and ecology can have similar 16S rRNA genes, as mentioned before. Cases like this indicate that evolution of the organism has diverged from evolution of the 16S rRNA gene. This can occur because of the conservative nature of this RNA gene; it changes only slowly over time. Two sequences differing by only 0.06% in this gene are thought to have diverged one million years ago, but there are many problems with this conversion factor (Kuo and Ochman, 2009). Also, two bacteria with similar 16S rRNA genes may not have a particular metabolic function in common because the genes for that function were acquired independently of the rest of the genome for one bacterium but not the other (Chapter 10).

One solution to some of these problems is to use other phylogenetic markers. Another marker is the intergenic spacer (ITS), the DNA between the 16S rRNA and 23S rRNA genes in bacteria and between other rRNA genes in eukaryotes. Archaea do not have an ITS region. The ITS region has been useful for examining closely related microbes. Other phylogenetic markers are genes for various proteins (Table 9.3). Bacteria typically have only one version or copy of these genes per cell, thus avoiding the multi-copy problem of the 16S rRNA gene. The phylogeny of these protein-encoding genes generally agrees with that of the 16S rRNA gene. However, while these alternative phylogenetic markers have proven useful for some specific questions, none has replaced the 16S rRNA gene for characterizing entire prokaryotic communities. There are a couple of practical problems with using the alternatives.

One is that the number of known sequences for these protein-encoding genes is small compared to the 16S rRNA data set. A small data base with few known sequences complicates identifying unknown sequences and hence sorting out the composition of microbial communities. This problem could be overcome by obtaining more data. The second problem, which cannot be solved so easily, is that protein-encoding genes have few conserved regions present in all bacteria. Gene sequences can be quite variable for these protein-encoding genes because the key function of the protein, which is conserved among microbes, takes up only a small portion of the entire gene. Another problem is that protein-coding genes can vary in the third base of the nucleotide triplet encoding some amino acids. In contrast, the 16S rRNA gene has several conserved regions because many parts of the product of this gene (a 16S rRNA molecule) participate in ribosome functions where changes arising from mutations cannot be tolerated. Consequently, it is difficult, often impossible, to devise PCR primers for retrieving all of a particular protein-encoding gene in a complex sample. Without PCR, many cultivation-independent approaches cannot be used.

Table 9.3 Protein-encoding genes that have been used in phylogenetic studies of bacteria. "% Detection" is the percentage of 11 bacteria that have the indicated gene (Santos and Ochman 2004). "Number of studies" is the number of publications found by Web of Science (6 December, 2010) using the gene abbreviation and (phylogen* or taxonom*) as search terms. The corresponding numbers for 16S rRNA genes are given for comparison.

Gene	Full name or function	% Detection	Number of Studies
rpoB	RNA polymerase subunit	55	323
gyrB	DNA gyrase	91	291
recA	general recombination and DNA repair	45	290
fusA	elongation factor G	45	16
ileS	isoleucine-tRNA synthetase	45	14
lepA	elongation factor EF4	55	9
leuS	leucyl-tRNA synthetase	82	7
pyrG	component of CTP synthase	73	7
rplB	50S ribosomal subunit protein L2	64	3
rrn	16S rRNA	100	8913

Box 9.3   Molecular clock

The difference between sequences of phylogenetic markers can be assigned an elapsed time, making a very powerful molecular clock for exploring the pace of evolution over geological time. For example, molecular clock data puts the divergence of humans from the great apes 4–8 million years ago. The clock according to 16S rRNA genes can be set by sequence differences among obligate endosymbionts found in host organisms whose evolutionary time line can be determined by other approaches. For most applications, however, microbial ecologists use just the sequence data alone and do not convert dissimilarities to time.

Community structure of protists and other eukaryotic microbes

There are several parallels between studying the structure of eukaryotic microbial communities and that of bacterial communities. Analogous to 16S rRNA genes for prokaryotes, the 18S rRNA gene is often used in cultivation-independent studies of eukaryotes. Many of the same methods for 16S rRNA genes can be used to examine the 18S rRNA gene and other phylogenetically informative genes from protists and other eukaryotic microbes. A big difference is that unlike prokaryotes, some protists can be identified by morphology—by their physical appearance. These microbes can have distinctive shapes and are propelled (or not) by various numbers of flagella or by cilia; they can be covered with scales, or extrude nets or other structures, and some have diagnostic pigments. These and other physical features enabled taxonomists to name and classify protists without sequence data of a phylogenetic marker gene. With some organisms, using gene sequences or morphology to identify an organism gives the same answer. But there are many cases in which they do not.

A common problem is that seemingly identical microbes may in fact have different sequences of a phylogenetic marker gene (McManus and Katz 2009). One example is the ciliates in Figure 9.14. These appear to belong to the same morphospecies of ciliates, being nearly the same size and having very similar shapes. However, sequencing of the small subunit ribosomal gene revealed differences of >6%, indicating that they are different species. Discrepancies in identifying organisms by morphology versus by gene sequences are especially common for the smallest protists (<5 µm) that have few distinguishing features. Organisms that appear to be morphologically similar but are in fact different are sometimes called cryptic species. Some microbial ecologists question whether two organisms with different rRNA genes but which otherwise are similar in appearance should be considered to be different species. Differences in these genes may not have any real biological consequences and may represent neutral variation within a species (Fenchel, 2005). However, most microbial ecologists think that this variation in gene sequences cannot be ignored (McManus and Katz, 2009).

Types of protists and other eukaryotic microbes in nature

We have already encountered the major groups of eukaryotic microbes living in soils, oceans, and lakes. These include algae (Chapter 4), protozoa, ciliates, and other protists (Chapter 7), photoheterotrophic eukaryotes that can carry out photosynthesis and predation (Chapter 4 and 7), and fungi, which are abundant in soils (Chapter 5). Cultivation-independent methods that examine various rRNA genes have revealed an even more diverse world of eukaryotic microbes. Pyrosequence data indicate that the diversity of these microbes equals that of bacteria in the oceans (Brown et al., 2009), and may even exceed bacterial diversity in soils, according to a study using a clone library approach (Fierer et al., 2007b). Adding to the 40 major subgroups of eukaryotes found by cultivation-dependent methods, studies using cultivation-independent methods have discovered at least another 10 and probably closer to 30 new subgroups.

Some of these new eukaryote subgroups belong to the phylum Alveolata. The microbes in this phylum identified by more traditional methods include ciliates and dinoflagellates. Many new alveolates, sometimes called novel alveolates (NA), have been discovered in the oceans (Worden and Not, 2008) and are not closely

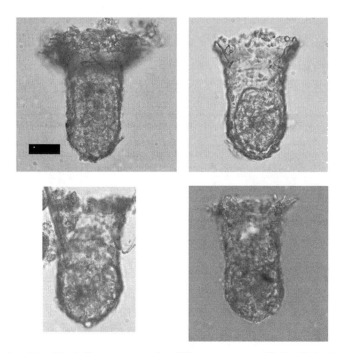

Figure 9.14 An example of protists with similar appearance but different sequences. Tintinnids in the genus *Tintinnopsis* can be classified by the size and shape of the lorica, a loose-fitting shell constructed around the cell. The top two individuals share identical 18S rRNA sequences as do the bottom two organisms. Sequences from the top individuals, however, are >6% different from the bottom two. Scale bar = 20 μm. Images taken by Luciana Santoferrara, which were provided and used with permission by George McManus.

related to the known alveolates, raising questions about their role in nature. Some may be parasites (Brown et al., 2009). Similarly, new marine stramenopiles have been found and placed in various MAST groups (MArine STramenopiles) only somewhat related to well-characterized stramenopiles. Also known as heterokonts, the well-characterized stramenopiles include diatoms and other chlorophyll *c*-containing algae. One defining trait of these organisms is their two flagella of unequal size. Diatoms fit into this group, because their gametes have flagella even though mature diatoms do not (Jürgens and Massana, 2008).

Fungi account for a large proportion of the rRNA genes recovered by cultivation-independent approaches from soils, with protists a close second in one study (2987 versus 1370 small subunit rRNA genes) (Urich et al., 2008). The Ascomycota accounted for over 60% of all fungi,

while Glomeromycota made up less than 25% (Fig. 9.15). Several cultivation-dependent studies have examined fungi, evident from the 100 000 species of fungi that have been described (McLaughlin et al., 2009). The number of described fungal species is still only a small fraction of the 0.7 to 1.5 million fungal species thought to be in the biosphere.

The protist community in the example given in Figure 9.15 was dominated by members of the Amoebozoa, although other types of protists, including green algae, protozoa, and ciliates, were also found. The most abundant protist group, the Mycetozoa, includes slime molds, which can spread over surfaces and attain large sizes (approaching a meter) under favorable conditions. Some of these soil protists are related to well-studied marine relatives. Cercozoa are closely related to marine foraminifera and radiolarians while Alveolata include ciliates and dinoflagellates as mentioned above.

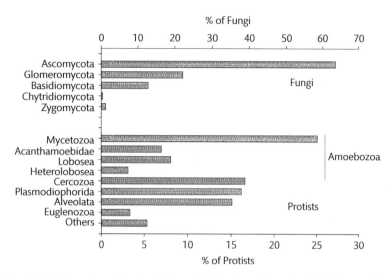

Figure 9.15 Eukaryotic microbes found by cultivation-independent analyses in sandy soil in Germany. Data from Urich et al. (2008).

Relevance of community structure to understanding processes

This chapter began by pointing out that biogeochemical processes can be examined without knowing anything about the types of microbes carrying out the process. This raises the question of what, if anything, we really need to know about microbial community structure if the goal is to understand processes. The answer will vary with the process and with what specifically we want to know. In fact, some answers have already been given in previous chapters. For example, the type of microbe carrying out photosynthesis has several ramifications for understanding food web dynamics (the cell size of the primary producer determines the size of the herbivore and so on) and the cycling of various elements (different primary producers require different elements). Likewise, it clearly matters whether fungi or bacteria are the main decomposers of organic material in soils. Several other processes, such as methanogenesis (Chapter 11) and nitrogen fixation (Chapter 12), depend on specific types of microbes. In all other areas of ecology, it is essential to know which organism is present and active in an

environment. It would be remarkable if microbial ecology were different.

In addition to learning about which microbe is carrying out which process, microbial ecologists are exploring general relationships among function, diversity, and stability of the community, analogous to questions first posed for larger organisms. One question is about the resilience of microbial communities to perturbation and the contribution of functional redundancy to that resilience. One hypothesis is that microbial communities are dominated by redundant organisms that carry out very similar processes, but which differ in how they respond to temperature, pH, or other environmental properties. This redundancy may become important when environments change. A diverse, functionally redundant community may respond quickly to perturbations and be resilient to environmental change, allowing critical processes to continue uninterrupted. It may also be essential for an ecosystem to be resilient to environmental changes caused by pollution from humans (Cardinale, 2011). In short, the diversity of microbial communities may be key to the maintenance of biogeochemical cycles in the biosphere.

Summary

1. To circumvent the problem of culturability, microbial ecologists use cultivation-independent methods to examine specific genes, usually 16S rRNA for prokaryotes and 18S rRNA for eukaryotes. Sequences of these genes are informative for deducing taxonomic and phylogenetic relationships among organisms.

2. The microbes isolated by standard cultivation approaches are quite different from those observed in nature by cultivation-independent methods.

3. Of the 50–100 phyla of bacteria found in the biosphere, only a few (<10) are abundant in any particular habitat. Microbial communities are usually dominated by a few phylotypes and clades while most are in low abundance, making up a rare biosphere.

4. The same bottom-up factors affecting bulk properties of microbes, such as biomass and growth, also shape the composition of microbial communities. These factors interact with top-down control especially by viruses in determining the presence and abundance of various taxa in microbial communities.

5. In contrast to prokaryotes, useful information about the community structure of eukaryotic microbes can be gained by microscopy and by traditional cultivation-dependent approaches. However, especially for small protists, cultivation-independent methods have revealed new types of eukaryotic microbes in many natural environments.

6. In every other field of ecology, identifying the organisms is essential for understanding the role of organisms in the environment. The ecology of microbes is likely to be no different, but the connection between community structure and biogeochemical processes is an open question.

CHAPTER 10

Genomes and metagenomes of microbes and viruses

Questions about the types of microbes present in natural environments are usually addressed by examining rRNA genes, as discussed in the last chapter. When interested in a specific biogeochemical process, however, microbial ecologists often turn to other genes. These other genes, called "functional genes", usually encode a key enzyme of the process being investigated, such as *rbcL* for CO_2 uptake, *pmoA* for methane oxidation, or *nah* for naphthalene degradation. While informative, there are several problems with this functional gene approach. These genes are usually examined by PCR-based methods using primers that target conserved regions of the genes. Of course, any genes too dissimilar to the primers will not be sampled. Often, several genes are important in a process, meaning several primer sets have to be used and several PCR reactions run. Also, functional genes often cannot be used to deduce which microbe is carrying out the process, for reasons to be discussed in this chapter.

This chapter will describe some genomic approaches to circumvent these problems. From the viewpoint of microbial ecology, genomics, metagenomics, and environmental genomics, all defined below, can be considered simply as a suite of approaches to examine the physiology and biogeochemical role of microbes in nature. However, genomic-based fields are much more than collections of methods. The genome (DNA), transcriptome (RNA), and proteome (protein) are all major, defining features of an organism. To know these features for a microbe is to know that microbe, a big step forward in understanding its roles in nature.

What are genomics and environmental genomics?

In contrast to the study of a single gene or even of several genes simultaneously, the field of genomics uses data about entire genomes of organisms: the complete sequence of all genes in the right order and organization in each chromosome (but see Box 10.1). The first genome, that of a bacteriophage (φX174), worked out in 1977, was extremely small, only 5386 nucleotides, just enough for only eight genes (Smith et al., 2003). The genomic field really got its start in 1995 with the complete sequencing of two bacteria, *Haemophilus influenzae* and *Mycoplasma genitalium* (Fraser et al., 1995, Fleischmann et al., 1995). These two bacteria were chosen for sequencing in part because they are pathogens but perhaps more importantly because they have small genomes—small for bacteria, but still much larger than the genome of φX174 and other viruses. The genome sizes are 1.8 and 0.58 Mb for *H. influenza* and *M. genitalium*, respectively, where "Mb" is a million base pairs of DNA. The bacteria were sequenced by a shotgun cloning approach, a radical idea when it first appeared (Fig. 10.1). The sequence of the first eukaryote, *Saccharomyces cerevisiae* (baker's yeast), was published in 1996, followed by publication of a draft of the human genome in 2000.

Now it is routine to sequence prokaryotic genomes, and soon it will be routine for eukaryotes, including humans. It has become routine because the cost of sequencing has greatly decreased since φX174 was sequenced. In the 1970s, sequencing a gene cost nearly

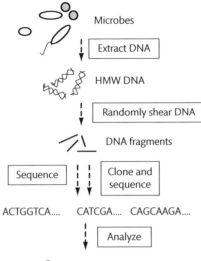

Genome sequence

Figure 10.1 Shotgun sequencing of genomes. The approach was first used for bacteria in pure culture, but now it is used for microbes in complex communities as implied here. DNA fragments used to be cloned before sequencing was possible, but next-generation sequencing techniques have eliminated the need for cloning. "HMW" is high molecular weight DNA.

Box 10.1 Complete or draft genomes?

Some sequencing projects may gather all of the sequences from the automated part of the sequencing process but not attempt to put them together into a "closed" genome with no spaces (gaps). Instead of being complete or closed, the genome is said to be a "draft". Nearly all of an organism's genome may be determined, but the remaining 1–10% is left undone. Assembling the sequences, filling in the gaps, and finding all of the missing pieces, are time-consuming and expensive steps. The time and money saved by not finishing the genome can go into doing more genomes, which is important for many ecological questions. The disadvantage is that a gene that is missing from a draft genome may in fact be in the unsequenced genetic material.

$1000 per base pair, and $2.3 billion was spent in the late 1990s sequencing the human genome. Today, costs are below $0.0001 per base pair, and they continue to decrease, thanks to the development of new high-throughput sequencing approaches like pyrosequencing mentioned in Chapter 9. Soon it will be possible to sequence a human genome for about $1000, and knowing your own genome will be the starting point of evaluating your health and physical well-being. New sequencing technology has been driven by the promise of high profits in biomedical applications, but the new approaches are being used in all areas of biology, including microbiology and microbial ecology.

The number of ecologically relevant organisms that have been sequenced is still small compared to the number of pathogens, but these numbers are increasing, nearly exponentially so. Now over 1500 genomes of microbes are available in one form or the other (Wu et al., 2009). "Environmental genomics" or "ecological genomics" are terms used when genomes of organisms important in the environment or in ecological processes are examined. Even though our main focus here is on natural communities of uncultivated microbes, there is much to be learned from the genomes of cultivated microbes, even if these microbes are often only distantly related to the most abundant uncultivated microbes in nature. Analyses of genomic data obtained directly from uncultivated microbes would not go very far without data from cultivated organisms.

Turning genomic sequences into genomic information

Once the sequence of a genome has been determined, the real work begins. While some analyses can use the raw sequence data, most questions require that the sequences be given some meaning, if not a name and a solid description of the function. The first step is to find open reading frames (ORFs), sequences of DNA that begin with an initial codon (usually ATG, which encodes methionine) and end with a stop codon (TAA, TGA, or TAG) and are possible genes. Next, bioinformaticians try to determine if the ORF is actually a gene and, if so, to assign a function to it by comparing the sequence with others in databases such as Genbank. The most frequently used tool for comparing sequences is "BLAST" or Basic Local Alignment Search Tool. BLAST analysis and other

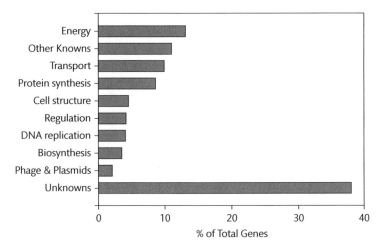

Figure 10.2 Annotation of genes in *E. coli* K12 when it was first sequenced. "Other Knowns" is the sum of all other known genes, which individually accounted for <2% of all genes. The number of unknown genes has decreased by about 50%, but it is still high. Data from Blattner et al. (1997).

steps in the annotation process are automated with most of the work done by sophisticated computer programs. People doing "manual annotation" are also necessary for some genes and parts of the genome.

The BLAST analysis can turn up significant "hits" or genes in Genbank that are similar to the unknown gene in the genome being examined. Ideally, the function of the known gene in Genbank has been established experimentally, in which case it is likely that the similar unknown gene has the same function. Often, however, the enzyme or other gene product from the known gene has not been characterized; the gene may have been identified by its similarity to a characterized gene in Genbank. Enzymes identified by sequence similarity alone are often called "putative", to emphasize their uncertain nature. The new gene may turn out to be most similar to another gene in Genbank without any known function. These are "conserved unknown" genes. In addition to known unknown genes, sequencing often turns up truly unknown ORFs, sometimes called ORFans, without a significant similarity (homology) to known genes. Typically, ORFans make up 10–20% of a prokaryotic genome (Koonin and Wolf, 2008).

In fact, even well-characterized organisms have large numbers of genes with unknown function (Fig. 10.2). When one of the best characterized organisms, *E. coli* K12, was first sequenced, nearly 40% of its 4288 protein-encoding genes could not be assigned a function

(Blattner et al., 1997). The fraction of unknown genes has dropped, but it is still surprisingly high (20% in 2009) even after years of work. Calling a gene "unknown" depends on one's definition and standards for the amount of data needed before concluding that the function of a gene is truly known. Regardless of semantics, there is still much to be learned about many genes in genomes even from *E. coli* and other intensively studied organisms. The number of unknown genes is even higher for microbes and other organisms which have not been examined extensively.

Lessons from cultivated microbes

The number of poorly characterized and completely unknown genes is one of the first lessons to be learned from examining cultivated microbes. There are several others, all useful for applying genomic approaches to and thinking about uncultivated microbes in nature.

Similar rRNA genes, dissimilar genomes

The first phase of genomic studies focused on microbes from different genera and even kingdoms; an archaeon (a methanogen) was sequenced soon after the first bacterium (Bult et al., 1996). The second phase focused on organisms that appeared to be very similar, and even strains of the same microbial species were compared. These organisms were usually pathogens or microbes such as *E. coli* whose physiology and molecular biology had been examined extensively over the years. It was startling when early comparative genomic studies found that closely related organisms can have quite different genomes. For example, three strains of *E. coli* have in common only about 40% of all of their protein-encoding genes (Welch et al., 2002), and strains of the plant pathogenic bacterium *Ralstonia solanacearum* are only 68% similar at the genome level; diversity within both bacterial species is greater than the difference between humans and pufferfish (Philippot et al., 2010). To put it another way, two bacteria sharing nearly identical 16S rRNA genes can have very different genomes. Another example is *Prochlorococcus*. Although the 16S rRNA genes from the high- and low-light ecotypes differ only slightly, whole genome sequencing revealed many differences between the two (Rocap et al., 2003), as mentioned before. The high-light ecotype has a smaller genome than the low-light ecotype (1716 versus 2275 genes) and is the smallest of any known oxygenic phototroph.

Genome size

Genome size varies greatly among prokaryotes and between prokaryotes and eukaryotes (Fig. 10.3). The genome size of sequenced bacteria ranges from about

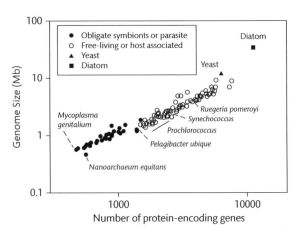

Figure 10.3 Genome size as a function of the number of protein-encoding genes in various bacteria and two eukaryotes. The data are from Giovannoni et al. (2005) and Armbrust et al. (2004).

0.18 Mb for the intracellular symbiont *Carsonella ruddii* to 13 Mb for the soil bacterium *Sorangium cellulosum* (Koonin and Wolf, 2008); there is about a tenfold variation in genome size among free-living bacteria, much less than the variation among eukaryotes—57 000 (Pellicer et al., 2010). As suggested by the genome of C. *ruddii*, obligate symbionts and parasites have very small genomes whereas some of the largest are found in soil bacteria. The current record (1.3 Mb) for the smallest genome of a free-living microbe belongs to the marine bacterium in the OM43 clade, followed closely by another marine bacterium, *Pelagibacter ubique*. Chapter 9 described how these bacteria were not isolated by the traditional agar plate method, and they grow very slowly for laboratory bacteria in unamended seawater. Other uncultivated bacteria in oligotrophic environments probably also have small genomes, as most oceanic bacteria are about the same size (both in terms of DNA and protein) as these bacteria (Straza et al., 2009). In contrast, other natural environments are likely to harbor uncultivated bacteria with large genomes, evident from the genome size of various soil bacteria (Konstantinidis and Tiedje, 2004).

Overall, there is a bimodal distribution in genome sizes for bacteria. That is, in graphs of the number of bacteria with a particular genome size versus genome size, there is one peak around 2 Mb and a second, smaller one around 5 Mb (Koonin and Wolf, 2008). This bimodal distribution is consistent with data from flow cytometry (Chapter 6) of bacterial communities in lakes and the oceans. In graphs of flow cytometry data comparing DNA content (fluorescence from a DNA stain) and side scatter (related to cell size), there are often two clouds of points, one termed low nucleic acid cells, the other high nucleic acid cells, although the fluorescence is due mainly to DNA. In contrast to bacteria, there is no bimodal distribution in the distribution of genome size for archaea, and the median is around 2 Mb (Koonin and Wolf, 2008).

The genomes of most eukaryotic microbes are much larger than those of prokaryotes. The genome of the diatom *Thalassiosira pseudonana*, for example, is 31.3 Mb (von Dassow et al., 2008), making it nearly 20-fold larger than that of the bacterium *P. ubique*. Another example is the fungus *Neurospora crassa* which has a 40 Mb genome with about 10 000 protein-encoding genes (Galagan et al., 2003). An exceptional eukaryote is the green alga *Ostreococcus tauri* (Derelle et al., 2006). Its genome of 12.6 Mb is actually smaller than the genome of the bacterium *Sorangium cellulosum*, although the alga has 20 chromosomes, while the soil bacterium has only one. Reminiscent of *P. ubique*, *O. tauri* has a large number of genes per total genome and minimal non-coding DNA. The genome of *O. tauri* may have some bacteria-like features because this eukaryote is very small, only about one micron. At the other extreme, the current record for the largest genome, held by the flowering plant *Paris japonica*, is 150 000 Mb (Pellicer et al., 2010).

Organization of eukaryotic versus prokaryotic genomes

Eukaryotic and prokaryotic genomes differ in many other aspects in addition to size (Table 10.1). Although eukaryotic genomes are much bigger than bacterial genomes, there is less of a difference in the number of protein-encoding genes. Taking *T. pseudonana* as an example again, this diatom has genes for about 11 000 proteins or only about tenfold more than that for the bacterium *P. ubique*, much smaller than the twenty-fold difference in genome size. Part of the difference is the larger number of regulatory genes in eukaryotes than in prokaryotes. It is these regulatory genes that help to explain differences among eukaryotes. Perhaps because of our inflated self-regard, it was a surprise to discover that *Homo sapiens*

has only about 22 000 genes, no more than many vegetables. We differ from turnips and other mammals because of how genes are regulated, as well as in the nature of those genes. Eukaryotic genomes are also large because they have large regions of non-coding DNA, which does not code for a protein or an RNA molecule and seems not to do anything. Once called "junk DNA", we are beginning to understand the essential functions of these large regions of eukaryotic genomes. Another difference is that eukaryotic genes are often interrupted by stretches of DNA (introns) that do not encode for any amino acid and after transcription are cut out of the resulting mRNA molecule before translation to a protein. All of these and other differences in genome structure have many implications for the regulation of metabolism and thus for the ecology of prokaryotes and eukaryotic microbes.

Related to the number of protein-coding genes per total genome is the space between genes ("intergenic space"). This space is much larger for eukaryotes in general than for prokaryotes. Even among bacteria, there is a large range in the length of intergenic space. The median intergenic space is only three base pairs for *P. ubique* (Giovannoni et al., 2005) versus 137 base pairs for *Photobacterium profundum*. Intracellular symbionts and parasites such as *Mycobacterium leprae* or *Rickettsia* have even larger intergenic spaces, some of which are filled with "clustered regularly interspaced short palindromic repeats" (CRISPRs) and pseudogenes.

Table 10.1 The structure of genomes in eukaryotic microbes and prokaryotes.

Property	Prokaryotes	Eukaryotes	Comments
Genome size (Mb)	0.18–13	10–150 000*	Invertebrates and plants
Organization	One circular chromosome	Several linear chromosomes	Bacteria can have several replicons
Related genes together?	Genes in operons	Few operon-like clusters	Operons are clusters of genes that are co-regulated
Introns?	Rare	Common	Introns are DNA fragments between sections of a real gene
"Junk" DNA	Little to none	Lots	"Junk" DNA probably have essential purposes
Repeated sequences	Rare	Common	Eukaryotes can have long stretches of two or more repeating nucleotides
Protein-encoding genes: total genome	High	Low	This is the result of the previous three characteristics
rRNA genes	1–10 (mean <5)	Hundreds	

* From Gregory (2010) and Pellicer et al. (2010).

CRISPRs are thought to enable bacteria to resist infection by phages (Horvath and Barrangou, 2010) and pseudogenes are genes that are no longer capable of being turned into proteins.

Growth rates and genomics

The presence or absence of individual genes will obviously determine a microbe's growth rate in a specific environment, but overall characteristics of a microbe's genome also set fundamental limits on how fast it can grow. Small genomes are thought to be one reason why bacteria grow faster than other organisms, at least in nutrient-rich conditions, perhaps because of the low energetic cost of replicating a small genome. However, there is no correlation between genome size and minimal growth rates among bacteria (Vieira-Silva and Rocha, 2010) as a result of various selection factors pushing against each other. Bacteria with large genomes may be able to take advantage of diverse and rich nutrient conditions to grow quickly where bacteria with small genomes may not have the necessary functional or regulatory genes. In contrast to genome size, there are several tight relationships between growth and various genomic traits connected to protein synthesis.

Perhaps of most interest to microbial ecologists is the 16S rRNA gene. There is a fairly tight correlation between the minimal growth rate of a bacterium and the number of 16S rRNA genes it has (Fig. 10.4). A bacterium capable of high growth rates can have several 16S rRNA genes (multiple "copies"), which differ only slightly from each other in sequence (<1%). Interestingly, other genes connected to protein synthesis are not necessarily present as multiple copies in fast-growing bacteria (Vieira-Silva and Rocha, 2010). Rather, these genes tend to be closer to the origin of replication; the genes include those for RNA polymerase, ribosomal proteins, and tRNA—all connected to protein synthesis. By being close to where the bacterial chromosome starts to replicate, these genes in effect are present in higher numbers in a rapidly dividing cell than genes further away from the origin of replication. This arrangement of protein synthesis genes helps explain growth rates under constant conditions as well as why some bacteria can respond more quickly than others to changes in growth conditions.

Another genomic feature related to growth is the preference for use of one codon for an amino acid over another, termed "codon usage bias". Remember that several amino acids can be encoded by more than one triplet of DNA bases. Isoleucine, for example, can be encoded by ATT, ATC, and ATA. Intrinsically slow-growing bacteria tend to use these various codons equally whereas fast-growing bacteria have high codon usage bias (Vieira-Silva and Rocha, 2010). The favoring of some codons over others enhances translation efficiency and thus protein synthesis and growth. Interestingly, psychrophilic bacteria tend to have higher codon usage bias than the average bacterium for a given growth

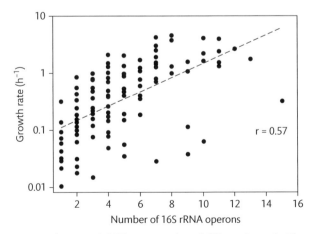

Figure 10.4 Relationship between growth rate and rRNA copy number of different bacteria. The correlation coefficient is from a linear regression analysis of log (growth rate) versus copy number. Data from Vieira-Silva and Rocha (2010).

rate while thermophilic bacteria tend to have lower bias. This feature of genomes may allow bacteria to compensate for temperature effects. Psychrophilic bacteria can grow at high rates in spite of cold temperatures slowing down chemical reactions, while thermophilic bacteria can maintain control of their metabolism in the face of high temperatures pushing reactions faster than can be sustained over the long run.

Chromosomes, plasmids, and replicons

The genetic material in bacteria is often depicted as a single circle of DNA, a single chromosome, although microbiologists certainly knew that many bacteria have non-chromosomal DNA, such as plasmids. These are usually circular molecules of DNA that are much smaller than a chromosome, consisting of only a few thousands of base pairs. Plasmids were thought to contain genes such as those for antibiotic resistance or for the degradation of odd organic compounds—genes that are essential for the bacterium's growth and survival only in a few, perhaps atypical

environments. Bacteria may lose plasmids if genes on them are no longer required. Plasmids bearing antibiotic resistance genes, for example, are selected against and disappear when the antibiotic is no longer present.

Whole genome sequencing of bacteria revealed a more complicated picture of how genomic material is physically organized (Moran, 2008). In addition to the main chromosome and the usual plasmids, bacteria were found to house several "extrachromosomal" pieces of DNA. Some of these approach the size of the main chromosome and contain many essential genes. The distinction between a plasmid and chromosome is based on size, on the type of genes directing replication, and on the presence of essential genes like an rRNA gene. Chromosomes have them while plasmids don't. The term "replicon" includes both plasmids and chromosomes. Many bacteria contain more than one replicon (Table 10.2).

The organization of a bacterial genome has evolutionary and ecological implications. It is not clear if large replicons are formed by splitting off from the main chro-

Table 10.2 Basic properties of the genomic material in some prokaryotes. "Replicons" refers to chromosomes and plasmids. "GC" refers to the DNA bases guanine and cytosine. GC content is a commonly used, overall characteristic of genomes. Data from Moran (2008) and http://cmr.jcvi.org/.

| Organism | Genome size (Mb) | Number of | | | |
		Replicons	ORFs	rRNA operons	GC %
Archaea					
Aeropyrum pernix K1	1.67	1	1841	1	56
Archaeoglobus fulgidus DSM 4304	2.18	1	2420	1	48
Methanocaldococcus jannaschii	1.74	3	1786	2	31
Methanococcus maripaludis S2	1.66	1	1722	3	33
Methanosarcina acetivorans C2A	5.75	1	4540	3	42
Bacteria					
Colwellia psychrerythraea 34H	5.37	1	4910	9	38
Bacillus anthracis Ames	5.23	1	5637	11	35
Desulfotalea psychrophila LSv54	3.66	3	3234	7	46
Geobacillus kaustophilus HTA426	3.59	2	3540	9	52
Nanoarchaeum equitans Kin4-M	0.49	1	536	1	31
Pelagibacter ubique HTCC106	1.31	1	1354	1	29
Photobacterium profundum SS9	6.4	3	5491	15	41
Rhodopirellula baltica SH1	7.15	1	7325	1	55
Ruegeria pomeroyi DSS-3	4.6	2	4284	3	64
Cyanobacteria					
Anabaena variabilis ATCC 29413	7.07	4	5697	12	41
Nostoc sp. PCC 7120	7.21	7	6127	11	41
Prochlorococcus marinus MED4	1.66	1	1713	1	30

mosome or by the coalescing of large plasmids ("mega-plasmids"). In an ecological context, genes on plasmids are likely to be more mobile than genes housed in the main chromosome or in other large replicons. As used here, "mobile" implies the potential loss of a gene if environmental conditions no longer select for it, but also the gain of genes through exchange among bacteria. "Mobile genetic elements" is the term used to describe DNA pieces that are capable of exchange among bacteria. This exchange can occur on ecological timescales—hours to days, the lifetime of a bacterium, short enough to affect the bacterium's success in a particular environment and time. Because of these mobile genetic elements, the genomes of bacteria and probably of all microbes in nature are much more dynamic than those of large organisms. These mobile genetic elements are potentially exchanged between distantly related organisms, as described next.

Lateral gene transfer

Genomic sequence data of microbes revealed a new mechanism in evolution not envisioned by Darwin and his successors. The traditional mode of evolution is that genes are handed down from generation to generation, from parent to offspring, with some genes persisting in offspring that survive while others disappear when offspring die before reproduction. This vertical exchange of genes is captured by traditional phylogenetic trees of rRNA gene sequences. Similar trees can be constructed with other genes. The problem comes when the trees don't agree. These discrepancies first became apparent when even just a few genes were compared, but they became even more evident as whole genome data accumulated. Discrepancies can arise when rates of evolution for various genes diverge. Another reason is lateral gene exchange, also called horizontal gene transfer.

Lateral gene transfer is the movement of genes from one organism to another, unrelated organism, in contrast to the vertical passing of genes from mother to daughter cells. This exchange can be effected by viruses (transduction—Chapter 8) or by the uptake of undegraded DNA ("transformation"). Genes have a greater chance of remaining in the recipient organism if the donor and recipient are related, but there are many

examples of genes being exchanged between unrelated organisms, even between bacteria and archaea, and between both prokaryotic kingdoms and eukaryotes, including humans. In addition to discrepancies between a gene's phylogenetic tree and a 16S rRNA tree, the effect of lateral gene transfer is seen in the relatedness of various genes next to each other in a microbial genome. While most of the genes are most similar to those from a close relative sharing the same ancestor, genes brought in by lateral transfer are most similar to those from a distantly related organism. The GC content and codon usage of the recently transferred gene may also differ from that of the rest of the genome.

Lateral gene transfer calls into question the idea that any single gene can be used to follow the evolution of an organism. It is certainly true that the tree metaphor or model for describing evolution must be modified to include many intertwining branches due to instances of lateral gene transfer, and it is now possible to use whole genome sequences ("phylogenomics") to explore microbial evolution. However, those genes having to do with information processing, such as DNA and protein synthesis, tend not to be subjected to lateral gene transfer (Daubin et al., 2003). In particular, the 16S rRNA gene appears to follow an organism's phylogeny when defined by whole genome sequences (Wu et al., 2009). This and similar genes probably do not undergo substantial lateral gene transfer because the gene products cannot be easily accommodated into the existing molecular machinery. In the case of 16S rRNA genes, a foreign rRNA molecule cannot easily fit into the complex structure of a ribosome with its several other rRNA molecules and more than 50 proteins.

In contrast, functional genes encoding enzymes often act alone and can tolerate more variation. The portion of the enzyme most directly catalyzing the reaction (the catalytic site) maybe very similar ("highly conserved") among enzymes from very diverse organisms, whereas the rest of the enzyme and thus the gene may vary substantially.

Lateral gene transfer has consequences for trying to determine which organism a particular gene came from, an issue that arises when sequence data for only that gene from natural microbial communities are available. In these cases, often it is impossible to link the gene to its source organism with any confidence. When sequences

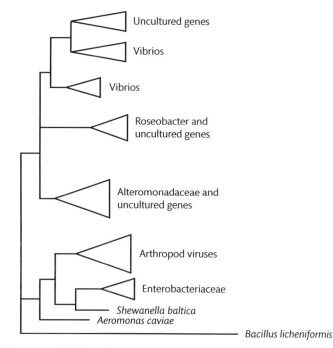

Figure 10.5 Neighbor-joining tree of chitinases from various bacteria and from viruses infecting insects. The wedges contain several related genes. "Uncultured genes" refers to chitinase genes retrieved by cultivation-independent methods; all genes with taxonomic names come from cultivated (cultured) organisms. There are several indications of lateral gene transfer in this tree, such as the presence of virus chitinases among these bacterial genes, and the *Roseobacter* sequences (members of the Alphaproteobacteria) sequences among the vibrios and Alteromonadaceae (both in the Gammaproteobacteria). Data from Cottrell et al. (2000).

of both the functional gene and the 16S rRNA gene are known, often the phylogeny of the 16S rRNA gene doesn't match the phylogeny of the functional gene. The enzyme used to hydrolyze chitin (chitinase) is one example (Fig. 10.5). In this case, chitinases from various types of bacteria do not fall into the same group as defined by their 16S rRNA genes, probably because of lateral gene transfer. The presence of chitinases in viruses infecting insects, which have chitin exoskeletons, is further evidence that these genes are exchanged laterally. The genes retrieved from marine waters in this example seem to come from vibrios, but we cannot say for sure. This is the problem in interpreting short stretches of DNA sequences, such as retrieved by PCR-dependent approaches, which have parts of functional genes without any sequences from a phylogenetic marker gene. Some cultivation-independent approaches get around this problem, as discussed next.

Genomes from uncultivated microbes: metagenomics

Genomes from cultivated microbes are invaluable for understanding the ecology of uncultivated microbes, but as mentioned repeatedly throughout this book, there are many differences between most of the cultivated microbes grown in the lab and uncultivated microbes in nature. Another problem is that there are too many microbes in nature to cultivate them all, even if we could. Fortunately, genomic information can be accessed directly from uncultivated microbes without growing them in the lab. Rarely is the end result, however, a complete genome of an uncultivated organism. Rather, the goal is to obtain sequence data for a large number of genes simultaneously without using PCR to retrieve the genes. This approach is called metagenomics, with "meta" emphasizing that several types of organisms are examined simultaneously.

Metagenomics can provide clues about the physiology and thus potential biogeochemical role of uncultivated microbes. The sequences of RNA genes and other phylogenetic markers say much about the types of microbes present in nature, but we cannot link those data easily to a particular biogeochemical function. Metagenomics is one way to make those links. While most metagenomic studies have focused on bacteria, a few have examined eukaryotic microbes and viruses.

Metagenomic approaches

The data and insights gained from a metagenomic approach depend on what type of metagenomic approach is actually used. Figure 10.6 presents some of the alternatives. One approach is to examine individual microbial cells, which arguably is not metagenomics, although it is certainly a part of ecological genomics. The single cell approach is easiest for larger eukaryotic microbes (see below), but it has been used for prokaryotes as well (Stepanauskas and Sieracki, 2007). The other metagenomic approaches can be distinguished by the size of the DNA fragment being targeted by cloning or by direct sequencing.

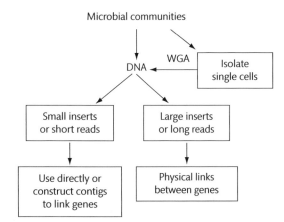

Figure 10.6 Summary of metagenomic approaches. WGA is "whole genome amplification", a method to generate more DNA from small or dilute samples, here a single cell. The original metagenomic approaches cloned either small inserts (<10 kb) or large inserts (>10 kb) of foreign DNA. New metagenomic approaches rely on direct sequencing without cloning and generate either short or long stretches of DNA sequences ("reads").

One metagenomic approach examines small DNA fragments that bear less than a gene's worth of genomic material. In the past, these small DNA fragments ("inserts") were first cloned into *E. coli*, resulting in a "small insert library" (see Box 10.1). Cloning and library construction were necessary for sequencing and other analyses. The cloning step, however, is not a part of next-generation sequencing methods. Another advantage of new sequencing methods is that they generate huge data sets and thousands to millions of sequences or "reads" per analysis. A disadvantage, however, is that the number of bases sequenced per read, the read length, for many of these new sequencing methods is short, only 100–1000 bp long, depending on the method. These reads are not usually long enough to cover an entire gene, much less several contiguous genes in a genome.

The sequence data from a small insert metagenomic library or from a next- generation approach can be used immediately to make inferences about potential biogeochemical functions without the potential biases of PCR. But those functions cannot be linked to a specific microbe, given problems with lateral gene transfer. A single insert or short read is too small to have both a phylogenetic marker, such as an rRNA gene, and a functional gene. However, the link can be made in another way. Given enough sequence data, it is possible to piece together ("assemble") sequences from two or more small sequences that overlap to form a "contig", One potential artifact is that two unrelated sequences may be joined artificially into a "chimera". If chimeras are removed, however, the reconstructed DNA fragment, the contig, is assumed to originate from a single microbe.

Another metagenomic approach analyzes large DNA fragments, ranging from about 40 000 bases (40 kb) for fosmid clone libraries to >100 kb for BAC clone libraries. Briefly, DNA is isolated from the sample, sheared to the appropriate size, inserted into either fosmids or BAC vectors, and then cloned in *E. coli*. The DNA fragments carried by the fosmids and BAC vectors can be examined by various methods and completely sequenced, now most efficiently by next-generation approaches. The end result is a collection of sequences 40 kb to >100 kb in length. To put that length in perspective, each sequence potentially represents roughly 40 to >100 genes from a single organism, if we assume that a "typical" gene is about 1000 base pairs (1 kb). Therein lays the power of this metagenomic approach. These cloned DNA fragments harbor not only entire genes but often whole operons or related sets of genes. Ideally, one of those genes is for rRNA or some other phylogenetic marker. The presence of such a marker gives the origin, the type of microbe, of the genes. In this case, the presence of both a phylogenetic marker and functional genes is a strong argument for that particular organism carrying out that particular biogeochemical function in nature.

Currently, the type of metagenomic information obtained from fosmids or BAC clones cannot be matched by next-generation sequencing approaches alone. The read length from these approaches is not long enough to cover more than a gene, maybe two. Although the small insert library approach is not used these days, there is no replacement today for large insert library construction. However, the technology is changing rapidly and what seemed impossible a few years ago is commonplace today.

The proteorhodopsin story and others

The discovery of proteorhodopsin is a good example to illustrate the power of various metagenomic approaches. The original studies on proteorhodopsin used various cloning-based approaches, before next-generation sequencing approaches were available.

Chapter 4 describes how proteorhodopsin harvests light energy and may fuel ATP synthesis in many bacteria in natural environments. Genes for this chromophore-protein complex were discovered in a metagenomic BAC library of DNA from coastal seawater (Béjà et al., 2000). Screening of the library for 16S rRNA genes turned up a 130 kb clone with a 16S rRNA gene most similar to one from the SAR86 clade of Gammaproteobacteria originally discovered in the Sargasso Sea. Sequencing the clone revealed several other genes, including one most similar to rhodopsin from archaea (Fig. 10.7). The presence of the 16S rRNA gene on the same BAC clone as the rhodopsin gene proved that the rhodopsin was in fact from a bacterium. It is an example of a particular function attributable to a particular uncultivated bacterium. The large size of the BAC clone also ensured that the entire proteorhodopsin could be retrieved from the original BAC clone and examined in greater detail. Having the entire protein was important for showing that its function, at least in laboratory experiments with *E. coli*, is similar to that of the archaeal rhodopsin as predicted by the sequence data (Béjà et al., 2000).

Small insert metagenomic libraries also played an important role in the early chapters of the proteorhodopsin story. Soon after its discovery in SAR86, other studies using PCR-based and large insert metagenomic approaches turned up evidence that rhodopsin may be in bacteria other than the SAR86 clade (de la Torre et al., 2003). However, it was a study using a small insert metagenomic approach that really demonstrated the diversity of bacteria bearing the proteorhodopsin gene in the oceans (Venter et al., 2004). That study showed

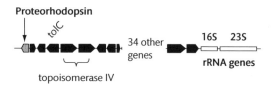

Figure 10.7 Map of a BAC clone with genes for proteorhodopsin and 16S rRNA, indicating the genomic material was from a SAR86 bacterium. The direction of transcription for each gene is indicated by an arrow. Only a few genes and names are given for simplicity. Figure used with permission from Oded Béjà, based on Béjà et al. (2000).

that bacteria other than those in the Proteobacteria phylum have this gene, making the "proteo" part of the name a misnomer, although it is still widely used. These metagenomic approaches can uncover much more diversity than a PCR-based approach because the genes recovered by metagenomic approaches are not necessarily just those matching the PCR primers. Although small insert clone libraries were used by these studies, today the work would be done without cloning, instead with a next-generation sequencing approach.

The story of proteorhodopsin illustrates how metagenomics can reveal previously unknown functions and metabolisms carried out by microbes in natural environments. These functions potentially change our ideas about how biogeochemical cycles work and how they are regulated. Other examples of novel metabolisms, or at least novel for where they were found, include the oxidization of sulfide and carbon monoxide in oxic environments (Moran and Miller, 2007). Sulfide and other reduced inorganic sulfur compounds are not expected in oxygen-rich habitats (Chapter 11), so it was a surprise to see genes using these compounds in surface environments. Carbon monoxide was known to be produced by photochemical reactions and some bacteria were known to use this highly oxidized, low energy-yielding compound, but microbial ecologists thought that bacteria capable of using higher energy-yielding organic compounds would not have carbon monoxide oxidization genes (*cox*). Another example of the power of metagenomics is ammonia oxidation by archaea (Schleper et al., 2005). Although ammonia oxidation by bacteria has been known for decades, it took metagenomic work in both the oceans and soils to reveal the role of archaea in carrying out this important reaction of the nitrogen cycle.

Metagenomics of a simple community in acid mine drainage

The microbial communities examined by the metagenomic studies discussed so far were complex, consisting of thousands of different bacteria. At the other extreme is the biofilm microbial community that forms on rocks covered by the waste stream from old mines. The water draining from mines is highly acidic and loaded with reduced sulfur and iron compounds. These compounds are used by a few prokaryotes for energy (chemolithotrophy) to support autotrophic fixation of carbon dioxide (chemoautotrophy), the only source of carbon in this dark environment, far from algae and higher plants. The acid mine drainage (AMD) microbial community is very simple. In one AMD community from the Richmond mine at Iron Mountain, California, a classic study found only three bacterial and three archaeal taxa (Tyson et al., 2004). The most abundant bacterium was *Leptospirillum* group II, followed by *Leptospirillum* group III in the Nitrospira phylum. The archaea were *Ferroplasma*, members of the Thermoplasmatales. *Ferroplasma acidarmanus* fer1 had been isolated from the Richmond mine and had been sequenced by previous studies. Eukaryotic microbes are rare in AMD communities and their genes were not detected by this metagenomic study.

Because the community is simple, it was possible to construct nearly complete genomes for *Leptospirillum* group II and for a very close relative of *F. acidarmanus* fer1, called *Ferroplasma* type II. The nearly complete genomic data became important for addressing several questions about the functions carried out by the bacteria and archaea in this consortium. The sequence data indicated that both *Leptospirillum* group II and III have the genes needed to fix CO_2 via the Calvin–Benson–Bassham cycle, although there was also evidence that some AMD microbes use the reductive acetyl-coenzyme A cycle for CO_2 fixation. The surprise was that the dominant bacterial taxon, *Leptospirillum* group II, does not have genes for nitrogen fixation, an essential process for this AMD community remote from sources of fixed nitrogen (nitrogen other than N_2). In contrast, *Leptospirillum* group III has nitrogen fixation genes. Since *Ferroplasma* type II also does not have these genes, *Leptospirillum* group III emerged as a key member of the AMD community even though it is not abundant. This is one example of a general problem in microbial ecology, that of figuring out the biogeochemical role of seemingly minor members of microbial communities in nature.

Useful compounds from metagenomics and activity screening

Chapter 1 pointed out that microbes produce many compounds useful in industry and in biomedical applications. The last category includes antibiotics, many of which

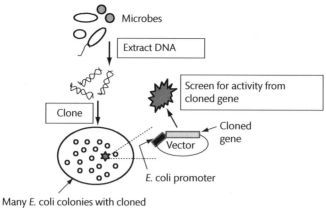

Many *E. coli* colonies with cloned
DNA fragments on a plate

Figure 10.8 Screening a metagenomic library for activity. The clone with the targeted gene is found by looking at the library for activity (represented here by a star), such as fluorescence from a cleaved by-product. This activity is due to a protein made from the targeted cloned gene. The protein must be synthesized by *E. coli*. The first step in the synthesis of this protein is the initiation of mRNA synthesis by an *E. coli* promoter, followed by translation of the mRNA by *E. coli* ribosomes and other aspects of protein synthesis.

come from microbes isolated from soil. Metagenomics is one way to discover these compounds from the vast majority of microbes that cannot be isolated and cultured in the lab. This practical application has been used most frequently to-date with soil communities (Daniel, 2005), but both the general idea and a key methodological step (activity screening) are applicable to any environment.

Many of the metagenomic clones producing useful compounds have been found by screening the metagenomic library for activity (Fig. 10.8). Although potentially quite powerful, it is not an easy approach. In addition to having the gene, an *E. coli* clone has to produce a functional protein from that gene, and that protein's activity has to be detectable using an approach that can be applied to many clones; a key is the capacity to screen thousands or even millions of clones rapidly. For example, in order to detect enzymes, the clone library can be exposed to analogs of the substrate attacked by the enzyme. These analogs produce color or fluorescence when hydrolyzed by the enzyme and are often the same as those used to examine exoenzyme activity (Chapter 5). Those clones that become colored or produce fluorescence harbor the gene of interest. A related approach is to modify the growth media for *E. coli* such that only those bacteria bearing the correct gene and producing the correct gene product survive and grow on the agar plate. This approach

has been used to find antibiotic resistance genes. Those *E. coli* bearing the targeted antibiotic resistance-encoding gene are able to grow in media with the antibiotic whereas all other *E. coli* do not grow. Instead of looking at thousands to millions of *E. coli* colonies, only a few resistant colonies will appear.

Activity screening has its weaknesses and strengths. The weakness of this approach is that many interesting proteins may not be produced by or be functional in *E. coli* at all or may not be produced under the conditions of the screening assay. The compounds may be toxic, or genes and synthesis of the target compound may be too foreign for *E. coli* to handle. Also, often huge libraries of thousands to millions of clones have to be screened to find one positive clone producing the targeted compound. The strength of this approach is that genes found by activity screening are often quite different from those found by other approaches. Screening clones by sequencing would find only those genes similar to what is known already whereas activity screening has the potential to find novel genes.

Metatranscriptomics and metaproteomics

The genes from a natural microbial community indicate the potential for a particular function to be carried out,

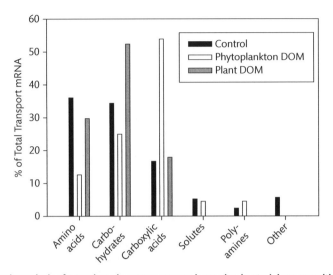

Figure 10.9 Metatranscriptomic analysis of organic carbon transporters in marine bacterial communities. The results are from bacterial communities incubated for one hour without any addition (control) or with the addition of organic material extracted from phytoplankton or a vascular salt marsh plant. "Solutes" refer to compounds used by bacteria as compatible solutes, such as proline and glycine betaine. "Other" includes known transporters for compound classes such as lipids and nucleic acids. The large number of transcripts for amino acid and carbohydrate transporters is consistent with other data indicating the importance of these organic compounds for supporting bacterial growth. Data from Poretsky et al. (2010).

but those genes may or may not be expressed. Biogeochemical and "black box" approaches (Chapter 1) sometimes can be used to determine if the function is actually occurring in the environment, but these approaches do not reveal which organisms are carrying out the process and do not detect at all some functions, such as light harvesting by proteorhodopsin. One way to address these problems is to examine expression of genes (mRNA synthesis) for metabolic functions connected to a biogeochemical process of interest. Especially for prokaryotes, the presence of mRNA for a particular process from a particular microbe is a good argument that that microbe is carrying out the process.

Transcriptomics is the study of all RNA molecules synthesized by an organism while metatranscriptomic approaches examine the RNA from an entire community of organisms. These RNA-based approaches are analogous to genomics and metagenomics, and in fact the RNA is turned back into DNA using reverse transcriptase before sequencing. However, there are several technical problems with examining RNA, one being that it is easily degraded. Also, rRNA accounts for about 80% of the total RNA pool (Chapter 2), complicating analysis of mRNA, usually the main target of metatranscriptomic

studies. The expression of a single gene can be followed by PCR-based approaches, but transcriptomic and metatranscriptomic approaches arose to avoid PCR and to explore questions involving expression of many genes simultaneously.

The types of mRNAs found by metatranscriptomic studies give some insights into metabolic functions being carried out by microbes in nature. Many of the mRNAs from soil and aquatic bacteria are for enzymes involved in RNA and protein synthesis, reflecting the amount of energy cells devoted to those activities. Expression of genes for protein folding and export and DNA repair is also high. Among functions related to more specific biogeochemical processes, metatranscriptomic studies have shown that genes for transport proteins are highly expressed by aquatic bacteria (Poretsky et al., 2009), with some transport processes favored over others, depending on the organic carbon source (Fig. 10.9). PCR-based assays indicated that expression of ammonia oxidation genes (amoA) by archaea is higher than that by bacteria in soils, but a metatranscriptomic analysis was also an important part of the story (Leininger et al., 2006).

Metatranscriptomic approaches are especially informative for examining eukaryotic microbes. In addition to gaining insights into the active microbes, it is easier to find the protein-encoding parts of eukaryotic genomes by looking at mRNA rather than the genome or metagenome directly. Remember that eukaryotic genomes are usually much larger than prokaryotic genomes because eukaryotic protein-encoding genes are often interrupted by introns, and eukaryotic genomes have a large number of regulatory elements and DNA with seemingly no function. The junk DNA is not transcribed and the introns are removed soon after transcription, leaving behind mRNA with only protein-encoding sequences. One study took advantage of dinoflagellate-specific spliced leader sequences to explore the metatranscriptomes of dinoflagellates in a lake and an estuary (Lin et al., 2010). Unexpectedly, this study found a proteorhodopsin-like gene in these aquatic eukaryotes.

One big problem remains, however: so little is known about the genomes and about even the protein-encoding genes of eukaryotic microbes. For example, a metatranscriptomic study of soil eukaryotes found a large number (32% of the total) of new hypothetical proteins, and the source of many of the genes could not be identified (Bailly et al., 2007). About 35% of the known genes came from fungi and metazoans, but few were from protists, even though protists as well as fungi accounted for most of the rRNA genes amplified from the soil sample. The low number of protist genes in the metatranscriptomic data is probably due to the paucity of protist sequences in genomic databases.

Proteomics and metaproteomics

The existence of an mRNA for a particular enzyme is some evidence that the process mediated by that enzyme is actually occurring, especially in prokaryotes. However, eukaryotic microbes and even prokaryotes have additional regulatory mechanisms that may prevent the mRNA from being translated into a protein and the process from occurring. Examining proteins is a way to get closer to the actual process. Analogous to genomics and transcriptomics, proteomic approaches examine all of the various proteins in a cell, and metaproteomics does the same for natural communities. These proteins are separated by chromatographic techniques, broken into pieces of peptides (the size varies with the method), and analyzed by mass spectrometry, often by more than one mass spectro-

metric technique strung together ("MS/MS") (VerBerkmoes et al., 2009). The mass or size of the peptide is sufficient for deducing its composition, but identifying the proteins requires information about the genes coding for them.

Metaproteomics has confirmed the conclusion from metatranscriptomic studies about the prevalence of transport proteins in aquatic microbial communities (Fig. 10.10). Metaproteomic studies also find unknown proteins, analogous to the unknown genes of metagenomes. Even data about seemingly uninformative proteins have revealed new insights, one example being biofilm development by an acid mine drainage (AMD) community (Denef et al., 2010). Although the metaproteomic dataset by itself was informative, it gained additional power when combined with metagenomic data for examining the closely related microbes making up the AMD biofilm. One important observation made by the study was that only a small fraction of the AMD genomes was actually expressed and appeared as proteins at a given point in time.

Metagenomics of viruses

Metagenomic approaches have been especially informative about viruses in natural environments. Remember that viruses have nothing analogous to the rRNA molecule used frequently for taxonomic and phylogenetic analyses (Chapter 8), although some studies have examined genes for proteins, such as those for capsids and DNA polymerase, common in viruses. Even the number of different viruses in a habitat is not well known, and many other basic questions have not been answered. Metagenomic approaches offer some solutions to this problem. In brief, viruses are collected by using filtration to remove large organisms, followed by ultrafiltration to concentrate the viral size fraction. Ultracentrifugation and other steps are then used to purify viral nucleic acids. Although single-strand DNA viruses have been examined in the Sargasso Sea (Angly et al., 2006), most of the work has focused on double-strand DNA viruses.

The most striking observation from viral metagenomic studies is the overwhelming diversity of viruses (Kristensen et al., 2010). The number of different viruses in the oceans alone has been estimated to be on the order of 10^{30}; soils and sediments are thought to be even

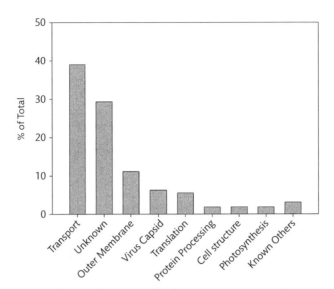

Figure 10.10 Example of metaproteomics, showing the proteins found in a marine bacterial community. The high number of transport proteins is consistent with metatranscriptomic studies of the oceans. The "Known Others" category includes proteins for redox reactions, proton pumps, transcription, and carbon and nitrogen metabolism. Data from Morris et al. (2010).

more diverse. One study estimated that surface waters of a coastal ocean had 7000 viral genomes (Edwards and Rohwer, 2005), roughly the same number of bacterial types in this habitat, assuming the metagenomic estimate for viruses is comparable to the 16S rRNA-based estimate for bacteria. But the evenness of the two communities differed. In contrast to a bacterial community with its few dominant phylotypes and many rare ones, viral communities are more even, with each virus present in low abundance and none dominating the community. This difference is evident in plots of abundance versus the rank in terms of abundance (organisms ranked from most to least abundant) of each phylotype in the community (Fig. 10.11).

The metagenome of these viruses is difficult to study because of its diversity but also because most of the viral genes do not match up with known genes. About 50% to 90% of viral genes from natural environments do not match genes from cultivated viruses and microbes. Among the known genes, there is a good correlation between the occurrence of genes in the viral metagenome and in the microbial metagenome (Kristensen et al., 2010). Carbohydrate metabolism genes, for example, are common in both the viral metagenome and the microbial metagenome while cell signaling genes are low in both metagenomes. Similarly, the phylogenetic

origin of the microbial genes in the viral metagenome reflects the phylogenetic composition of the microbial community. That is, the microbial genes are mainly from bacteria, not archaea or eukaryotes, in environments where bacteria are the most abundant organism, and the microbial genes in the viral metagenome are predominantly from groups, such as Proteobacteria, that dominate the bacterial community. These correlations reflect the intimate relationship between viruses and their hosts.

Some of the known genes in the viral metagenome also give clues about the types of viruses in the habitat being examined and also about the prevalence of virulent versus temperate viral lifestyles. The bacteriophage order Caudovirales is the most abundant type of virus, but the families within this order vary in abundance. In the Chesapeake Bay, for example, Myoviridae and Podoviridae families accounted for most (>80%) of Caudovirales while the Siphophage family was not abundant (Bench et al., 2007). In contrast, Siphophage were the most abundant form in sediments and terrestrial subsurface environments examined to-date (Edwards and Rohwer, 2005). These names are important because cultivated representatives of Myoviridae and Podoviridae are virulent viruses while Siphophage are temperate viruses. The metagenomic data are

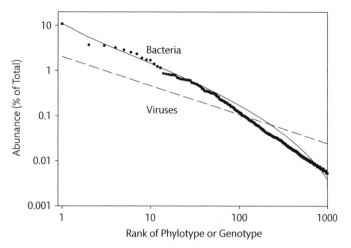

Figure 10.11 Rank abundance curves for viruses and bacteria in coastal oceans. The bacterial rank abundance curve is best described by a lognormal distribution while the viruses curve is a negative exponential. The bacterial curve is based on the frequency of 16S rRNA genes (actual data given as filled-in circles) whereas the viral curve is generated from metagenomic data. Only the most abundant 1000 phylotypes (bacteria) or genomes (viruses) are shown. Much of the apparent difference between the two curves is likely to be due to the low phylogenetic resolution of 16S rRNA genes. Data from Gilbert et al. (2009) and Angly et al. (2005).

consistent with other data about the extent of lysogeny in aquatic, sediment, and terrestrial environments (see Chapter 8). Lysogeny, perhaps due to high abundance of Siphophage, appears to be more common in sediments and terrestrial systems than in the water column of aquatic habitats.

RNA viruses

In addition to the double-strand DNA viruses just discussed, a variety of RNA viruses are known to infect mainly eukaryotes in nature. Much less is known about natural RNA viruses because these viruses are smaller than DNA viruses and because of the difficulties in working with RNA. Similar to metatranscriptomic approaches, the metagenome of RNA viruses is examined by reverse-transcribing the RNA into cDNA and then sequencing the cDNA. The few relevant studies completed so far found that RNA viruses are less diverse than DNA viruses and have a high fraction of genes similar to those in known RNA viruses (Kristensen et al.,

2010). The low diversity of RNA viruses may reflect the diversity of their hosts, eukaryotic microbes, whereas the diversity of DNA viruses and their prokaryotic hosts is high.

Many negative-strand and double-strand RNA viruses are known to attack plants and animals, but there is no sign of these viruses in the viral metagenomic data collected so far. Instead, RNA viruses in natural habitats seem to have positive-strand RNA and are in the order Picornavirales. This order accounted for nearly all of the RNA viruses in one study of coastal waters of British Columbia (Culley et al., 2006). Other viruses in Picornavirales are known to infect animals and higher plants, but the marine picorna-like viruses appear to attack protists, specifically those in the supergroup Chromalveolata (Kristensen et al., 2010). This supergroup is a diverse collection of eukaryotic microbes, including diatoms, Raphidophyte, and dinoflagellates. This picture of RNA viruses may change as more data are collected.

Summary

1. Genomic studies of microbes in pure cultures have revealed new insights into regulation and growth strategies, though a large fraction of genes remains unknown even for well-studied organisms.

2. Metagenomic approaches have been used to identify organisms carrying out specific functions in biogeochemical processes while also suggesting new functions that were not obvious by studies using microbiological or biogeochemical approaches.

3. Complete genomes can be reconstructed by metagenomic approaches of simple communities with only a few members, allowing the ecological role of each member to be defined.

4. Useful compounds, such as antibiotics, can be retrieved by activity screening metagenomic libraries in which entire genes are captured and expressed by *E. coli* hosts.

5. Gene expression and proteins examined by metatranscriptomic and metaproteomic approaches are closer to actual biogeochemical processes and provide insights into microbial communities not gleaned by metagenomic or biogeochemical techniques.

6. Metagenomic approaches have been especially important in revealing the high diversity and large number of unknown genes in viruses and have given clues about which hosts are attacked by viruses and about the extent of lysogeny.

CHAPTER 11

Processes in anoxic environments

All of the organisms discussed in detail so far in this book live in an oxygen-rich world. The production of oxygen by light-driven primary producers is enough for use by heterotrophic organisms not only in sunlit environments, but also in many habitats without direct exposure to the sun. In these dark habitats, the supply of oxygen by diffusion and other physical mechanisms matches or exceeds the supply of organic material, meaning that there is more than enough oxygen to degrade and oxidize organic carbon coming from sunlit worlds. But there are times and places when the supply is not enough and oxygen runs out. What happens then? This chapter provides some answers.

Oxygen-deficient habitats vary in size and shape. Some anoxic habitats are small and are close to oxic worlds. Intense aerobic heterotrophy in the top few millimeters of sediments stops oxygen from penetrating far, creating anoxic mud only millimeters away from oxic waters. Likewise, aerobic heterotrophy can use up the oxygen in the middle of organic-rich particles, resulting in anoxic microhabitats in otherwise oxic soils and waters. These anoxic microhabitats explain the presence of anaerobic by-products, such as methane, in oxic environments. Other anoxic worlds are huge and distant from sunlight oxic environments. The subsurface environment below soils and ocean sediments is largely devoid of oxygen, as are a few oceanic water bodies where physical exchange with oxygen-rich waters is restricted. Fueled by organic material raining down from sunlit surface layers, the bottom waters of hypolimnions in lakes are often anoxic. Whether large or small, anoxic worlds are dominated by bacteria and archaea. Only a few eukaryotic microbes are anaerobes.

While earth's surface is now oxic, that has not been always the case. The entire planet was anoxic for the first half of its existence (Fig. 11.1), with oxygen becoming abundant in the atmosphere only about 2.5 billion years ago after the invention by cyanobacteria of oxygenic photosynthesis. Oxygen was the first pollutant of the planet, potentially lethal for anaerobic bacteria without antioxidants that are ubiquitous in aerobic organisms. Of course oxygen is essential for many other organisms. Only after atmospheric oxygen became sufficient was evolution of larger, eukaryotic organisms possible. Atmospheric oxygen increased during the Carboniferous period when massive forests on land were buried without being decomposed, eventually turning into coal. Burning of coal and other fossil fuels is now causing a slight but significant decrease in atmospheric oxygen. Current oxygen concentrations are still optimal for life as we know it. Any higher and fires would easily start; any lower and large organisms could not survive. But life started when the earth was anoxic.

Introduction to anaerobic respiration

In terms of the carbon cycle, the most important process occurring in anoxic environments is the mineralization of organic material synthesized by land plants and algae in oxic environments. Much of this organic material is degraded by organisms carrying out aerobic respiration, but some of it escapes that fate and is deposited or transported to regions without sufficient oxygen where it is mineralized by anaerobic processes, including anaerobic respiration. To understand anaerobic respiration, let us go back to aerobic mineralization and break down the

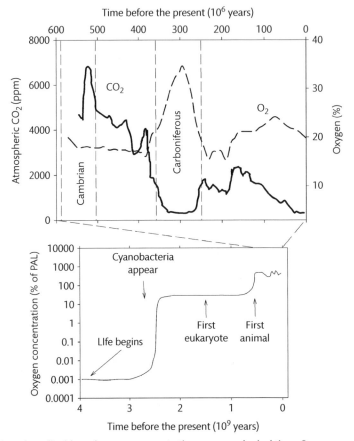

Figure 11.1 Atmospheric carbon dioxide and oxygen concentrations over geological time. Oxygen concentrations are given in either percent of total atmospheric gases (top panel) or percent of present atmospheric levels (PAL, in the bottom panel). Note that the top graph is the last 600×10^6 years before the present while the bottom graph covers the last 4.0×10^9 years before the present, nearly the entire geological history of the earth. Two important geological periods are also given in the top panel. The Cambrian saw a huge explosion of metazoan diversity while lignin-rich plants were abundant during the Carboniferous period, leading eventually to the formation of many coal beds. Data from Berner (1999), Berner and Kothavala (2001), Donoghue and Antcliffe (2010), and Kump (2008).

familiar equation describing this process. As seen before, the equation for aerobic oxidation of organic material is:

$$CH_2O + O_2 \rightarrow CO_2 + H_2O \qquad (11.1)$$

where CH_2O again symbolizes generic organic material, not a specific compound. Equation 11.1 describes a redox reaction that can be broken down into two half-reactions. One reaction generates electrons (e^-):

$$CH_2O + OH^- \rightarrow CO_2 + 4e^- + 3H^+ \qquad (11.2)$$

while the other half-reaction accepts electrons:

$$O_2 + 4e^- + 4H^+ \rightarrow 2H_2O \qquad (11.3).$$

Combining Equations 11.2 and 11.3 yields Equation 11.1. In words, organic material is the electron donor, while oxygen is the electron acceptor.

We can write a more general form of Equation 11.1,

$$CH_2O + 1/2X_2 \rightarrow CO_2 + H_2X \qquad (11.4),$$

to illustrate that organic material can be oxidized to carbon dioxide with an electron acceptor, X_2, other than

Box 11.1 Balancing equations

A chemical equation balanced in terms of electrons and elements is a succinct and powerful description of a biogeochemical process potentially occurring in an environment. To balance a chemical equation, the starting point is to make sure the number of electrons from the electron donor matches the electrons being received by the electron acceptor. These are set by the valence of the elements being oxidized and reduced. Elemental hydrogen is always +1 or 0 (H_2). Elemental oxygen can be –2 or 0 while carbon can take on anything between –4 (CH_4), when it is in its most reduced form, and +4 (CO_2) when it is most oxidized. The main elements other than hydrogen and oxygen should be balanced and equal in number on both sides of the equation. To balance hydrogen and oxygen atoms, H^+ or OH^- (but not O_2) can be added to either side as needed, because the reaction is in an aqueous solution, even in soils, where H^+ and OH^- are plentiful. If done correctly, at this point everything should be in balance: electrons, elements, and charge. Among many resources, *Brock Biology of Microorganisms* gives a primer on how to balance chemical equations and to calculate energy yields.

oxygen. Anaerobic respiration uses various electron acceptors symbolized by X_2 in Equation 11.4. Of the many elements and compounds that can take the place of X_2, all have the characteristic of being oxidized, meaning they can take on more electrons. For example, nitrate (NO_3^-) is one possibility because of the oxidized state of its nitrogen (+5) whereas ammonium (NH_4^+) is not because its nitrogen is highly reduced (-3).

The order of electron acceptors

The equations illustrate in theory how organic material can be oxidized by electron acceptors other than oxygen, but they are not of much use in predicting which acceptor is most important in nature. Answering this question is likely to help us understand

variation in three common electron acceptors, O_2, nitrate (NO_3^-), and sulfate (SO_4^{2-}), in a typical sediment profile (Fig. 11.2). Invariably, oxygen disappears before sulfate begins to decline. Nitrate also disappears quickly soon after oxygen but before sulfate. Figure 11.2 illustrates variation in space, but it also illustrates how these compounds would vary through time. If enough organic material were placed in a bottle along with possible electron acceptors, oxygen (O_2) would disappear first, followed by nitrate (NO_3^-), and then sulfate (SO_4^{2-}). Why this order?

The order is explained by the tendency of these compounds to accept electrons. This tendency is measured in relationship to the reduction of H^+ to H_2 which is set at 0 mV, by definition. Possible electron acceptors are put in the "electron tower" of half-reactions (Fig. 11.3), with oxygen at the top at +1.27 V and CO_2 at the bottom with +0.21 V. So, oxygen is the strongest electron acceptor while CO_2 is the weakest. The strength of an electron acceptor is an important characteristic in explaining the contribution of various elements and compounds to anaerobic respiration and to the mineralization of organic material in anoxic environments.

The electron tower is enough to explain the order of electron acceptors used up over time and down a depth profile, but it is insufficient for exploring the benefit to an organism of using one acceptor over another and for predicting which would be most important in oxidizing organic material in the absence of oxygen. To explore these issues, it is useful to calculate a theoretical energy yield for an electron acceptor oxidizing an organic compound. This energy yield is the Gibb's change in free energy (ΔG°) where the superscripts indicate that standard biochemical conditions are assumed: pH = 7, the temperature is 25 °C, and each of the compounds other than H^+ in the reaction occurs in equal molar amounts. To compare the electron acceptors, one approach is to assume that the same electron donor, here an organic compound, is being oxidized for all electron acceptors; in Table 11.1 the "compound" is in fact a hypothetical one with the main elements (C, N, and P) occurring in Redfield ratios (Chapter 5). As we will soon see, sulfate reducers and carbon dioxide reducers do not use the same electron donors. But the calculations and the theoretical energy yields are still useful in thinking about these various electron acceptors.

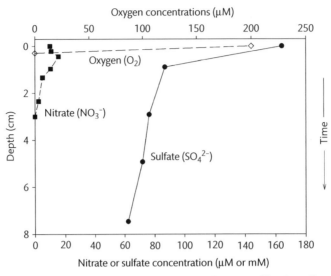

Figure 11.2 Concentrations of three important electron acceptors in a sediment profile. The sulfate concentrations are in mM, but divided by 10 (the highest concentration was about 17 mM in this example) whereas nitrate concentrations are in μM. Going down a profile is equivalent to the passing of time. In closed incubations, these three electron acceptors are used up over time in the following order: oxygen, nitrate, and sulfate. Data from Sørensen et al. (1979).

The order of the electron acceptors defined in Table 11.1 by energy yield is nearly the same as seen in the electron tower figure (Fig. 11.3) with only iron and man-

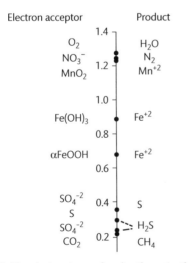

Figure 11.3 The electron tower showing the potential (volts) for some half-reactions involving electron acceptors commonly used by microbes. The form of Mn(IV) given here is pyrolusite (MnO_2) and $Fe(OH)_3$ is an amorphous iron oxide. Data from Canfield et al. (2005) and Stumm and Morgan (1996).

ganese flipped. The electron donor side of the reaction is the same in Table 11.1 for all of the electron acceptors while the differences in energy yield are due to the electron acceptors, which is the data given in Figure 11.3. But a couple of new points are illustrated by the energy yield calculations. Note the small difference in energy yield between oxygen and the next electron acceptors in Table 11.1. The implication is that these other electron acceptors should be nearly as commonly used by microbes as oxygen. In contrast, using sulfate or carbon dioxide yields very little energy, nearly tenfold less, to oxidize the same hypothetical organic carbon. The implication of these calculations is that both sulfate and carbon dioxide are even less desirable as electron acceptors and perhaps even less important than suggested by the electron tower.

The energy yield also explains why eukaryotic microbes do not use the electron acceptors at the bottom of the list, such as sulfate and carbon dioxide. The energy yield with these electron acceptors is too small to support the high energy requirements of the eukaryotic lifestyle. Some of the other electron acceptors are not as easily ruled out. There is a seemingly small drop-off in the theoretical energy yield in switching from oxygen to manganese or nitrate. In fact, the only electron acceptor

Table 11.1 Theoretical yield of energy from organic material oxidation using various electron acceptors. The organic material (C_o) has Redfield ratios for C, N, and P. The equations often include more than one reaction. For example, the first equation for oxygen includes both aerobic respiration by heterotrophs and nitrification (Chapter 12). The oxidized form of manganese given here is birnessite and the oxidized iron is goethite. Other calculations indicate that nitrate yields more energy than manganese and is second only to oxygen. Data from Nealson and Saffarini (1994).

Electron Acceptor	Reaction	Energy yield (kJ mol^{-1})
Oxygen	$C_o + 138\ O_2 \rightarrow 106\ CO_2 + 16\ HNO_3 + H_3PO_4 + 122\ H_2O$	−3190
Manganese	$C_o + 236\ MnO_2 + 472\ H^+ \rightarrow 106\ CO_2 + 236\ Mn^{+2} + 8N_2 + H_3PO_4 + 366\ H_2O$	−3090
Nitrate	$C_o + 94.4\ HNO_3 \rightarrow 106\ CO_2 + 55.2\ N_2 + H_3PO_4 + 177.2\ H_2O$	−3030
Iron	$C_o + 424\ FeOOH + 848\ H^+ \rightarrow 106\ CO_2 + 424Fe^{+2} + 16\ NH_3 + H_3PO_4 + 742\ H_2O$	−1330
Sulfate	$C_o + 53SO_4^{2-} \rightarrow 106\ CO_2 + 53S^{2-} + 16\ NH_3 + H_3PO_4 + 106\ H_2O$	−380
CO_2	$C_o \rightarrow 53CO_2 + 53\ CH_4 + 16NH_3 + H_3PO_4$	−350

used by eukaryotic microbes other than oxygen is nitrate (Risgaard-Petersen et al., 2006, Kamp et al., 2011), although some protists thrive in anoxic environments using fermentation, which does not involve external electron acceptors. Along with their lower energy yield, manganese and iron are probably not used by eukaryotic microbes for the same reason (their mineral form) that limits their use by bacteria and archaea, a topic returned to below.

Oxidation of organic carbon by various electron acceptors

So far, oxygen and other electron acceptors have been evaluated on theoretical grounds using basic thermodynamics under standard biochemical conditions. How do the predictions based on the electron tower and the energy yields compare to the real world? Answering this question experimentally is not easy. It is often difficult to measure the use of a particular electron acceptor and to calculate its contribution to organic carbon oxidation. The technical difficulties and workload explain why few studies have examined more than a couple of electron acceptors simultaneously.

However, enough data sets have been collected over the years to discuss which electron acceptor is most important in oxidizing organic material back to carbon dioxide (Table 11.2). Globally, the answer is oxygen. This should be no surprise after seeing its high energy yield in oxidizing organic material and after realizing that an entire chapter (Chapter 5) was focused on aerobic oxida-

tion of organic material. It may also not be a surprise to see that iron and manganese may be important in some environments, given their high energy yields. Next on the energy yield list is nitrate. This electron acceptor is responsible for relatively little organic material oxidation, except for polluted waters and some water-saturated soils with high nitrate concentrations. In the oceans, nitrate reduction has been examined mainly because of its role in denitrification and the N cycle (Chapter 12), but it may account for as much as 50% of organic carbon mineralization in low-oxygen basins in the Pacific Ocean (Liu and Kaplan, 1984). Still, nitrate reduction accounts for little organic carbon oxidation on a global scale.

The two electron acceptors lowest on the electron tower are often next to oxygen in importance in oxidizing organic material (Table 11.2). Sulfate reduction is crucial in marine environments while carbon dioxide reduction fills that role in freshwater environments, such as wetlands and rice paddies. So, energetic yield only partially explains why some electron acceptors are more important than others. The two other factors that help to explain Table 11.2 are concentrations and the chemical form of the electron acceptors.

Limited by concentrations and supply

The thermodynamic calculations for energy yields given in Table 11.1 assume equal concentrations of everything. However, that is far from being true. One reason why oxygen is so important is that it is often readily available and concentrations are high. In fact,

Table 11.2 Contribution of various electron acceptors to organic carbon mineralization in marine and freshwater sediments and soils. The global importance of oxygen is under-represented in this table because several studies focused only on anoxic environments. Oxygen is the predominant electron acceptor in the oceans and in unsaturated soils. Data from references: 1 = Capone and Kiene (1988); 2 = Canfield et al. (2005); 3 = Keller and Bridgham (2007); 4 = Yavitt and Lang (1990); 5 = Roden and Wetzel (1996); 6 = Thomsen et al. (2004). nd = not determined

Location	Total oxidation (mmol C m^{-2} d^{-1})	% of total carbon flow						
		O_2	NO_3^-	Fe (III)	Mn (IV)	SO_4^{2-}	CO_2	Ref
Marine								
Sippewissett saltmarsh	458	10	nd	nd	nd	90	nd	1
Sapelo saltmarsh	80	0	0	95	0	5	0	1
Chile margin	60	0	0	0	0	100	nd	2
Laurentian trough	5	17	4	18	2	59	nd	2
Deep sea	<0.1	80	20	0	0	0	nd	2
Freshwater and soils								
Michigan bog*	175	nd	<1	<1	<1	13	35	3
Sedge meadow*	120	nd	nd	nd	nd	<1	9	4
Alabama wetlands*	117	nd	nd	55	nd	7	38	5
Michigan fen*	110	nd	<1	<1	<1	10	19	3
Wintergreen Lake*	14.4	nd	nd	nd	nd	13	87	2
Lake Michigan	6.8–8	37	<3	44	0	19	0	6

* Percentages refer to anaerobic respiration only.

the production of oxygen is as high as the production of organic material. The other electron acceptors become important only when oxygen transport by physical processes is impeded somehow. It is no accident that the supply of oxygen and organic material is so evenly matched. The dominant form of primary production by far is oxygenic photosynthesis which produces enough oxygen to oxidize all of the organic material made by photosynthesis.

This is not the case for nitrate. In contrast to oxygen, nitrate production is not intimately linked to organic carbon production. For this and other reasons, concentrations and supply rates of nitrate are low, explaining why nitrate respiration does not consume more organic material even with its high theoretical energy yield. Nitrate formation starts with nitrogen fixation ($N_2 \rightarrow NH_3$), a slow process, carried out by only a few organisms (Chapter 12), unlike the widespread capacity of oxygen formation by oxygenic photosynthesis. By contrast, nitrate reduction is a rapid process that can lead to nitrogen gases (N_2 and N_2O) and the loss of fixed nitrogen from the system. Another sink for nitrate is its use as a nitrogen source for biomass synthesis by higher plants, algae, and heterotrophic bacteria.

Concentrations and supply explain much about the contribution of the other electron acceptors to organic material oxidation. The concentration of sulfate is the reason why organic material oxidation by sulfate reduction is so high in marine environments. Sulfate is not lost from these environments because the end product of sulfate reduction (H_2S) is usually easily converted back to sulfate. Sulfate concentrations are low in freshwaters and soils, explaining why sulfate reduction is usually not important in those environments. Carbon dioxide reduction is never limited by carbon dioxide concentrations which are usually high enough everywhere. Likewise, iron and manganese are often abundant, explaining why these two elements are often important in organic material oxidation. But use of oxidized iron and manganese as electron acceptors is complicated by their chemical form.

Effect of chemical form

The chemical form or state of the electron acceptor has an effect on how it is used by microbes. The forms of the acceptors commonly used by microbes vary from a gas to a solid. Oxygen once again has a form most conducive

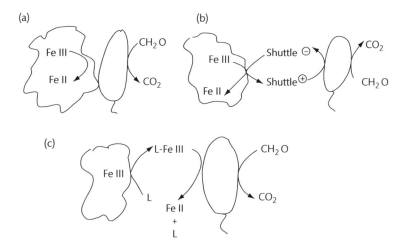

Figure 11.4 Three strategies for using insoluble iron oxides by iron-reducing bacteria. The first (A) is to be in direct physical contact with the iron oxide. In the second strategy (B), the bacterium relies on an external electron shuttle, either produced by the microbe or supplied by the environment, to bring electrons from the electron donor (e.g. an organic compound) and to Fe(III). The third strategy (C) is for the bacterium to produce a complexing ligand (L) that leads to the dissolution of the iron oxide. Adapted from Weber et al. (2006).

to use by microbes. It has the lowest molecular weight of all compounds in the electron tower, resulting in its diffusion rate being highest of all electron acceptors. Because it is a gas, it can be transported to microbes without water. Finally, because it is uncharged and small, oxygen easily enters into cells without special transport mechanisms. The only other common electron acceptor with any of these traits is carbon dioxide, but its chemical form (low molecular weight, gas, and uncharged) and high concentrations are not enough to offset its low energy yield.

Concentration is the main reason why nitrate reduction contributes so little to organic carbon oxidation, but its chemical form doesn't help. Its charge and thus non-gaseous state means that it is transported to microbes only via water. The charge also means that specialized transport mechanisms and energy are required to bring it across membranes and into cells.

Chemical state has a big impact on the use of ferric iron (FeIII) as an electron acceptor. Perhaps most importantly, because ferric iron is insoluble at the pH of most environments, it occurs as particulate oxides which are much too large to transport across membranes. Consequently, iron-reducing bacteria may have to be in physical contact with iron oxides and somehow transport electrons from organic

material oxidation to the iron oxide (Fig. 11.4). This contact may be via "nanowires" (Roden et al., 2010). Alternatively, direct physical contact may not be necessary if electrons can be shuttled from the bacterium to the insoluble iron oxide. Another complication is the type of crystal form iron takes on ("crystallinity"). This crystallinity, which ranges from amorphous oxides to highly crystallized ones, affects the access of ferric iron to iron-reducing bacteria. Iron in amorphous oxides is more easily reduced and thus supports more organic material oxidation than highly crystallized iron. So, even though concentrations of iron are high in soils and sediments as is the energy yield using ferric iron as an electron acceptor, the chemical form of iron can limit its contribution to oxidizing organic material.

The anaerobic food chain

So far, we have assumed that organisms using electron acceptors other than oxygen can use the same organic compounds, the same electron donors. In fact, this is far from being the case. It is true that the suite of organic compounds used by nitrate reducers is about the same as oxygen reducers (aerobic respiration), and geochemists often treat the two processes as being

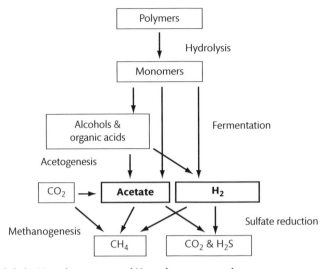

Figure 11.5 Anaerobic food chain. Note that acetate and H_2 are key compounds.

nearly equivalent. With these electron acceptors, in theory, any labile organic material could be degraded, oxidized, and mineralized by a single organism. In contrast, most of the anaerobic bacteria and archaea cannot use many organic compounds. Consequently, an entire consortium of organisms is needed to mineralize organic material when sulfate or carbon dioxide is the terminal electron acceptor in the system. The consortium is called the anaerobic food chain. Although members of this consortium are not eating one another, "food chain" does convey the correct idea of organic carbon being passed from one organism to another (Fig. 11.5).

Let us follow plant detritus as it is degraded and eventually oxidized back to carbon dioxide. After being broken up into smaller fragments by larger organisms (Chapter 5), many microbes have hydrolyases (cellulase, in the case of cellulose) that cleave macromolecules in plant detritus to various monomers (glucose, in the case of cellulose). The monomers and other by-products from macromolecule hydrolysis are not used directly by sulfate- or carbon dioxide-reducers. Instead, these by-products are used by bacteria carrying out fermentation. The fermenting bacteria in turn produce several compounds, most importantly acetate and hydrogen gas (H_2). These two compounds and a few others are then used by sulfate- or carbon dioxide-reducers, thus completing the anaerobic food chain.

The anaerobic food chain model assumes that either carbon dioxide or sulfate is the last or "terminal" electron acceptor. How other electron acceptors, such as iron and manganese, fit into this model is not clear.

Fermentation

This form of catabolism is an important intermediate step between biopolymer hydrolysis and oxidation by the terminal electron acceptors. Bacteria using fermentation are key members of the anaerobic food chain, but not much is known about their ecology. It is known that fermentation is common among microbes and even among eukaryotes. Even muscle cells of mammals have the capacity for a type of fermentation. When the supply of oxygen is insufficient, our muscles carry out lactic acid fermentation:

$$Glucose \rightarrow 2\,lactate + 2H^+ + 2ATP \qquad (11.5)$$

which yields 196 kJ mol^{-1}, much lower than the nearly 3000 kJ mol^{-1} released by aerobic respiration. Muscle cells are forced to do lactic acid fermentation when the supply of oxygen is insufficient. There is no external electron acceptor. In this form of fermentation, no carbon

Box 11.2 Useful waste

Many types of fermentation are carried out by microbes. These pathways take their name from their end product, excreted as waste. One example is lactic acid fermentation used for centuries to make yogurt, cheeses, and other food. Another example is ethanol fermentation, a key process in making wine and beer. Many commercially valuable products are made by fermentation, including enzymes, vitamins, and antibiotics. Compounds, such as insulin, are produced by fermenting microbes with cloned genes from other organisms. Several hydrocarbons can be produced by fermentation, including butane and oil suitable for use as fuel.

dioxide is released because there is no net oxidation of the glucose carbon; there is no place for electrons from glucose oxidation to go. Instead of respiration, muscle cells and fermenting organisms in general gain energy from substrate-level phosphorylation. This type of ATP production does not involve membranes, in contrast to respiration. Our muscle cells cannot go for long without oxygen, but many microbes can grow using energy only from fermentation.

Fermentation is widespread among bacteria and archaea, and even a few anaerobic eukaryotes carry out this form of metabolism. Some organisms make their living only on fermentation, such as strains in the *Lactobacillus* genus which are well-known lactic acid fermenters. Stable isotope experiments have suggested bacteria related to *Acidobacterium capsulatum* in the Acidobacteria phylum, which is very abundant in soils (Chapter 9), are important in fermenting sugars in fens and peatlands (Hamberger et al., 2008). Similar to our muscle cells, many fermenting microbes switch back to oxygen when it becomes available due to the higher energy yield of aerobic respiration. Microbes capable of making this switch are called facultative anaerobes. In addition to using fermentation, some facultative anaerobes also use electron acceptors other than oxygen.

In terms of the anaerobic food chain, a key feature of fermentation is the release of organic compounds. These compounds are not as energy-rich as the starting material but they still are reduced enough to yield energy when

oxidized by sulfate- and carbon dioxide-reducers. These organisms use many of the compounds released by various fermentation pathways, but the main fluxes of carbon and energy are through acetate and hydrogen gas. This conclusion was reached by studies examining concentrations and fluxes through various compounds, such as lactate and propionate, in sediments (Parkes et al., 1989). Why are acetate and hydrogen key compounds in the anaerobic food chain? Both acetate and hydrogen gas can be produced directly by fermentation pathways, but in addition, they are produced by another group of microbes in another step in the anaerobic food chain.

Interspecies hydrogen transfer and syntrophy

The next step in the anaerobic food chain is the production of acetate and hydrogen gas by acetogenic bacteria, using another metabolic pathway, acetogenesis. There are about 20 genera of acetogenic, mainly Gram-positive bacteria, with most known strains in the *Acetobacterium* and *Clostridium* genera (Drake et al., 2008). These organisms have been isolated from a wide variety of environments, including soils, animal guts, and sediments.

Acetogenic microbes can use several organic compounds, such as ethanol, a common end product of fermentation. The reaction describing use of ethanol by acetogens is:

$$Ethanol + H_2O \rightarrow acetate + H^+ + 2H_2$$
$$\Delta G^{\circ\prime} = 9.6 \text{ kJ mol}^{-1} \quad (11.6).$$

The reaction has a serious problem as now written; the Gibb's change in free energy ($\Delta G^{\circ\prime}$) is positive, implying that the reaction is thermodynamically impossible. But, experimental studies have demonstrated it occurs and that organisms grow with energy from it. How is this possible? This is one example of many where biology seems to break the laws of thermodynamics. What actually happens is less dramatic. Although no thermodynamic laws are broken, biology does not necessarily operate under standard biochemical conditions.

Note that the energy yield given for Equation 11.6 assumes that the reaction occurs under standard biochemical conditions, most importantly, equal concentrations of the reactants and by-products. In fact, the reaction does go forward when hydrogen gas is removed and its concentration drops far below the hydrogen to

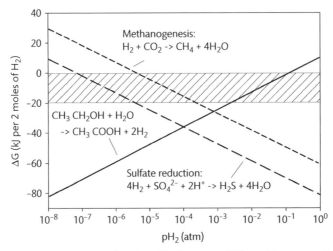

Figure 11.6 Energetic yield for three reactions as a function of hydrogen gas (H_2) partial pressure. Calculated changes in Gibb's free energy increase with increasing pH_2 for acetogenesis (production of CH_3COOH, the unlabeled reaction in the middle) because hydrogen gas is a product in this reaction (right side of the equation). The top line of the hatched area is at Gibb's free energy equal to zero, below which reactions are thermodynamically favorable, whereas the bottom line indicates the minimal negative change thought to support microbial metabolism. Data from Canfield et al. (2005).

ethanol ratio implied by Equation 11.6. According to theoretical calculations, acetate formation from ethanol becomes thermodynamically feasible when the partial pressure of hydrogen gas drops below one atmosphere and it becomes energetically profitable for growth when it is less than about 0.01 atmosphere (Fig. 11.6). In theory, removing the other end product, acetate, could also "pull" the reaction to the right and make it thermodynamically possible, but that is easier to accomplish with a gas (here hydrogen) which can diffuse away more quickly than a charged, larger compound like acetate.

In addition to diffusion, hydrogen gas concentrations are reduced by its use by other organisms. This connection between a hydrogen gas producer (the acetogenic bacterium) and a hydrogen gas user (sulfate or carbon dioxide reducer) is referred to as interspecies hydrogen transfer. It is greatly facilitated when the two organisms are close together physically in a mutually beneficial arrangement called syntrophy (Stams and Plugge, 2009).

A famous example of syntrophy is the isolation of "*Methanobacillus omelianskii*" described by H.A.Barker in 1940. This seemed to be one organism that used ethanol and carbon dioxide to produce acetate and methane. The reaction is thermodynamically favorable ($\Delta G^{0'} = -116.4$ kJ/reaction) and thus seemed possible. However,

later it was shown that this reaction was actually carried out by two organisms, one an acetogenic bacterium (*Acetobacterium woodii*) that produces hydrogen gas and acetate from ethanol, and another, a methanogen (*Methanobacterium bryantii*), that uses hydrogen gas and carbon dioxide (but not ethanol) and produces methane. That these microbes were isolated and maintained together for years is indicative of the tight physical relationship between the two.

Sulfate reduction

The next step in the anaerobic food chain in marine systems is sulfate reduction, which oxidizes acetate, hydrogen gas, and other by-products of fermentation and acetogenesis, using sulfate as the terminal electron acceptor. The various electron donors for sulfate reduction are discussed below. When acetate (CH_3COO^-) is the electron donor, the reaction is:

$$CH_3COO^- + SO_4^{2-} \rightarrow 2HCO_3^- + HS^- \quad (11.7).$$

This process and the rest of the sulfur cycle are worthy of more discussion because sulfate-reducing bacteria oxidize a large amount of organic material in the biosphere and because these microbes and others involved in the sulfur

Table 11.3 Comparison of assimilatory and dissimilatory sulfate reduction.

Characteristic	Assimilatory	Dissimilatory
Purpose	Biosynthesis	Energy production
Fate of reduced sulfur	Assimilated into organic compounds	Excreted
Requires energy?	Yes	No
Membrane-associated	No	Yes
Key enzyme (gene)	ATP sulfurylase	Dissimilatory sulfite reductase (*dsr*)
Organisms	Widespread	Deltaproteobacteria

cycle are abundant in many natural ecosystems, not just marine ones. Sulfur biogeochemistry also plays a big part in figuring out the history of early life on the planet.

The focus here is on those organisms using sulfur in generating energy, but all organisms need sulfur for biosynthesis of protein and other macromolecules. Since it is the most available form of sulfur in all environments, sulfate is the main sulfur source for many organisms, ranging from microbes to higher plants. Sulfate needs to be reduced before its sulfur can be assimilated and used in biosynthetic pathways. Assimilatory sulfate reduction is quite different from dissimilatory sulfate reduction (Table 11.3), the main topic of this section. There is a similar difference between dissimilatory nitrate reduction, mentioned above as just "nitrate reduction", and assimilatory nitrate reduction.

The capacity to carry out dissimilatory sulfate reduction is not as common as assimilatory sulfate reduction, dissimilatory nitrate reduction, and oxygen reduction (aerobic respiration). Most known sulfate reducers are in the Deltaproteobacteria (Barton and Fauque, 2009), although there are some Gram-positive sulfate reducers in the genus *Desulfotomaculum* in the Firmicutes phylum, and there are several other interesting thermophilic Gram-positive sulfate-reducing bacteria. Sulfate reduction in the Archaea is restricted, as far as is known, to the genus *Archaeoglobus*. This is an interesting organism because it grows at high temperatures, reaching maximum growth at over 90 °C. Probably other, uncultivated sulfate reducers grow at even higher temperatures because sulfate reduction by natural communities has been observed at temperatures hotter than 90 °C. Another interesting aspect of this archaeon is that it is most closely related to methanogenic archaea, suggesting *Archaeoglobus* lost its methanogenic pathway and gained sulfate reduction later during its evolution. Support for this hypothesis comes from comparing phylogenetic trees of 16S rRNA genes and of genes for dissimilatory sulfite reductase (*dsr*), a key enzyme and gene in sulfate reduction (Pereyra et al., 2010, Wagner et al., 2005).

Box 11.3 Examining sulfate reduction and sulfate reducers

Rates of sulfate reduction can be estimated by following the radioactive isotope ^{35}S, added as ^{35}S-sulfate, into reduced by-products such as HS⁻. This is a technically difficult approach because of problems in maintaining the original environmental conditions during the experiment and because the sulfur cycle is complex. Several cultivation-independent approaches based on *dsr* genes are available for examining the microbes carrying out sulfate reduction.

Electron donors for sulfate reduction

An individual sulfate-reducing bacterium may not be able to use many organic compounds, but the entire collection of sulfate reducers can use a great variety of electron donors, ranging from hydrogen gas to many organic compounds. In fact, the list is quite long: hydrocarbons, organic acids, alcohols, amino acids, sugars, and aromatic compounds, to name just the broad classes. Some of these electron donors are given in Table 11.4. Still, acetate is the most important organic compound for natural communities of sulfate reducers. This finding was surprising to microbiologists because many sulfate reducers were isolated and grown in the laboratory with

Table 11.4 Some sulfate reducers and some of the electron donors they use. Data from Itoh et al. (1998), Itoh et al. (1999), Klenk et al. (1997), Muyzer and Stams (2008), and Widdel and Hansen (1992).

Phylum or subphylum	Order	Genus	Electron donors					
			H$_2$	Acetate	Lactate	Propionate	Larger fatty acids	Ethanol
Deltaproteobacteria	Desulfovibrionales	*Desulfovibrio*	+	–	+	–	–	+
Deltaproteobacteria	Desulfobacteriales	*Desulfobulbus*	+	–	+	+	–	+
Deltaproteobacteria	Desulfobacteriales	*Desulfobacter*	+	+	–	–	+	+
Deltaproteobacteria	Desulfobacteriales	*Desulfococcus*	–	(+)	+	+	+	+
Firmicutes	Clostridiales	*Desulfotomaculum*	+	+	+	+	+	+
Thermodesulfobacteria	Thermodesulfobacteriales	*Thermodesulfobacterium*	+	–	+	+	–	–
Nitrospirae	Nitrospirales	*Thermodesulfovibrio*	+	–	+	–	–	–
Euryarchaeota	Archaeoglobales	*Archaeoglobus*	+	–	+	–	–	+
Crenarchaeota	Thermoproteales	*Thermocladium*	+	–	–	–	–	–
Crenarchaeota	Thermoproteales	*Caldivirga*	+	–	–	–	–	–

lactate. In spite of acetate and lactate being quite similar, both being organic acids that differ by only one carbon, several sulfate reducers able to use lactate cannot grow on acetate (Table 11.4). This apparent preference for lactate over acetate may be a bias introduced by cultivating organisms and growing them in the laboratory. Another aspect of sulfate reduction, incomplete oxidation of organic compounds, may be more important in the laboratory than in natural environments.

Some sulfate reducers only partially oxidize the electron donor and release organic compounds back into the environment. An example is the oxidation of valerate, a six carbon organic acid, to lactate and acetate. Another, perhaps even more surprising example is the incomplete oxidation of lactate to acetate. Why would an organism "throw away" organic compounds that seem quite "good" and are used by others? Organisms carrying out incomplete oxidation are able to grow fast, which is the real objective of the organism, not efficient use of substrates. In organic-poor environments, complete oxidation is probably selected for over incomplete use of valuable electron donors. However, in sediments and other environments receiving pulses of rich organic material, rapid growth fueled by incomplete oxidation may be advantageous.

Sulfur oxidation and the rest of the sulfur cycle

Sulfate reduction, one prominent end of the anaerobic food chain, produces hydrogen sulfide and several other reduced sulfur compounds. Concentrations of these compounds are high in sulfate-rich environments, but they do not build up indefinitely because both biotic and abiotic processes oxidize the reduced sulfur compounds back to less reduced forms, eventually all the way to sulfate. Sulfide reacts abiotically with amorphous oxides of iron and manganese very quickly. The half-life of sulfide in the presence of colloidal manganese oxides can be as short as 50 seconds, for example. The speed at which sulfide is abiotically oxidized by oxygen varies by 10^6 depending on the conditions (Jørgensen, 1982). However, calculations based on only abiotic processes indicate that sulfide should persist in the presence of oxygen much longer than it actually does, suggesting that biotic oxidation dominates. The sulfur cycle is complicated because both biotic and abiotic reactions are important and because sulfur can take on many oxidation states (Fig. 11.7).

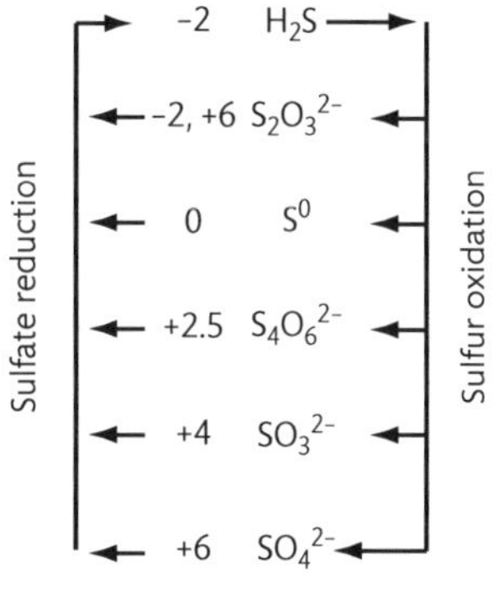

Box 11.4 Biocorrosion by sulfate-reducing bacteria

In addition to their roles in natural environments, sulfate-reducing bacteria (often abbreviated as SRB) are big contributors to microbially influenced corrosion of ferrous metals. This problem costs hundreds of millions of dollars in the USA alone. The problem starts when aerobic bacteria colonize metal surfaces, creating anoxic micro-environments for sulfate-reducing bacteria. This results in an uneven distribution of microbes and biofilm along the metal surface, a key feature of biocorrosion. Sulfate-reducing bacteria accelerate corrosion by producing sulfides which combine with Fe^{+2} from ferrous metals, forming iron sulfide, and eventually iron oxides, better known as rust. As these bacteria grow and excrete extracellular polymers, other microbes join the biofilm, further exacerbating the corrosion problem.

Figure 11.7 The sulfur cycle, including the major transformations mediated by microbes. The numbers refer to the oxidation state of sulfur in each compound. Thiosulfate ($S_2O_3^{2-}$) can be pictured as a sulfate molecule with one of the oxygens replaced with sulfide (S^{2-}), resulting in the outer sulfur having an oxidation state of -2 and the inner one $+6$. Other formulations indicate the oxidation states to be -1 and $+5$. Compiled with input from George Luther.

Table 11.5 Major characteristics of the two main types of sulfur (S) oxidizing bacteria and archaea in nature. Non-phototrophic sulfur oxidizers are also referred to as being colorless.

Characteristic	Non-phototrophic S oxidizers	Phototrophic S oxidizers
Pigments	None	Bacteriochlorophyll a and others
Role of light	None	Energy source
Role of reduced S	Source of energy and reducing power	Source of reducing power and energy
Role of oxygen	Electron acceptor for S oxidation	Represses photosynthesis, used as electron acceptor for oxidation of reduced S (chemolithotrophy) or organic carbon (heterotrophy), or kills cells*
Carbon source	CO_2	CO_2 (when not growing heterotrophically)

* Phototrophic sulfur-oxidizing bacteria are anaerobes that vary in their response to oxygen, depending on the species. Some are strict anaerobes that are killed by oxygen.

There are two types of microbial metabolisms that oxidize sulfide and other reduced sulfur compounds (Table 11.5). One depends on light and is a form of phototrophy, including both photoautotrophy and photoheterotrophy. The other sulfur-oxidizing metabolism does not depend on light ("non-phototrophic sulfur oxidation") and is a form of chemolithoautotrophy. The two types of metabolisms are carried out by very different organisms and are quite different from the phototrophic and heterotrophic metabolisms discussed so far in this book.

Non-phototrophic sulfur oxidation

A wide variety of organisms, often called colorless sulfur bacteria (the phototrophic sulfur oxidizers have color), obtain energy from oxidizing sulfide and other reduced sulfur compounds in the dark. Sulfide oxidation is carried out by bacteria in the Alpha-, Beta-, Gamma-, Delta-, and Epsilonproteobacteria and by archaea in the *Sulfolobales* family. These organisms also oxidize other reduced sulfur compounds, such as elemental sulfur and thiosulfate. A common reaction is:

$$H_2S + O_2 \rightarrow SO_4^{2-} + 2H^+ \quad \Delta G^{\circ\prime} = -796 \, kJ \, mol^{-1} \quad (11.8).$$

The reaction can stop at elemental sulfur which is deposited within the cell where it serves as an energy store. *Thiothrix nivea* oxidizes it further to sulfate while *Beggiatoa alba* uses elemental sulfur as an electron acceptor and reduces it back to hydrogen sulfide.

Sulfide oxidation is an example of chemolithotrophy, meaning that these microbes gain energy from the oxidation of inorganic compounds, hydrogen sulfide in this case. In essence, hydrogen sulfide takes the place of organic compounds (CH_2O) in Equation 11.2 and is the electron donor:

$$H_2S + 4OH^- \rightarrow SO_4^{2-} + 6H^+ + 8e^- \quad (11.9).$$

The electron acceptor is often oxygen, as described in Equation 11.3. Putting Equation 11.3 and 11.9 together yields Equation 11.8 and energy for the sulfur-oxidizing microbe.

Optimal conditions for sulfide oxidation are at the interface between the oxic world where oxygen concentrations are high and the anoxic world which produces sulfides (H_2S) via sulfate reduction (Fig. 11.8). At this interface, oxygen and hydrogen sulfide may overlap only for 50 µm. One gammaproteobacterial genus, *Beggiotoa*, is well known to reside at this interface and to glide away from high concentrations of either oxygen or sulfide; it seeks the interface, not the extremes of either compound. It also demonstrates negative taxis against light as part of its strategy to avoid high oxygen concentrations produced by oxygenic photosynthesis. *Beggiotoa* may migrate several millimeters over a day as oxygen concentrations vary due to photosynthesis and aerobic respiration. This bacterium and other sulfide oxidizers are microaerophilic, meaning they prefer low oxygen concentrations, about 5–10% of atmospheric levels.

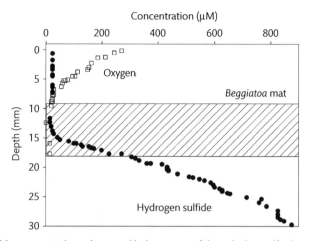

Figure 11.8 Oxygen and sulfide concentrations above and below a mat of the colorless sulfur bacterium, *Beggiatoa* (hatched area). Data from Kamp et al. (2006).

In addition to oxygen, nitrate can act as an electron acceptor for sulfide oxidation, yielding 785 kJ per reaction, only slightly less than with oxygen as an electron acceptor. The nitrogenous end product is ammonium, although some bacteria, such as *Thiobacillus denitrificans*, produce nitrogen gas. Some well-studied examples of nitrate-reducing, sulfide-oxidizing bacteria are *Thioploca* and *Thiomargarita*. These bacteria are interesting because their cells are huge, taken up mostly by a nitrate-filled vacuole (Schulz and Jørgensen, 2001). The bacteria fill the vacuole at the sediment–water interface where nitrate concentrations are high and then migrate deeper into sediments where hydrogen sulfide is available.

Perhaps the most fascinating example of sulfide-oxidizing bacteria is the symbiotic relationship between these bacteria with select marine invertebrates, including those found at hydrothermal vents. Vent animals are bathed in waters with high sulfide concentrations and also sufficient oxygen to support sulfide oxidation by symbiotic chemolithoautotrophic bacteria. This topic is discussed in Chapter 14.

Sulfide oxidation by anoxygenic photosynthesis
The other biotic mechanism for oxidizing reduced sulfur is carried out by anaerobic anoxygenic photosynthetic bacteria, abbreviated as AnAP, also as AAnP or AnAnP bacteria; no archaeon or eukaryote is known to have this form of metabolism. Although both oxidize sulfide, AnAP bacteria and colorless sulfide oxidizers are quite different in phylogeny and in use of the reduced sulfur. The phototrophic sulfur oxidizers use the reduced sulfur as an electron source for synthesis of the NADH needed for carbon dioxide reduction. Since the reduced sulfur replaces the water used by oxygenic phototrophic organisms (Chapter 4), AnAP bacteria do not evolve oxygen (they are anoxygenic). Unlike many colorless sulfide oxidizers, AnAP bacteria oxidize sulfide without oxygen (they are anaerobic) without gaining any energy from the process. ATP synthesis in AnAP bacteria is driven by light when growing photosynthetically. In contrast, light has no role in the metabolism of colorless sulfide oxidizers. Similar to colorless sulfide oxidizers, however, AnAP bacteria are found at interfaces where light and hydrogen sulfide are both present. They are common in waterlogged soils, salt marshes and stagnant pools where their unusual pigments can color the water brilliant purples and reds. Nearly all AnAP bacteria have bacteriochlorophyll *a*, but in addition they have several other types of bacteriochlorophyll and carotenoids (Table 11.6).

The five main groups of AnAP bacteria differ in their potential for heterotrophy and tolerance of oxygen, among other characteristics. The purple sulfur bacteria are mainly obligate anaerobes relying on photolithoau-

Table 11.6 Summary of sulfur-oxidizing anoxygenic phototrophs. "S" and "nonS" refer to sulfur and nonsulfur. Bacteria in this table incapable of using sulfide for photolithoautotrophic growth may oxidize other reduced sulfur compounds, such as thiosulfate and elemental sulfur. Alpha, Beta, and Gamma refer to subdivisions of the Proteobacteria. CBB = Calvin-Benson-Bassham (CBB), rTCA = reductive TCA cycle, and 3HPP = 3-hydroxypropionate/4-hydroxybutyrate. Data mainly from Canfield et al. (2005) and Sattley et al. (2008).

Microbe	Phylum or subphylum	Photolithoautotrophy with sulfide	Pigments	Aerobic heterotrophy	C fixation pathway
Purple S	Gamma	Yes	Bchl a,b	Yes	CBB
Purple nonS	Alpha, Beta	No	Bchl a,b	Yes	CBB
Green S	Chlorobi	Yes	Bchl a,c,d,e	No	rTCA
Green nonS	Chloroflexi	No	Bchl a,c,d	Yes	3HPP, CBB
Heliobacteria	Firmicutes	No	Bchl g	No	None

totrophy. The purple nonsulfur bacteria, including the intensively studied *Rhodobacter sphaeroides* (Choudhary et al., 2007), use a great variety of electron donors, including organic compounds in place of reduced sulfur for photosynthesis; photo-organotrophy is the group's preferred mode of metabolism. Some purple nonsulfur bacteria can withstand low oxygen concentrations while others thrive in oxic environments. In the dark, many even grow heterotrophically on organic compounds and oxygen. Purple nonsulfur bacteria do best in organic-rich environments with low light. The green sulfur bacteria, including the well-known *Chlorobaculum* (formally *Chlorobium*) *tepidum*, are obligate anaerobic phototrophs and can grow with low light levels at rates observed for purple sulfur bacteria with much higher light intensities. One genus in the green nonsulfur bacteria group, *Chloroflexus*, has strains that grow chemolithoautotrophically on H_2S or H_2 in addition to photolithoautotrophy, while others can carry out aerobic respiration in the dark, although are inhibited by atmospheric levels of oxygen. Finally, the heliobacteria are strict anaerobes that use photo-organotrophy or fermentation. These bacteria do not appear capable of carbon dioxide fixation and autotrophic growth (Madigan and Ormerod, 2004).

The carbon source for sulfur oxidizers
Sulfide oxidizers, other chemolithotrophs, and AnAP bacteria when not growing heterotrophically use carbon dioxide as their carbon source, making them autotrophs. The full name of the metabolism carried out by the colorless sulfide-oxidizing bacteria is chemolithoautotrophy

whereas it is photolithoautotrophy for the AnAP bacteria. For colorless sulfide oxidizers, the carbon dioxide fixation pathway is the same as for higher plants, eukaryotic algae, and cyanobacteria: the Calvin-Benson-Bassham (CBB) cycle. Bacteria capable of oxidizing elemental sulfur also use the reverse trichloroacetic acid cycle (rTCA) and the 3-hydroxypropionate/4-hydroxybutyrate (3-HPP) pathway. Depending on the species, AnAP bacteria fix carbon dioxide by the rTCA cycle or by the 3-HPP pathway, in addition to the CBB cycle (Hanson et al., 2012).

Methane and methanogenesis

Carbon dioxide reduction is another branch of the anaerobic food chain that is common in freshwaters and waterlogged soils where concentrations of sulfate and of all other electron acceptors are low. What gives this process added importance and global significance is the end product of carbon dioxide reduction, methane. Although its concentration is one hundredfold less than that of carbon dioxide in the atmosphere, methane is a strong greenhouse gas over twentyfold more effective in trapping heat than carbon dioxide as mentioned in Chapter 1. Both have been increasing, albeit at different rates, since the nineteenth century (Fig. 11.9). Unlike carbon dioxide, the anthropogenic inputs of methane now exceed natural ones (Chen and Prinn, 2006). Again unlike carbon dioxide, the main anthropogenic inputs of methane are agricultural, with emissions via belches and farts by cows and other ruminants high on the list. Rice paddies and other anoxic habitats on land are also major sources of methane. Some methane escapes into the

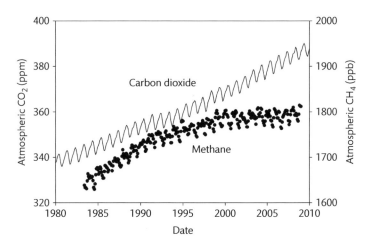

Figure 11.9 Methane and carbon dioxide concentrations in the atmosphere at the Mauna Loa Observatory, Hawaii. The data sets are used with permission from Pieter Tans (carbon dioxide) and Edward J. Dlugokencky (methane) at the NOAA Earth System Research Laboratory. The carbon dioxide data are presented as a trend line, while monthly means are given for the methane data. It is not clear why methane concentrations stopped increasing after 2000, but there are some indications that concentrations have begun to rise again (Rigby et al., 2008).

atmosphere during mining or transport of natural gas and other fossil fuels. Natural gas is mostly methane, of which some is directly from methanogens while the rest is from geothermal reactions working on preserved organic material.

Methanogenesis is carried out exclusively by strict anaerobes in the Euryarchaeota phylum of archaea. There are five well-defined orders of methanogens. Methanobacteriales, Methanomicrobiales, and Methanosarcinales are common in anoxic environments. Methanopyrales is a deeply branching order of archaean hyperthermophiles that produce methane from carbon dioxide and hydrogen gas. Its position in phylogenetic trees and ecophysiology suggests that members of Methanopyrales arose early in the history of life. The first cell on the planet has been hypothesized to use carbon dioxide as an electron acceptor and hydrogen gas as the electron donor (Lane et al., 2010). There is evidence against as well as for methanogens being among the first life forms on earth (House et al., 2003, Cameron et al., 2009).

Carbon dioxide and several other single carbon compounds (C_1), including methanol, formate, carbon monoxide (CO), and methanol, are used by various methangens to make methane. As indicated in Figure 11.5 of the anaerobic food chain, acetate is also a common substrate.

Methylamines, which have the general formula of $(CH_3)_xNH_3^+$ where x can be 1, 2, or 3, are said to be "noncompetitive" substrates for methanogens because they are not used by other bacteria, most importantly, sulfate reducers. When carbon dioxide is used, the reductant is hydrogen gas (Thauer et al., 2008). The other compounds used in methanogenesis undergo a disproportionation reaction, which means that a reductant is not needed. The following equation with formate (HCOO⁻) is one example of this type of reaction:

$$4HCOO^- + 4H^+ \rightarrow CH_4 + 3CO_2 + 2H_2O \quad (11.10).$$

The above reaction yields slightly more energy than the reduction of CO_2 with hydrogen gas ($\Delta G^{0'} = -144$ kJ versus -131 kJ) (Buckel, 1999). Methanogens are autotrophic and use the reductive acetyl-CoA pathway to fix carbon dioxide (Berg et al., 2010).

Thermodynamics seems to explain why methanogens are not abundant and why methanogenesis does not operate where sulfate concentrations are high and sulfate reduction is prevalent. Sulfate reduction is energetically more favorable than methanogenesis. However, thermodynamics does not explain how this energetic advantage is manifested at the physiological level of the microbes. The energetic advantage shows up in uptake kinetics for two key compounds used by both methanogens and

Box 11.5 Disproportionation

Formally, this type of reaction (also called dismutation) can be described as 2A → A' + A" where A, A', and A" are different chemicals but having the same main element, such as carbon in Equation 11.10. Fermentation can be considered as a type of disproportionation. Other examples of disproportionation are those involving sulfur compounds of intermediate oxidation state. One reaction is:

$$4S^\circ + 4H_2O \rightarrow 3H_2S + SO_4^{2-} + 2H^+$$

A variety of obligate anaerobic bacteria carries out disproportionation reactions of sulfur compounds. These reactions were only relatively recently discovered (Bak and Cypionka, 1987), long after fermentation was well-understood.

sulfate reducers. When sufficient sulfate is available, sulfate reducers outcompete methanogens because the half-saturation constant (K_m) for both acetate and hydrogen gas use by sulfate reducers is much lower than that for methanogens (Lovley and Klug, 1983, Muyzer and Stams, 2008). Consequently, uptake by sulfate reducers and other microbes reduce acetate and hydrogen concentrations to levels too low for methanogenesis to operate. The lack of methanogenesis in environments with active sulfate reduction implies that the concentrations and fluxes of noncompetitive substrates used only by methanogens are also low.

Methanotrophy

The flux of methane to the atmosphere is the net outcome of methanogenesis and of methane oxidation, or methanotrophy. Atmospheric methane concentrations would be even higher if not for methanotrophy. Environments such as rice paddies with high rates of methanogenesis can have equally high rates of methanotrophy. About 20% of the methane produced in rice paddies is oxidized before reaching the atmosphere (Conrad, 2009). Globally, soils are the dominant biological sink for atmospheric methane. Once in the atmosphere, methane can be oxidized by OH radicals, resulting in a residence time of about eight years.

Aerobic methane degradation

Methane oxidation with oxygen as the electron acceptor is described by:

$$CH_4 + 2O_2 \rightarrow HCO_3^- + H^+ + H_2O \quad \Delta G^{\circ\prime} = -814 \text{ kJ mol}^{-1} (11.11).$$

The known aerobic methanotrophs mostly are Alpha- and Gammaproteobacteria and differ in key steps in the methane oxidation pathway (Fig. 11.10). Type I and Type X use the RuMP pathway while the serine pathway is found in Type II. All methanogens have internal membranes presumed to be involved in methane oxidation, although the arrangement of these membranes differs among the types. All known methanotrophs also have the same first step in the pathway, the oxidation of methane to methanol by particulate methane mono-oxygenase (pMMO); "particulate" here refers to membranes. Only one methanotroph (*Methylocella*) does not have pMMO while many also have a soluble methane mono-oxygenase.

Methanotrophs are generally thought to be obligate methane oxidizers. Many can also use methanol (CH_3OH)

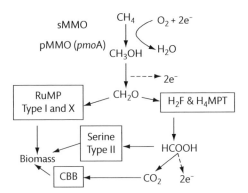

Figure 11.10 Pathway for aerobic methane oxidation. Two key genes are pMMO = particulate methane mono-oxygenase (and the gene *pmo*A for one of the pMMO subunits, a common target for studying uncultivated aerobic methanotrophs) and sMMO = soluble methane mono-oxygenase, RuMP = ribulose monophosphate, H_4F = tetrahydrofolate, H_4MPT = tetrahydromethanopterin, and CBB = Calvin-Benson-Bassham cycle. Type 1 and X methanotrophs use the RuMP pathway while Type II uses the serine pathway. The steps with electrons (e⁻) indicate connections to the electron transfer system and ATP production. The boxes indicate other pathways connected to methane oxidation. Modified from Chistoserdova et al. (2005).

but few other single carbon ("C₁") compounds. While some methanotrophs can oxidize other compounds, they cannot grow on these. There is some evidence of a methanotroph using acetate (Conrad, 2009). Methanotrophs are often included in a broader group, "methylotrophs", defined by the capacity to oxidize and grow on C_1 compounds, such as formate and carbon monoxide, in addition to methane and methanol.

Anaerobic methane oxidation

Microbiologists once thought that oxygen was required for methane oxidation, but geochemists long had evidence that methane was consumed in anoxic sediments by the following reaction:

$$CH_4 + SO_4^{2-} \rightarrow HS^- + HCO_3^- + 2H_2O$$
$$\Delta G^{\circ\prime} = -16.7 \ kJ \, mol^{-1} \quad (11.12).$$

Microbiologists were not convinced that microbes mediated this reaction because they were unable to isolate an organism capable of carrying it out. Early work turned up isolates of methanogens that oxidized methane anaerobically (Zehnder and Brock, 1979), but the rates did not seem high enough to explain the geochemical evidence. Nevertheless, Zehnder and Brock proposed that in nature methane was oxidized by methanogens carrying out "reverse methanogenesis" that produced H_2 or acetate which was subsequently used by a sulfate reducer.

An important piece of the puzzle was more geochemical evidence indicating that the lipids of archaea were highly deleted in ^{13}C in sediments near a methane seep (Canfield et al., 2005). The ^{13}C data made sense only if the lipid carbon came from methane. Analysis of the 16S rRNA genes turned up a new cluster of archaea, called ANME-1 (ANaerobic MEthane), which was related to but was not exactly the same as methanogens. Soon after the geochemical study, microbial ecologists who applied fluorescence in situ hybridization (FISH) to samples from methane seeps found aggregates of archaeal cells surrounded by sulfate-reducing bacteria (Boetius et al., 2000). More 16S rRNA gene sequencing revealed that the archaeal cells were in another cluster, ANME-2, related to ANME-1. Data showing that ANME-2 cells carry out methane oxidation came from FISH-type experiments using secondary ion mass spectrometry (SIMS) which demonstrated that organic carbon in the archaeal cells was highly depleted in ^{13}C (Orphan et al., 2001).

Now more is known about the diversity of the archaea carrying out anaerobic oxidation of methane (AOM) from additional studies of the 16S rRNA gene and of a key functional gene, the alpha subunit of methyl-coenzyme M reductase (*mcrA*) (Knittel and Boetius, 2009). These studies have revealed more about ANME-1 and ANME-2 and have found a third group, ANME-3, which all differ in morphology, aggregate formation, and association with bacterial partners. It appears that these organisms grow very slowly (one estimate of the doubling time is seven months!) and convert only 1% of methane into biomass with the other 99% lost as CO_2. Although AOM may occur without a bacterial partner, the best-known model is the coupling of a methane-oxidizing archaean with sulfate-reducing bacteria as originally hypothesized by Zehnder and Brock (Fig. 11.11). Figure 11.11 assumes that hydrogen gas is the compound released by the archaean and used by the sulfate-reducing bacteria, but this has not been demonstrated (Knittel and Boetius, 2009).

Still another form of anaerobic methane degradation uses nitrite (NO_2^-) as an electron acceptor, and is carried out by bacteria, not archaea, at least as known so far (Ettwig et al., 2010). This reaction is:

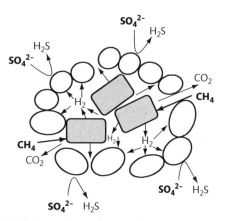

Figure 11.11 Anaerobic methane oxidation by archaea (shaded rectangles) surrounded by sulfate-reducing bacteria (open circles). Hydrogen is thought to be one possible compound exchanged between the microbes, but this has not been demonstrated. The cells are much closer together than actually depicted here.

$$3CH_4 + 8NO_2^- + 8H \rightarrow 3CO_2 + 4N_2 + 10H_2O$$
$$\Delta G^{\circ\prime} = -928 \text{ kJ mol}^{-1}$$

$$(11.13).$$

These bacteria belong to a poorly characterized phylum known only as "NC10". Although the process occurs in anoxic environments, genomic data and experimental evidence indicate that molecular oxygen is generated from nitrite (NO_2^-) which then goes on to oxidize methane using pMMO as in aerobic methane oxidation. Another important feature of this reaction is that it results in the loss from the system of nitrogen as nitrogen gas. Other electron acceptors for methane oxidation include Mn(IV) and Fe(III) (Beal et al., 2009).

Anaerobic eukaryotes

While bacteria and archaea dominate anoxic environments, some eukaryotes are present. Yeasts are well known for their fermentation pathways and end products, and there are studies of their natural distribution in vineyards, fruits, and soils, but little else is known about their role in anoxic environments. Other anaerobic eukaryotes are mainly flagellates and ciliates, although some metazoans can survive in the absence of oxygen, occasionally for extended periods of time (Fenchel and Finlay, 1995). In marine environments and other habitats with sufficient sulfate, sulfides as well as the lack of oxygen shut nearly all eukaryotes out of anoxic environments.

The main ecological role of anaerobic protists is similar to that in oxic environments: to graze on bacteria and archaea. Few studies have examined grazing in anoxic environments (First and Hollibaugh, 2008), but these indicate that rates may be low, implying that viruses are the main form of top-down control of bacterial and archaeal communities in anoxic environments. The abundance of these protists is lower in anoxic environments than oxic ones because of low growth efficiencies inherent with the energy-generating metabolism, fermentation, used by anaerobic protists. A few eukaryotes, such as the flagellate *Loxodes*, some fungi, and a few diatoms (Kamp et al., 2011), appear to use dissimilatory nitrate reduction. Fungi may be particularly important in denitrification (Chapter 12) in some soils (Laughlin and Stevens, 2002).

Overall, however, the metabolic and phylogenetic diversity of eukaryotes in anoxic environments is low. Except for assimilation of sulfate and organic sulfur, eukaryotes are not directly involved in the sulfur cycle, nor do they produce or consume methane.

While their diversity and abundance may be low, some anaerobic protists are important and interesting for other reasons. *Giardia* is an anaerobic flagellate that lives in the small intestine of humans and other animals, causing diarrhea when it attaches to epithelial cells of its hosts. It survives outside its hosts as a cyst and is transmitted by ingestion of fecal-contaminated water. Aside from public health concerns, *Giardia* and related organisms, including those in the trichomonad order, are interesting because they provide clues about the early evolution of primitive eukaryotes. *Giardia* and other anaerobic protozoa occupy deep branches in the Tree of Life, suggesting that they are primitive eukaryotes (Fig. 11.12).

Further support for that hypothesis came from the observation that these protozoa lack mitochondria. At

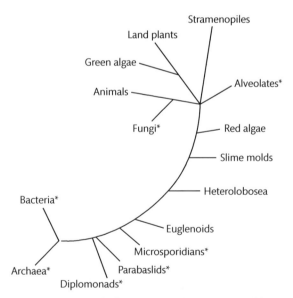

Figure 11.12 Tree of Life, with an emphasis on anaerobic organisms. All taxa with a * have representatives that can grow in anoxic habitats. The protistan groups without mitochondria are Diplomonads (which includes *Giardia*), Parabasalids, and Microsporidians. Protists related to dinoflagellates (alveolates) were recently found to be abundant in anaerobic waters (Stoeck et al., 2010). This tree was modified from one in Dacks and Doolittle (2001).

first, the microbes appeared to be missing links between prokaryotes and fully equipped eukaryotes, but subsequent work demonstrated that amitochondriate protozoa retain some mitochondria-like proteins even while losing their mitochondria during evolution (Bui et al., 1996, Hjort et al., 2010). In some protozoa, the mitochondrion evolved into a hydrogenosome which generates ATP by oxidizing pyruvate and producing hydrogen gas. Other anaerobic protozoa, including *Giardia*, have a mitochondrion-like organelle, called a mitosome, whose function remains unknown (Hjort et al., 2010). Mitosomes are much smaller than mitochondria and definitely are not involved in ATP generation. They may synthesize Fe-S proteins needed elsewhere in anaerobic protists. Some anaerobic protozoa have symbiotic bacteria or methanogenic archaea (Fenchel and Finlay, 1991). One cellulose-degrading protozoa is lined with a "wriggling fringe" of motile spirochetes, a type of bacteria, that has been hypothesized to be the predecessor of cilia in eukaryotes (Wier et al., 2010).

Summary

1. Thermodynamics explains why oxygen and nitrate are preferred electron acceptors, whereas sulfate and carbon dioxide are major terminal electron acceptors in anoxic environments because of high concentrations and their chemical form.

2. In the absence of oxygen, organic material is mineralized by a complex consortium of microbes in the anaerobic food chain. Two key compounds include acetate and hydrogen gas produced by fermenting bacteria and acetogens.

3. Hydrogen sulfide and other reduced sulfur compounds produced by sulfate reduction are oxidized in the dark by colorless sulfur-oxidizing bacteria and in the light by anaerobic anoxygenic phototrophic bacteria.

4. Colorless sulfur-oxidizing bacteria obtain carbon by fixing carbon dioxide, making them a type of chemolithoautotroph.

5. Methane, an important greenhouse gas, is produced only by strict anaerobic archaea from the reduction of carbon dioxide coupled to the oxidation of hydrogen gas or from the disproportionation of acetate, methanol, methylamines, and a few other compounds. Methanogens are outcompeted by sulfate reducers who use many of these same compounds, especially acetate and hydrogen gas.

6. Methane is degraded aerobically by specific methanotrophic bacteria or anaerobically by a consortium of archaea carrying out reverse methanogenesis and of sulfate reducers consuming an unknown reduced intermediate, perhaps hydrogen gas.

7. Although a few protists are capable of generating ATP from dissimilatory nitrate reduction, most anaerobic protists gain energy from fermentation. In many of these microbes, the mitochondrion has evolved into hydrogenosomes or mitosomes, providing fascinating examples of evolution in progress.

CHAPTER 12

The nitrogen cycle

Nitrogen is the only element with its own chapter in this book. It deserves special treatment for several reasons. Because microbes need so much nitrogen (Chapter 2), the supply of fixed nitrogen compounds, such as ammonium and nitrate, often limits growth and biomass of all organisms in terrestrial and aquatic ecosystems. Unlike phosphorus, which also is often a limiting element, nitrogen is involved in several important redox reactions because it can take on many oxidation states (Fig. 12.1), ranging from -3 in ammonium (NH_4^+) to +5 in nitrate (NO_3^-). Consequently, many nitrogenous compounds are involved in catabolic, energy-generating reactions, either as electron donors or acceptors, as well as being necessary for the biosynthesis of cellular components. In contrast, phosphorus occurs only as phosphate (PO_4^{3-}) and in organic forms, primarily in the oxidation state of +5 for all of these compounds. It is used only for biosynthesis. Phosphorus-rich compounds like ATP are important in catabolic reactions, but the oxidation state of phosphorus does not change in these reactions. Nitrogen is an interesting and an important element.

Another reason for paying special attention to nitrogen is that one nitrogenous compound, nitrous oxide (N_2O), is a potent greenhouse gas, being about 270-fold more effective in trapping heat than carbon dioxide. It is third behind carbon dioxide and methane in contributing to the overall greenhouse effect. Nitrous oxide is also now the most potent destroyer of ozone (Ravishankara et al., 2009), because concentrations of it have increased in the atmosphere, along with carbon dioxide and methane, over the last 100 years. Similar to methane, increases in nitrous oxide concentrations are due to increases in agriculture and to a lesser extent certain chemical industries. Nitrous oxide figures in both nitrification and denitrification as discussed below.

Humans have several other impacts on the nitrogen cycle (Galloway et al., 2008), starting with nitrogen fixation. The Haber-Bosch process, used to make fertilizers, produces about 120×10^{12} g N y^{-1} of ammonium from nitrogen gas while nitrogen fixation by human-managed legumes (Chapter 14) adds another 40×10^{12} g N y^{-1}. Together these anthropogenic rates are starting to approach natural nitrogen fixation rates estimated to be about 260×10^{12} g N y^{-1}. The extra nitrogen from anthropogenic sources has increased plant production in both terrestrial and aquatic ecosystems. This fertilization effect

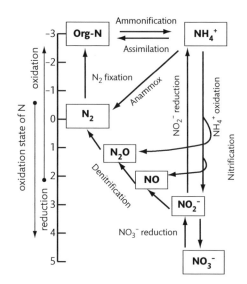

Figure 12.1 The microbial reactions in the N cycle. "Org-N" refers to organic nitrogen. Adapted from Capone (2000).

leads to higher agricultural yields, feeding a burgeoning world population, but it also contributes to more noxious algal blooms in lakes and coastal marine waters and to contamination of groundwater used for drinking. Humans add more nitrogen in the form of nitric oxide (NO) to the biosphere by burning fossil fuels and forests. Nitric oxide is a major component of acid rain. Over 80% of all NO emissions are thought to be from human activities, and in some regions the anthropogenic sources exceed natural inputs by tenfold.

Nitrogen fixation

The capacity for nitrogen fixation is widespread among bacteria and archaea, but it is a rather specialized process. Not all prokaryotes fix nitrogen, and even if two microbes are closely related, one may be a diazotroph and the other not. Although nitrogen fixation is not carried out by any eukaryotes, except for humans using the Haber-Bosch process, some eukaryotic microbes and higher plants form symbiotic relationships with diazotrophs (Chapter 14). Nitrogen fixation is found in prokaryotes carrying out every energy-generating form

---

Box 12.1 The moral balance of a scientist

Fritz Haber (1868–1934) was a German chemist who won the Nobel Prize in Chemistry in 1918 for inventing the process now bearing his name. In addition to synthesizing ammonium for fertilizer, the Haber-Bosch process was also important in manufacturing explosives for warfare. Haber's moral standing has been even more severely questioned because of his role as scientific director of the Kaiser-Wilhelm Institute in developing poisonous gases used during World War I (Szollosi-Janze, 2001). His institute went on to develop the cyanide gas Zyklon B, which was used first as a pesticide and then later against humans during the Holocaust. Members of Haber's extended family died in Nazi concentration camps during World War II. Haber left Germany in 1933 after losing his position at the institute because he was a Jew.

---

of metabolism: aerobic and anaerobic heterotrophy (chemoorganotrophy), oxygenic phototrophy (cyanobacteria only), anaerobic anoxygenic phototrophy, and chemolithotrophy. The one exception is aerobic anoxygenic phototrophy (AAP); no known AAP bacterium is capable of nitrogen fixation.

It is remarkable that diazotrophs can accomplish at room temperature (20 °C) under normal atmospheric pressure what the Haber-Bosch process can do only at high temperature (300–550 °C) and pressure (15–26 MPa).

Nitrogenase, the nitrogen-fixing enzyme

Nitrogen fixation is the reduction of nitrogen gas to ammonia and can be described by:

$$N_2 + 8H^+ + 8e^- + 16ATP \rightarrow 2NH_3 + H_2 + 16ADP \quad (12.1)$$

where the reducing power ($8e^-$) is supplied by NAD(P)H. The fate of hydrogen gas produced by nitrogen fixation is not completely clear. Although Equation 12.1 indicates that nitrogen fixation needs 16 ATPs, under natural conditions energetic costs may be even higher, up to 30 ATPs for some microbes (Hill, 1976). The huge energetic cost of nitrogen fixation helps to explain the limited distribution of the process among organisms and environments. It also helps to explain why every prokaryote does not routinely carry out nitrogen fixation even in nitrogen-limited environments. Finally, it explains why microbes turn off nitrogen fixation when fixed nitrogen, especially ammonium, is available. Energetic costs are high due to the difficulty of breaking the triple bond of nitrogen gas ($N \equiv N$).

The key enzyme carrying out nitrogen fixation is nitrogenase, a huge enzyme, as large as 300 kDa, that can make up as much as 30% of cellular protein. Nitrogenase is actually a complex of two proteins, dinitrogenase and dinitrogenase reductase (Fig. 12.2). The first contains both iron (Fe) and either molybdenum (Mo) or vanadium (V), while the second contains only Fe. Dinitrogenase reductase, the Fe protein coded for by the *nifH* gene, has two identical subunits with about four Fe atoms. The Mo-Fe protein has many more Fe (21–35 atoms) and two Mo atoms in its two pairs of subunits. These are encoded by the *nifD* and the *nifK* genes. Nitrogenase is thought to have evolved early in the history of life. The diversity of

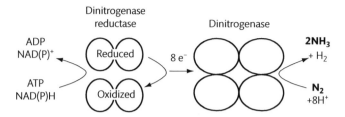

Figure 12.2 Nitrogen fixation by the nitrogenase complex. Dinitrogenase reductase, the Fe protein, has two subunits encoded by the *nifH* gene. Dinitrogenase is a Mo-Fe protein with four subunits, two encoded by *nifD* and another two encoded by *nifK*. Dinitrogenase reductase is about 60 000 Da while dinitrogenase is about 240 000 Da.

diazotrophs is one argument for it being an ancient process, potentially evolving soon after the first cell came into existence (Zehr and Paerl, 2008). There is some speculation that nitrogenase had another main function, such as reduction of cyanide or carbon monoxide, when it first appeared in the biosphere (Lee et al., 2010).

Solving the oxygen problem

Nitrogenase is irreversibly damaged by oxygen, creating a huge problem for the many diazotrophs that live in oxic environments, particularly those that evolve oxygen during photosynthesis. Reviewing the strategies for solving this oxygen problem serves to introduce some of the best-known nitrogen fixers and to illustrate their diversity (Table 12.1). The oxygen protection strategy taken by filamentous cyanobacteria, such as *Anabaena*, was mentioned in Chapter 4. These microbes house nitrogenase in a specialized cell, the heterocyst. Its thick cell walls physically limit oxygen diffusion and help to keep oxygen concentrations around the nitrogenase low. Unique among the cells of a cyanobacterial filament, heterocysts lack the oxygen-producing part of photosynthesis, photosystem II (PS II). A marine single-cell coccoid cyanobacterium also does not have PS II apparently to avoid poisoning its nitrogenase with oxygen (Tripp et al., 2010). For heterocyst cyanobacteria, the other, vegetative cells of the filament feed sugars and organic acids to hetero-

Table 12.1 Some types of nitrogen fixing bacteria, their main environment, and strategies to protect nitrogenase from oxygen poisoning. The O_2 protection strategy for *Rhizobium* is labeled as "physical" because the host plant limits the flow of oxygen to the bacterium. Several nitrogen fixers use more than one strategy. *Oscillatoria* partially avoids oxygen by living in a microaerophilic environment as well as separating oxygen-producing photosynthesis from nitrogen fixation over time. The soil bacterium *Azotobacter* is famous for its high respiration rates, which lower oxygen to tolerable levels, but some *Azotobacter* species also produce protective proteins that bind to nitrogenase in the presence of oxygen.

O_2 protection	Microbe	Type	Environment
Heterocyst	*Anabaena*	Cyanobacteria	Freshwater
Heterocyst	*Nostoc*	Cyanobacteria	Microbial mats
Heterocyst	*Nodularia*	Cyanobacteria	Microbial mats
Heterocyst	*Richelia*	Cyanobacteria	Marine endosymbiosis
Physical	*Rhizobium*	Alphaproteobacteria	Soil endosymbiosis
Time and spatial separation	*Trichodesmium*	Cyanobacteria	Marine waters
Time separation	*Oscillatoria*	Cyanobacteria	Microbial mats
Respiration	*Azorhizobium*	Alphaproteobacteria	Soils
Respiration	*Azotobacter*	Gammaproteobacteria	Soils
Respiration	*Azospirillum*	Alphaproteobacteria	Soils
Avoidance	*Methanosarcina*	Archaea	Various
Avoidance	*Clostridium*	Firmicutes	Various
Avoidance	*Chlorobium*	Chlorobi	Freshwaters

cysts in exchange for fixed nitrogen in the form of gluta-mate. Organic carbon-fueled respiration helps also to keep oxygen concentrations low in heterocysts. Analogous to the heterocyst strategy, the soil actinomyc-ete, *Frankia*, produces a vesicle to help protect its nitro-genase. Other microbes fix nitrogen when coated in a polysaccharide-rich slime which minimizes oxygen diffusion.

Another strategy for photosynthetic microbes is to separate nitrogen fixation in time from oxygen produc-tion by photosynthesis. This is the strategy used by one of the main nitrogen-fixing cyanobacteria in the oceans, *Trichodesmium*. This filamentous cyanobacterium forms aggregates large enough to be seen by the naked eye. It was once hypothesized that nitrogen fixation was high-est in the middle of the aggregate where oxygen concen-trations were presumed to be low. Actual micro-electrode measurements showed low oxygen concentrations at the center of aggregates, but the concentrations are not low enough to protect nitrogenase. Instead, this cyano-bacterium separates nitrogen fixation over space and time, as different cells take turns carrying out the two processes (Berman-Frank et al., 2001).

Soil bacteria in the genus *Azobacteria* use a couple of oxygen protection strategies. One is to maintain high respiration rates to consume oxygen. This respiration may even be uncoupled from ATP synthesis. Some spe-cies in this genus produce proteins that bind to nitroge-nase and protect it from oxygen.

N2 fixation in nature
On a global scale, rates of nitrogen fixation are roughly equal on land and in the oceans. Ignoring anthropo-genic sources, diazotrophs fix 110 × 10^{12} g N y^{-1} on land and 140 × 10^{12} g N y^{-1} for the oceans (Galloway et al., 2008). In soils, most studies have focused on agriculturally important symbiotic diazotrophs such as the bacterium *Rhizobium* and its host, the legumes (Chapter 14). Other symbiotic diazotrophs in terrestrial systems include the actinomycete *Frankia* and its angiosperm hosts and the cyanobacterium *Anabaena* and its host, the aquatic fern *Azolla*. Cyanobacteria fix nitrogen at the surface of soils, and heterotrophic dia-zotrophs can be important in organic carbon-rich soil habitats. Nitrogen fixation in freshwaters is very small

compared to rates in soils by symbiotic and free-living microbes. Rates per square meter are high in salt marshes and estuaries, but most nitrogen fixation in marine systems is in the open oceans because of their vast area and volume. It was once thought that *Trichodesmium* was the main diazotroph in the oceans until a small coccoid cyanobacterium ("UCYN-A") capable of N_2 fixation was discovered by a combina-tion of ^{15}N rate measurements and cultivation-inde-pendent studies of *nifH*. This cyanobacterium is the one without PS II and appears to be a type of photo-heterotroph (Zehr et al., 2008).

Regardless of the diazotroph, rates of nitrogen fixation are large compared to other external sources of nitrogen but small compared to internal nitrogen fluxes and exchanges. The nitrogen fixed by diazotrophs in the sur-face layer of the oceans can be roughly equal to the sup-ply rate of nitrate from deep waters, together referred to as "new" nitrogen (Carpenter and Capone, 2008). This nitrogen is "new" to the ecosystem, in contrast to nitro-gen recycled internally. Any biological production sup-ported by new nitrogen is likewise referred to as "new production". In the open oceans, new production is usu-ally small (about 10% of the total) compared to total pri-

---

Box 12.2 Measuring N₂ fixation

Nitrogen fixation is difficult to measure because of the lack of a convenient radioactive isotope of N, the difficulty of working with the stable nitrogen isotopes (^{15}N), and the importance of maintaining the physical structure of the nitrogen fixing micro-environment; introduction of oxygen could lead to underestimates of rates. Ideally, nitrogen fixation is estimated by following $^{15}N_2$ into biomass or ammonium. An easier and more sensitive assay is acetylene reduction. The technique relies on nitro-genase reducing acetylene (C_2H_2) to ethylene (C_2H_4), which is easily measured by gas chromatog-raphy. Nitrogenase works on acetylene because it has a triple bond like that in N_2. The problem is that the ratio of ethylene production to actual N_2 fixa-tion can vary. The acetylene reduction assay needs to be calibrated with ^{15}N assays for each habitat.

mary production, implying that rates of internal cycling of nitrogen (ammonium excretion and uptake) are about tenfold higher than rates of nitrogen fixation. In coastal waters, the fraction of new production is much higher, about 30% of total production. In soils, nitrogen fixation is about 10% of internal nitrogen fluxes (Schlesinger, 1997).

Limitation of N_2 fixation

Microbial ecologists have wondered why rates of nitrogen fixation are not higher and diazotrophs more abundant in nitrogen-limited systems, such as the oligotrophic oceans. There are several answers to these questions and several factors affecting nitrogen fixation rates in nature (Fig. 12.3). The relative importance of each varies with the diazotroph and the environment, but the high energetic cost of nitrogen fixation explains much. Because of energetic costs, light limitation would lead to low nitrogen fixation by phototrophs and a low supply of organic material would have the same effect on heterotrophic diazotrophs. The latter explains why heterotrophic bacteria are not very important in fixing nitrogen in the oceans, whereas they account for much of the nitrogen fixation in soils supplied with organic carbon from higher plants. Energetic costs also account for why nitrogen fix-

ation is shut down by high concentrations of ammonium and nitrate. Microbes can save energy by using these inorganic sources of nitrogen rather than fixing it. It is not totally clear why diazotrophs can be abundant in nitrogen-rich environments like estuaries.

Several inorganic nutrients in addition to ammonium and nitrate also affect nitrogen fixation. The very low concentration of iron is a big reason why nitrogen fixation is not higher in open-ocean regions. An alternative hypothesis is that diazotrophs in the open oceans are limited by low phosphate concentrations (Sañudo-Wilhelmy et al., 2001). A high input of phosphate can switch a lake from phosphorus limitation to nitrogen limitation and selects for cyanobacterial diazotrophs; nasty cyanobacterial blooms have been caused by phosphorus in detergents and other phosphorus-rich contaminants polluting reservoirs and lakes. Concentrations of the other trace element used by nitrogenase, molybdenum, are also low but are apparently sufficient for nitrogen fixation in aquatic and most terrestrial habitats. The exceptions may include some freshwaters and highly weathered acidic soils such as those found in the tropics (Barron et al., 2009, Glass et al., 2010).

Temperature affects nitrogen fixation and diazotroph abundance, as it does all microbial processes and all microbes. Temperature may explain why cyanobacterial diazotrophs in the oceans such as *Trichodesmium* occur as filaments without heterocysts, or are unicellular. One hypothesis is that solubility and diffusion of oxygen into regular cells is limited enough by temperature and salinity such that heterocysts are not needed to protect nitrogenase from oxygen (Stal, 2009). Still, why heterocystous cyanobacteria are not more common in marine habitats remains a mystery.

Ammonium assimilation, regeneration, and fluxes

Once synthesized by the nitrogenase complex, ammonium is assimilated into amino acids and then converted into other nitrogen-containing compounds within the diazotroph. The pathways for ammonium assimilation are the same as carried out by all other microbes, both prokaryotic and eukaryotic. There are two pathways. The first, designed for high ammonium concentrations, relies on the enzyme glutamate dehydrogenase (GDH)

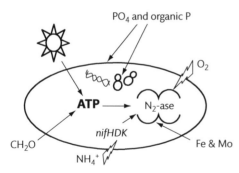

Figure 12.3 Factors affecting N_2 fixation by diazotrophs. The positive factors are indicated by simple arrows while the two negative factors are depicted by lightning bolts. Because this process is so energetically expensive, fixation by heterotrophs and phototrophs may be limited by organic material (CH_2O) and by light, respectively. Nitrogenase ("N_2-ase") requires iron (Fe) and molybdenum (Mo), whose concentrations may be too low. This enzyme is inactivated by oxygen. Diazotrophs need phosphorus supplied as phosphate ("PO_4") or in organic compounds. Expression of nitrogenase genes (*nifHDK*) is negatively regulated by ammonium.

and uses only one NAD(P)H. The second is a high affinity system for low ammonium concentrations. The first step of the second pathway consists of the synthesis of glutamine from glutamate, catalyzed by glutamine synthetase (GS):

$$\text{glutamate} + NH_4^+ + ATP \rightarrow \text{glutamine} + ADP \quad (12.2)$$

followed by the second step that yields the net production of one glutamate:

$$\text{glutamine} + \alpha\text{-oxoglutarate} + NADPH \rightarrow$$
$$2 \text{ glutamates} + NADP^+ \quad (12.3).$$

The enzyme for the second step is glutamate synthase, also called glutamine-α-oxoglutarate transferase (GOGAT). The entire two-step pathway is called the GS-GOGAT pathway. This pathway is used by diazotrophs to ensure that the newly synthesized ammonium does not inhibit further nitrogen fixation. It is also used by microbes in natural environments where ammonium concentrations are very low. With either pathway, the resulting glutamate supplies the nitrogen needed for all other biochemicals within a cell.

The nitrogen now in the diazotroph enters the microbial food web by all of the mechanisms discussed in previous chapters: grazing, excretion, and viral lysis (Fig. 12.4). The exchanges of nitrogen depicted in Figure 12.4 are dominated by organisms other than the diazotrophs, but all of the nitrogen came from nitrogen fixation at some point in time. The nitrogen may be transferred from prey to predator as amino acids or other building blocks for macromolecules without being released as dissolved inorganic or organic nitrogen.

When the nitrogen enters the detritus pool, it is mineralized eventually to ammonium by bacteria, fungi, and to a lesser extent other organisms, as discussed for overall organic material mineralization (Chapter 5). This production of ammonium is also referred to as regeneration. Mineralization of detrital protein to ammonium occurs after protein has been hydrolyzed to amino acids (Chapter 5), by deamination of amino acids (R-C(NH$_2$) COOH, where "R" represents the various side chains of amino acids):

$$\text{R-CHCOOH} + NAD^+ + H_2O \rightarrow \text{R-CHCOOH} + NH_4^+ + NADH$$
| |
NH$_2$ O   (12.4)

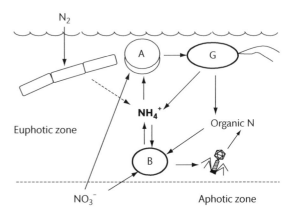

Figure 12.4 Internal cycling of nitrogen via ammonium. The diagram is most applicable to aquatic habitats, but all of these interactions occur in soils as well. "New" nitrogen is brought into the system by nitrogen fixation or, in the case of aquatic habitats, by advection and diffusion of nitrate from deep waters, the aphotic zone. Ammonium is released from diazotrophs (depicted here as a filamentous microbe) by several processes and is taken up by algae (A) and bacteria (B) in the surface layer, the euphotic zone. Grazers ("G") and bacteria are sources of regenerated ammonium while organic nitrogen comes from grazers and viral lysis.

forming an alpha-keto acid, in addition to ammonium.

The reactions producing ammonium from other detrital organic nitrogenous compounds are more complicated and not as well understood. These reactions collectively are referred to as ammonification. Some reactions produce urea $(CO(NH_2)_2)$, a major component of urine, which is released by many higher animals. Zooplankton and other primitive animals can release urea, but most release ammonium. Fluxes of urea are generally not as big as ammonium fluxes in most environments, most of the time.

Ammonium is an important nitrogen source for heterotrophic microbes, eukaryotic phototrophs, cyanobacteria, and higher plants. It is preferred over nitrate because the oxidation state of nitrogen in ammonium is the same as that in amino acids and other nitrogenous biochemicals in cells. Because of this preference, rates of ammonium uptake are often faster than nitrate uptake rates in both soils and aquatic habitats even when nitrate concentrations are higher; nitrate uptake exceeds ammonium uptake only when ammonium concentrations are substantially lower than nitrate. In aquatic systems, a large fraction of primary production is based on ammonium. Aquatic ecologists call primary production based on ammonium and urea "regenerated primary production". Rates of this ammonium-based production may be as much as 90% of gross primary production in the open oceans. Fluxes through the ammonium pool are high even though concentrations are low because of rapid uptake and adsorption and desorption of ammonium to detritus and clays.

Ammonium release in anoxic systems
In the absence of oxygen, ammonium is produced by anaerobic bacteria using other electron acceptors, such as nitrate, iron oxides, and sulfate. It is also produced by fermenting bacteria acting on the amino acids and purine and pyrimidine bases released by hydrolysis of nitrogenous macromolecules. Some of the best-known bacteria carrying out this type of fermentation are in the genus *Clostridium*. One example is fermentation of glycine:

$$4H_2N-CH_2-COOH+2H_2O \rightarrow 4NH_3 + 2CO_2 + 3CH_3-COOH \quad (12.5).$$

Note that all of the nitrogen in glycine is released as ammonium while much of the carbon remains as acetate, implying that the terminal electron acceptor for carbon mineralization has little impact on nitrogen mineralization, since acetate could be mineralized by microbes using one of several possible electron acceptors. Equation 12.5 also implies that rates of nitrogen mineralization would not be tightly correlated with carbon mineralization rates because it seems that fate of acetate is independent of ammonium production. In fact, carbon and nitrogen mineralization rates usually are coupled in anoxic habitats, since carbon dioxide and ammonium production often co-vary in sediment pore waters and incubation experiments (Canfield et al., 2005).

Ammonium uptake versus excretion, immobilization versus mobilization
Grazers, heterotrophic bacteria, and fungi consume organic nitrogen and potentially excrete ammonium. Heterotrophic bacteria and fungi can also take up ammonium. What determines the net direction of the ammonium going into or out of these microbes?

The short answer is the C:N ratio of microbes versus that of the organic material, plus how much carbon is lost during respiration. The long answer involves several simple equations. Based on the definition of growth efficiency (Y, see Chapter 6), the amount of C used by microbes for growth is $U_c \cdot Y$ where U_c is total uptake of the material in carbon units. To calculate N uptake, we convert C uptake to N uptake with the C:N ratio of the original organic material and of microbial biomass; here it is more convenient to use $N:C_s$ for the organic material (the substrate) and $N:C_b$ for microbial biomass. So, the total amount of N taken up is $U_c \cdot N:C_s$ and the amount of N used for growth is $U_c \cdot Y \cdot N:C_b$. At steady-state, these two are equal when balanced by the uptake or release of ammonium (F_N)

$$U_C \cdot N:C_S + F_N = U_C \cdot Y \cdot N:C_b \quad (12.6).$$

For net ammonium excretion, $F_N < 0$ whereas for net uptake, $F_N > 0$.

To explore Equation 12.6 in more detail, we can calculate conditions when there is no net excretion or uptake of ammonium, and then use these conditions as a dividing line between the two opposing processes (Fig. 12.5). For this example, let us use two extreme values for $C:N_b$, which are averages for bacteria ($C:N_b = 5.5$) and fungi ($C:N_b = 8$; Chapter 2). Growth efficiency varies even more, but it is likely to be between 0.1 and 0.5 in various environments (Chapter 5), which again can represent general averages for bacteria and fungi. Based on these values, bacteria should generally release ammonium except when using very nitrogen-poor organic material with $C:N_s$ exceeding 60. That number comes from seeing where the curve for $C:N_b = 5.5$ crosses the horizontal line set by a growth efficiency of 0.1 (Fig. 12.5). Fungi, which need less nitrogen than bacteria (they have a higher $C:N_b$), should also release nitrogen except for extremely high $C:N_s$. These results lead to the prediction that degradation of protein-rich detritus, such as from algae, should lead to net release of ammonium whereas microbes need to assimilate ammonium when growing on plant litter rich in carbohydrates and other components with high C:N ratios.

However, heterotrophic bacteria seem to assimilate more ammonium and nitrate than expected from

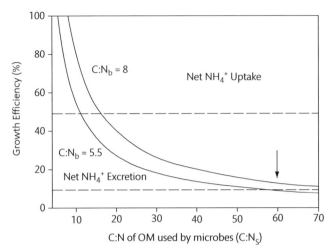

Figure 12.5 Uptake or excretion of ammonium as a function of C:N ratios of the microbial biomass (C:N$_b$) and of the organic material (OM) (C:N$_s$) and the growth efficiency. To determine whether there is net ammonium uptake or excretion, pick a particular growth efficiency and C:N$_s$ and find where the point is in relationship to the solid curves. If it is to the left or below the curve, then there is net ammonium excretion. If it is to the right or above the curve, then there is net uptake. The dashed horizontal lines indicate likely extremes of growth efficiencies. The arrow indicates the C:N$_s$ value (60) above which net ammonium assimilation occurs for growth efficiency equal to 10% and C:N$_b$ = 5.5.

Equation 12.4. These microbes account for about 30% of the uptake of ammonium and nitrate and effectively compete with larger phytoplankton for these important N sources in aquatic habitats (Mulholland and Lomas, 2008). Heterotrophic bacteria and fungi also use ammonium in soils (Inselsbacher et al., 2010). A problem with Equation 12.6 is that the C:N ratio of the organic material actually used by microbes is poorly known, in part because we do not know much about the composition of naturally occurring organic material (Chapter 5). Also, the equation does not take into account differences among microbes, and micro-environments with C:N ratios different from the bulk. But Equation 12.6 is still a useful way of thinking how ammonium fluxes are affected by nitrogen content (C:N ratios) and microbial energetics (Y).

The equation is also useful for exploring the balance between nitrogen mobilization and nitrogen immobilization in soils and sediments. Soil ecologists and biogeochemists consider ammonium and nitrate uptake by microbes and plants as part of an "immobilization" process. Immobilization also includes abiotic adsorption to soil constituents. These processes take nitrogen in the two inorganic nitrogen compounds that move more easily by diffusion and advective flow and put it into more

immobile forms, such as microbes, plants, and large soil particles.

Ammonia oxidation, nitrate production, and nitrification

We have seen that mineralization of organic material yields ammonium, yet concentrations of ammonium are often quite low in many natural environments. The more common inorganic form of fixed nitrogen is nitrate, one of the largest pools of nitrogen in the biosphere. In the deep ocean, for example, nitrate concentrations reach 40 µM whereas all other forms of nitrogen, except N$_2$, are low. Nitrate concentrations vary greatly in soils, depending on water content and fertilization, but usually there is more nitrate than ammonium. Where does this nitrate come from?

The answer is nitrification, a two-step process involving at least two types of microbes. No single microbe alone appears capable of oxidizing ammonia all the way to nitrate, perhaps because energetic constraints select against such a microbe (Costa et al., 2006). Here we will use "ammonia" (NH$_3$) when referring to the oxidation step and to the microbes because it is the actual substrate for this process even though concentrations of ammonium (NH_4^+) are higher in most environments. The switch from one to the other is set by pH and the pKa of the reaction (Chapter 3).

The first step in nitrification is usually considered to be the rate-limiting one that sets the overall pace of the process. This first step is the oxidation of ammonia to nitrite (NO_2^-):

$$NH_4^+ + 1.5\,O_2 \rightarrow NO_2^- + H_2O + 2H^+ \qquad (12.7)$$
$$\Delta G^{\circ\prime} = -272 \text{ kJ mol}^{-1}$$

while the second step, the oxidation of nitrite to nitrate, completes the process:

$$NO_2^- + 0.5\,O_2 \rightarrow NO_3^- \quad \Delta G^{\circ\prime} = -76 \text{ kJ mol}^{-1} \quad (12.8).$$

Both of these reactions are carried out mainly by chemolithotrophic microbes. As is typical of chemolithotrophy, neither of these reactions yields much energy. Note also that nitrification or the "making" of nitrate is an aerobic process. The microbes being discussed here depend on oxygen. Later we consider the anaerobic oxidation of ammonia, which shares very little in common with aerobic ammonia oxidation except that both processes oxidize ammonia.

In addition to the chemolithoautotrophic process, there is some production of nitrate during the aerobic oxidization of organic material by bacteria and fungi (Laughlin et al., 2008) by a process called heterotrophic nitrification. However, rates by this type of nitrification are 10^3 to 10^4 slower than rates for chemolithotrophic

nitrification. The mechanisms of heterotrophic nitrification are not well understood, but it is thought that heterotrophic nitrifiers do not gain energy from nitrogen oxidation, in contrast to chemolithotrophic nitrification. This type of nitrification has been examined mostly in soils, but it is thought to occur in aquatic habitats as well.

Aerobic ammonia oxidation by bacteria

Ammonia oxidizers appeared to make up a tight cluster of closely related organisms mostly in the Betaproteobacteria, along with a few in the Gammaproteobacteria. Classically, bacteria in the betaproteobacterial genera *Nitrosomonas* and *Nitrosospira* were considered to be the main microbes oxidizing ammonia in oxic environments. The classic picture is still largely correct, with some important additions, as will soon become apparent. The cultivated ammonia oxidizers are strict chemoautolithotrophs, relying solely on ammonia as an energy source. As with all microbes and microbial processes, however, many ammonia oxidizers cannot be cultivated and grown in the laboratory. To examine these uncultivated ammonia oxidizers, microbial ecologists use PCR-based approaches to examine the gene (*amoA*) for a subunit of a key enzyme, ammonia monooxygenase, catalyzing ammonia oxidation. Genes for other subunits of ammonia monooxygenase (*amoB* and *amoC*) are likely to be good markers for ammonia-oxidizing bacteria (Junier et al., 2008), but these have not been used extensively so far.

The *amoA* gene has proven to be a very powerful tool for exploring ammonia oxidation in nature. It can be used to identify these chemolithotrophic microbes because the phylogeny implied by *amoA* appears to match the 16S rRNA gene phylogeny of cultivated ammonia oxidizers (Fig. 12.6). Using *amoA* gene sequences and abundance, microbial ecologists have mapped out the biogeography of ammonia oxidizers and have estimated their abundance in various habitats. These PCR-based surveys have found different clades of *amoA* genes in different habitats. Not surprisingly, marine ammonia oxidizers differ from those in soil and both differ from those in freshwaters. Betaproteobacterial ammonia oxidizers greatly outnumber gammaproteobacterial ammonia oxidizers in the habitats examined so far.

Box 12.3 Measuring nitrification

Often, the individual steps of nitrification (ammonia and nitrite oxidation) are not measured separately, but the entire process is examined. One direct method consists of following the ^{15}N from added $^{15}NH_4^+$ into nitrate. Alternatively, either ammonia or nitrite oxidation is inhibited by the addition of nitrapyrin, allylthiourea (inhibitors of ammonia oxidation) or chlorate (inhibitor of nitrite oxidation). The build-up of ammonium or nitrite is then measured over time.

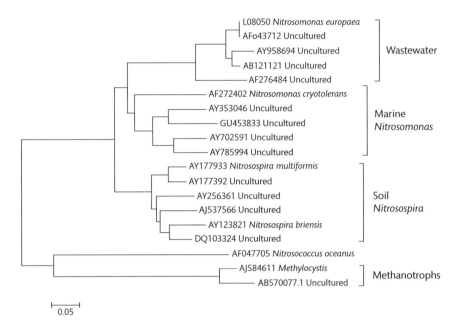

Figure 12.6 A phylogenetic tree of *amoA*, a key gene for ammonia oxidizers. The label with letters and numbers, such as LO8050, is a unique identifying number ("the accession number") for the sequence in gene databases such as Genbank. All are in the Betaproteobacteria, except for *Nitrosococcus oceanus* which is in the Gammaproteobacteria. The methanotrophs are represented by genes for particulate methane mono-oxygenases (*pmoA*). Ammonia oxidizers can oxidize methane and methane oxidizers can oxidize ammonia, albeit at tenfold lower rates than the microbes specialized for the substrate. The division between environments is not as clear cut as implied by this tree. Tree provided by Glenn Christman and used with permission.

Archaeal ammonia oxidation

Chapter 10 briefly mentioned how metagenomic research opened up a new chapter in the study of ammonia oxidation and of archaea in natural environments. A metagenomic survey of the Sargasso Sea found an *amoA* gene linked to a phylogenetic marker from the Crenarchaeota phylum (Venter et al., 2004). Ammonia oxidation by archaea in culture had not been observed before. Although the archaeal *amoA* gene was similar enough to the bacterial one to be recognized as an *amoA* gene, there are enough differences so that archaeal and bacterial versions of this gene can be distinguished by standard PCR methods. The Sargasso Sea finding was soon applied to soil communities (Schleper et al., 2005) and many more.

The abundance of ammonia-oxidizing archaea (AOA) and ammonia-oxidizing bacteria (AOB) has been assessed with quantitative PCR (QPCR), also called real-time PCR. QPCR studies have found that AOA abundance

exceeds AOB abundance in many environments, ranging from soils (Leininger et al., 2006) to the water column of the oceans (De Corte et al., 2009, Beman et al., 2008), although there are exceptions as always. The implication is that AOA are carrying out more ammonia oxidation than AOB. This is a hard question to address because actual rate measurements by current methods cannot distinguish oxidation by archaea from that by bacteria. Other approaches are needed.

Some of these other approaches were used by a study examining nitrification in soils of a maize field (Jia and Conrad, 2009). Similar to other studies, copies of AOA *amoA* genes outnumber AOB *amoA* genes in these soils, yet two lines of evidence suggested that most of the ammonia oxidation was by bacteria, not archaea. First, abundance of AOB *amoA* genes changed with nitrification rates in treatments that inhibited nitrification (acetylene) or stimulated nitrification (addition of ammonium), whereas AOA *amoA* genes did not, implying that the archaea were inactive in ammonia oxidation. The second piece of evidence came from using $^{13}CO_2$ in "stable isotope probing" (SIP) assays. Being chemoautotrophs, the active ammonia oxidizers incorporate $^{13}CO_2$ into DNA which can be separated by density gradient centrifugation from unlabeled, "light" DNA in organisms not active in assimilating the added $^{13}CO_2$. Jia and Conrad found that the ^{13}C-rich DNA contained AOB *amoA* genes but not those from AOA, suggesting that ammonia-oxidizing bacteria incorporated CO_2 whereas ammonia-oxidizing archaea did not. The two lines of evidence provided by Jia and Conrad (2009) argue strongly that bacteria dominate ammonia oxidation in these soils.

The abundance of both of AOA and AOB is low relative to total prokaryotic abundance in nearly all environments (<1% of total abundance), for reasons discussed below. However, the abundance of AOA, as a fraction of total prokaryotic abundance, may be quite high in the deep ocean. In this vast habitat, fluorescence in situ hybridization (FISH) studies showed that crenarchaea are abundant and account for about half of all microbes, the rest being bacteria, as discussed in Chapter 9. Subsequent work with QPCR confirmed the FISH results (Fig. 12.7a). Two lines of evidence point to most, if not all, of the crenarchaea being ammonia oxidizers.

First, QPCR studies found that the abundance of crenarchaeal *amoA* genes is high in deep waters, relative to archaeal 16S rRNA genes (Fig. 12.7b). The ratio of crenarchaeal *amoA* genes to crenarchaeal 16S rRNA genes is about one or higher, suggesting that most of the crenarchaea have *amoA* genes and carry out ammonia oxidation. A ratio of one is expected based on the number of *amoA* and 16S rRNA genes found in cultivated bacterial ammonia oxidizers and the few crenarchaeal ammonia oxidizers isolated to-date. Studies of other environments, including sediments, soils, and shallow waters, also found high ratios. The ratio data from natural environments leave some room for doubt because the ratio is calculated from two independent QPCR assays of the entire assemblage. Still, the data provide a strong argument for crenarchaea being ammonia oxidizers.

The second line of evidence for crenarchaeal chemoautotrophy in the deep ocean came from a study that examined the natural abundance of ^{14}C in unique archaeal lipids in the Pacific Gyre (Hansman et al., 2009). The investigators had to filter over 200 000 liters of water for their analyses, using a unique shore-based facility on the Big Island of Hawaii where the Mauna Loa Observatory is also located (Chapter 1). It was known that organic material at these depths is relatively young and thus has high amounts of ^{14}C (produced by cosmic bombardment of nitrogen in the upper atmosphere), whereas inorganic carbon in deep waters is much older and has much lower ^{14}C levels. The archaeal lipids also had low ^{14}C levels, sug-

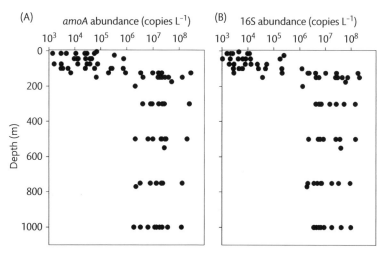

Figure 12.7 Abundance of the ammonia-oxidization gene, *amoA* (panel A), and 16S rRNA gene (panel B) for crenarchaea in the North Pacific Ocean. Data from Church et al. (2010).

gesting that the carbon came from inorganic pools. Since photosynthesis is not possible at these depths, CO_2 must have been fixed by a chemoautotrophic process, most likely ammonia oxidation. The ^{14}C data suggest that >80% of the carbon in archaeal lipids came from chemoautotrophic fixation.

While crenarchaeal *amoA* and 16S rRNA genes were being counted in the oceans and other environments, irrefutable evidence of ammonia oxidation by archaea was still missing. It finally came from a saltwater aquarium (Könneke et al., 2005). From there, an ammonia-oxidizing crenarchaeon was isolated and its metabolism experimentally confirmed. Even though its source was far from the oceans and other natural environments, the *amoA* gene from this crenarchaeon, subsequently called *Nitrosopumilus maritimus*, was very similar to *amoA* genes in the oceans and elsewhere, providing a strong link between the laboratory experiments showing ammonia oxidation and the environmental gene studies. Genomic analysis of *N. maritimus* and another archaeon, *Cenarchaeum symbiosum*, have provided valuable information about the role of archaea in ammonia oxidation (Hallam et al., 2006, Walker et al., 2010).

The distribution of *amoA* genes in various environments suggests that AOA and AOB respond differently to various environmental factors. One important factor is ammonium concentrations. Some evidence suggests that AOA have a higher affinity (lower K_m) for ammonium than do AOB (Martens-Habbena et al., 2009). This would explain the high numbers of AOA in the deep ocean and their response to ammonium additions in soils (Schleper, 2010). Archaeal and bacterial ammonia oxidizers appear to occupy different niches in nature.

Controls of aerobic ammonia oxidation
Equation 12.7 gives hints about three factors limiting rates of ammonia oxidation in natural environments. The low energetic yield of ammonia oxidation is one reason why these organisms are not more abundant. As with other chemolithotrophs, the low energetic yield of ammonia oxidation explains why cell yields of these organisms (Fig. 12.8) and thus abundances are low. The low energetic yield also explains why ammonia oxidizers cannot compete with algae, heterotrophic bacteria, and higher plants in soils for ammonium, although there may be exceptions (Inselsbacher et al., 2010). Exacerbating the competition problem is the low concentration of ammonium in most oxic environments, another impor-

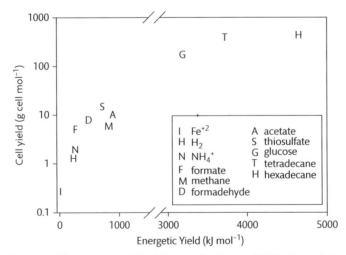

Figure 12.8 Cell yield as a function of the energetic yield of the growth substrate (Gibb's change in free energy). The energetic yield of ammonia oxidation and other chemolithotrophic reactions is low. Data from Bongers (1970), Candy et al. (2009), Farmer and Jones (1976), Goldberg et al. (1976), Jetten (2001), Kelly (1999), and Winkelmann et al. (2009).

tant factor limiting oxidation rates. Considering that the deep ocean has very low concentrations, it may be surprising that crenarchaeal ammonia oxidizers are so abundant. Although they are abundant relative to total prokaryotic abundance, their absolute abundance is low, only about 10^4 cells ml^{-1}, orders of magnitude less than total prokaryotes in surface waters and in soils. Ammonia oxidizers are relatively abundant because all other energy sources, such as organic material, are also very sparse in the deep ocean.

Oxygen is another factor affecting aerobic ammonia oxidation. The lack of oxygen prevents nitrification from occurring in sediments below the oxic surface layer and in waterlogged soils. Conditions for ammonia oxidizers are best at the interface between oxic environments with high oxygen concentrations and anoxic environments with high ammonium concentrations. The well-studied ammonia oxidizer *Nitrosomonas europaea* was found to be abundant at such an interface in a stratified lake (Voytek and Ward, 1995).

In addition to ammonium and oxygen concentrations, ammonia oxidation is influenced by two other physical properties of the environment. Inhibition by light is thought to be one reason why ammonia oxidation is not high in the upper surface layer of aquatic habitats, although ammonia-oxidizing bacteria and ammonia oxidation do occur in the euphotic zone. It is well known that extremes in pH select against ammonia-oxidizing bacteria, but more worrisome is the sensitivity of these organisms to ocean acidification (Chapter 4) (Beman et al., 2011). This part of the nitrogen cycle is especially sensitive to pH. Acidification forces a proton onto ammonia, producing ammonium (NH_4^+) and resulting in lower concentrations of ammonia (NH_3), the actual substrate used by ammonia oxidizers. Lower ammonia concentrations could mean slower rates of ammonia oxidation in the oceans even if the sum of ammonia and ammonium stays constant. This is one example of many complex effects on the biosphere caused by increasing greenhouse gases.

Nitrite oxidation and the second step in nitrification

Ammonia oxidation produces nitrite, yet the ultimate end product of nitrification is nitrate, so a second step is

necessary. It is carried out by a separate group of microbes that oxidize nitrite to nitrate. Much less is known about the ecology of nitrite oxidation, apart from the complete nitrification process. Arguably we do not need to know much, because usually ammonia oxidation is thought to be the rate-limiting step in nitrification, as mentioned before. Once ammonia is oxidized to nitrite, the next step, nitrite oxidation, appears to precede quickly most of the time, as nitrite rarely builds up in the environment. Still, the reaction is a critical and essential component of nitrification. We don't know, for example, whether archaea are involved or whether only bacteria oxidize nitrite.

We know the most about nitrite oxidation by cultivated bacteria. Of the cultivated nitrite oxidizers, four genera with diverse evolutionary pathways have been examined. These include the alphaproteobacterial genus *Nitrobacter* which has aquatic and soil species that are facultative nitrite oxidizers, in contrast to all known ammonia oxidizers being obligate. In addition to growing chemolithoautotrophically, *Nitrobacter* can grow heterotrophically on simple organic compounds. *Nitrococcus*, *Nitrospina*, and *Nitrospira*, in contrast, are all obligate nitrite oxidizers. *Nitrococcus* and *Nitrospina* are in the Gamma- and Deltaproteobacteria, respectively while *Nitrospira* makes up its own phylum.

Anaerobic ammonia oxidation

In 1965, the chemical oceanographer F.A. Richards pointed out the lack of ammonium accumulation in oxygen minimum zones of the oceans and speculated that ammonium was being oxidized with nitrate as the electron acceptor (Strous and Jetten, 2004). Over 10 years later, a microbiologist suggested the reaction:

$$NH_4^+ + NO_2^- \rightarrow N_2 + 2H_2O \qquad (12.9)$$

but it took about another 20 years before experimental evidence was found in a wastewater reactor for what is now called anaerobic ammonia oxidization (anammox). The stoichiometry of the reactants and products indicated that nitrite rather than nitrate was the electron acceptor in anammox (Equation 12.9). After the initial work with wastewater, anammox has been found in natural anoxic aquatic environments, and 16S rRNA genes

Table 12.2 Comparison of aerobic and anaerobic ammonia oxidation. Aerobic ammonia oxidation is carried out by both bacteria and archaea while only bacteria in the Planctomycetales are capable of anaerobic ammonia oxidation. Data from Jetten (2001).

Property	Aerobic	Anaerobic
Energy yield (kJ mol^{-1})	272	357
Oxidation rate (nmol mg^{-1} min^{-1})	400	60
Generation time (day)	1	10
Organisms	Several	Planctomycetes
Carbon source	CO_2	CO_2
End product	Nitrite	N_2 gas
Ecosystem role	Fuels denitrification	Removes fixed nitrogen

very similar to anammox bacterial 16S rRNA genes have been found in soils (Humbert et al., 2010). While aerobic ammonia oxidation and anammox both act on ammonia, they are very different processes involving very different organisms (Table 12.2).

Unlike aerobic ammonia oxidation, anammox is carried out by a very limited group of bacteria belonging to the phylum Planctomycetes (Jetten et al., 2009). So far, the known anammox bacteria belong to five genera within the order Brocadiales. No anammox has been grown in pure culture to-date, but much has been learned from enrichments dominated by anammox bacteria. A metagenomic approach applied to an enrichment culture was used to deduce the genome of the wastewater anammox bacterium, *Candidatus* Kuenenia stuttgartiensis (Strous et al., 2006). Among several unusual features, these bacteria have an intracellular compartment, the "anammoxosome", where ammonia oxidation takes place. The membranes of anammox bacteria contain an unusual lipid, ladderane, thought to be important in compartmentalizing a potent intermediate, hydrazine (N_2H_2), formed during ammonia oxidation. Hydrazine is in rocket fuel and is highly unstable. Anammox bacteria grow very slowly with generation times exceeding 10 days even under optimal conditions.

A key question concerns the extent of anaerobic ammonia oxidation and thus the release of N_2 gas and a potentially limiting element from ecosystems. The answer is that anammox often accounts for a high fraction of total N_2 production. The harder and more interesting problem is to compare release by anammox with the traditional mechanism of N_2 production,

denitrification. Before we address that problem, we need to learn more about dissimilatory nitrate reduction and other aspects of denitrification.

Dissimilatory nitrate reduction and denitrification

Denitrification ultimately produces nitrogen gas or nitrous oxide through a series of redox reactions. One equation describing denitrification is:

$$5\,\text{glucose} + 24NO_3^- + 24H^+ \rightarrow 30CO_2$$
$$+12N_2 + 42H_2O \; \Delta G^{\circ\prime} = -2657 \text{ kJ mol}^{-1} \quad (12.10).$$

The equation has glucose, but in fact many other organic compounds can serve as carbon and electron (energy) sources for this anaerobic heterotrophic reaction. The process requires four enzymes, starting with nitrate reductase encoded by the *nar* genes (Fig. 12.9). All of these redox-mediating enzymes require iron. In addition, nitrate reductase has a molybdenum co-factor, and both nitrite reductase (*nir*) and nitrous oxide reductase (*nos*) contain copper. Studies using cultivation-independent methods have examined the genes *nirS*, *nirK*, and *nosK* to explore questions about the potential for denitrification and the diversity of denitrifiers in natural environments (Thamdrup and Dalsgaard, 2008). Not surprisingly, the microbes carrying out denitrification in nature are quite different from those studied in laboratory cultures.

Denitrification starts with the dissimilatory reduction of nitrate to nitrite, a reaction carried out by many prokaryotes and even some eukaryotic microbes. Some fungi may contribute substantially to the process in some soils (Hayatsu et al., 2008). Even some protists such as a

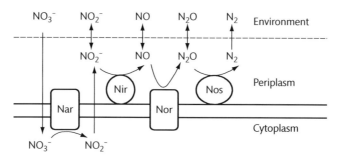

Figure 12.9 Pathway for denitrification. The enzymes are nitrate reductase (Nar), nitrite reductase (Nir), nitric oxide reductase (Nor), and nitrous oxide reductase (Nos), encoded by *nar, nir, nor,* and *nos,* respectively. Modified from Ye et al. (1994) and Zumft (1997). A subunit of nitrite reductase (*nirS*) is often examined to explore the potential for denitrification in natural environments.

marine foraminifera and diatoms can do it (Risgaard-Petersen et al., 2006, Kamp et al., 2011). Many of these microbes are facultative anaerobes with the capacity to switch to oxygen when concentrations are favorable. Once produced by dissimilatory nitrate reduction, nitrite is usually reduced further to N_2 gas, nitrous oxide, or ammonium. However, the process can stop at nitrite, which is excreted, resulting in a smaller energetic yield compared to reduction to N_2 gas: $\Delta G^{o'}$ = -2657 kJ/mol glucose for N_2 gas versus $\Delta G^{o'}$ = -1926 kJ/mol glucose for nitrite (Buckel, 1999). Some chemolithotrophs reduce nitrate while oxidizing hydrogen sulfide (a process called chemoautotrophic denitrification or lithotrophic denitrification), but heterotrophic organisms that use nitrate as a terminal electron acceptor and oxidize organic material are much more important in denitrification.

The environmental factors affecting denitrification include organic carbon and nitrate concentrations as discussed in Chapter 11. The obvious dependence on nitrate means that denitrification depends on nitrification. These two processes are often said to be "coupled", here meaning that the end product of one reaction, nitrate from nitrification, is used by another, denitrification. But these processes have to be separated in time or more commonly in space because nitrification is an aerobic process while denitrification occurs mainly in anoxic environments. A classic example occurs in sediments. In these environments, nitrification in the top, oxic layer supplies nitrate used by denitrifying microbes in lower, anoxic sediment layers. However, there are exceptions to the general rule that denitrification occurs only in anoxic

environments. Oxic (unsaturated) soils may have some denitrification activity because of anoxic microhabitats. More interestingly, some denitrification can occur when oxygen concentrations are low but still measurable (Fig. 12.10). It seems that the small advantage in energy yield in using oxygen over nitrate (Chapter 11) is not enough to prevent dissimilatory nitrate reduction when oxygen concentrations are low.

Dissimilatory nitrate reduction to ammonium

In addition to nitrogen gas and nitrous oxide, nitrate reduction can also produce ammonium by a process called dissimilatory nitrate reduction to ammonium (DNRA). Since bacteria in laboratory cultures do not gain energy from this reaction, other explanations have been suggested over the years for why bacteria carry out DNRA. Some environments may select for DNRA, as suggested by the stoichiometry of the reaction:

$$\text{glucose} + 3NO_3^- + 6H^+ \rightarrow 6CO_2 + 3NH_4^+ \\ + 3H_2O \ \Delta G^{o'} = -1767 \text{ kJ mol}^{-1} \quad (12.11).$$

Even assuming bacteria can derive energy from this reaction, it yields substantially less than reduction to N_2, putting DNRA bacteria at a disadvantage when competing for organic substrates and nitrate. However, note that the C:N ratio for the organic carbon and nitrate required by DNRA is 2:1 (Equation 12.11) whereas it is 1.25:1 for bacteria denitrifying nitrate to N_2 gas (Equation 12.10). That is, DNRA requires less nitrate than denitrification, leading to

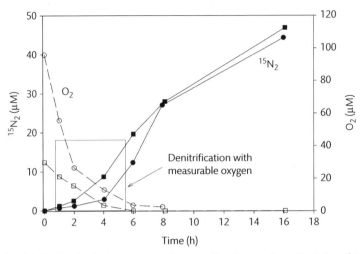

Figure 12.10 An example of denitrification in the presence of oxygen. The data are from the 0–2 cm (circles) and 2–4 cm (squares) layers in a sand flat of the German Wadden Sea. Denitrification was measured by following $^{15}NO_3^-$ into N_2 (solid symbols) while oxygen (open symbols) was measured by micro-electrodes. Data from Gao et al. (2010).

the hypothesis that DNRA is advantageous in organic-rich but nitrate-poor habitats. Evidence from studies in natural environments supports this hypothesis (Thamdrup and Dalsgaard, 2008). In some environments, more nitrate goes through DNRA than through denitrification (Koop-Jakobsen and Giblin, 2010).

Denitrification versus anaerobic ammonia oxidation

After anammox was discovered, an obvious question was about its importance in producing N_2 compared with traditional denitrification. The short answer is, it varies. More interesting is the question why the contribution of anammox and denitrification to N_2 production differs among various ecosystems. An important factor explaining this variation in marine sediments is the supply of organic material and its mineralization to ammonium.

The contribution of anammox to total N_2 production in marine sediments increases with increasing water depth (Fig. 12.11). The mechanism giving rise to this correlation probably involves the amount of organic material reaching the ocean floor. Denitrifying bacteria are favored in shallow sediments receiving high inputs of organic material. In these sediments, any ammonium

produced in the surface layer is oxidized by aerobic ammonia oxidizers and never reaches anammox bacteria deeper in the sediments. The nitrate produced by aerobic ammonia oxidizers would also favor denitrification. By contrast, in deep water columns, much of the organic material produced by primary producers in the surface layer is decomposed and worked over before reaching the sediments. These conditions are not favorable for denitrification and could support relatively more anammox activity. Also, denitrifying bacteria are apparently selected against in Mn-rich sediments where Mn can serve as a terminal electron acceptor in place of nitrate.

The supply of organic material may also explain why anammox is important in some sub-oxic marine waters and not others. Denitrification accounts for much of N_2 production in the oxygen minimum zone of the highly productive Arabian Sea but anammox is more important in the less organic material-rich waters of the Eastern Tropical South Pacific oxygen minimum zone (Ward et al., 2009). Other factors are the growth rates and corresponding response times of these microbes. The slow growth of anammox bacteria would limit their response and capacity to use episodic pulses of ammonium from organic material degradation in the water column, in contrast to the potential for rapid growth and response of denitrifying bacteria.

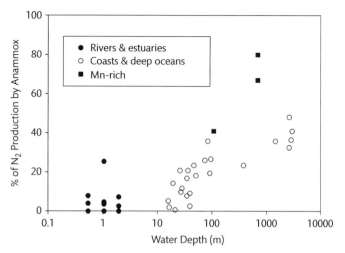

Figure 12.11 Contribution of anaerobic ammonia oxidation (anammox) to N$_2$ production in sediments as a function of water depth. Data from Thamdrup and Dalsgaard (2008).

But denitrification and anammox are not really competing against each, given the great differences between the two processes. In fact, anammox bacteria most likely depend on nitrate-reducing bacteria, if they produce nitrite or ammonium via DNRA. The supply of nitrite from aerobic ammonia oxidation is likely to be low because nitrite would have to diffuse from oxic habitats to anoxic ones. More work and data are needed to understand the relative importance of denitrification versus anammox.

Sources and sinks of nitrous oxide

Nitrogen gas is by far the main gas produced by denitrification and the only one produced by anammox, but another by-product of denitrification, nitrous oxide (N$_2$O), deserves a closer look because it is a potent greenhouse gas as well as being a route by which nitrogen exits ecosystems. Terrestrial ecosystems account for about twice as much to natural N$_2$O emissions as the oceans, and both are considered to be net producers of the gas (Gruber and Galloway, 2008). Denitrifying bacteria are possible sources of N$_2$O in both terrestrial and aquatic habitats, but more work has focused on N$_2$O production by nitrifying bacteria. N$_2$O is consumed by denitrifying bacteria in the oceans and soils. In the oceans and soils, nitrification is probably the main source of N$_2$O, although the question is still debated (Opdyke et al., 2009, Yu et al., 2010a). The "isotopologues" of N$_2$O (^{15}N-N-O versus

N-^{15}N-O) have been examined to explore sources of this greenhouse gas.

Nitrification can produce N$_2$O by two quite different mechanisms (Fig. 12.12), both involving ammonia oxidizers (Stein and Yung, 2003). The first is the production of N$_2$O as a by-product of hydroxylamine (NH$_2$OH) oxidation along the NH$_3 \rightarrow$ NO$_2^-$ pathway mediated by ammonia-oxidizing microbes. The second pathway, which is thought to be more important, is reduction of NO$_2^-$ to N$_2$O again by ammonia-oxidizing bacteria. Several terms have been used to describe this second pathway, including nitrifier denitrification, lithotrophic denitrification, and aerobic denitrification. While the production of N$_2$O by ammonia oxidizers results in nitrogen being lost from the system and thus warrants the "denitrification" label, the process differs in many ways from the denitrification pathway that begins with dissimilatory nitrate reduction and organic carbon oxidation. Of all the terms, perhaps most confusing is "aerobic denitrification" to describe the NO$_2^-$ to N$_2$O pathway. This term has also been used to describe N$_2$ production by heterotrophic denitrification in the presence of measurable oxygen.

One piece of evidence for the importance of the NO$_2^- \rightarrow$ N$_2$O pathway is the effect of oxygen. When oxygen concentrations are high (near saturation) N$_2$O production by ammonia-oxidizing bacteria is minimal, arguing against the hydroxylamine pathway. As concentrations decrease, N$_2$O production by nitrifying bacteria

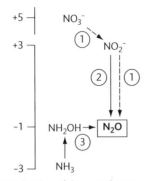

Figure 12.12 Production of the greenhouse gas nitrous oxide (N_2O). Pathway 1 is anaerobic and starts with the dissimilatory reduction of nitrate. Pathways 2 and 3 are carried out by ammonia-oxidizing bacteria. The numbers indicate the oxidation state of nitrogen in the five nitrogenous compounds. Based on Stein and Yung (2003).

increases such that production rates are about twenty-fold higher at 1% of saturated oxygen concentrations than at 100% oxygen (Goreau et al., 1980). Presumably, ammonia oxidizers switch to NO_2^- as an electron acceptor when oxygen concentrations are low. This mechanism helps to explain variation in N_2O production in the oceans and is part of the explanation in soils as well.

In terrestrial environments, N_2O production is also affected by temperature, pH, and soil moisture. During droughts, N_2O concentrations in the soil surface layer are below atmospheric levels, which leads to net N_2O diffusion into soils and net N_2O degradation (Billings, 2008). Soils wetted again return to being net producers of N_2O. Too much water, however, reduces diffusion of oxygen into soils and promotes N_2O consumption by denitrifying bacteria (Chapuis-Lardy et al., 2007). These bacteria use N_2O as an electron acceptor especially when nitrate concentrations are low.

Balancing N loss and N_2 fixation

Biogeochemists take the rate measurements from microbial ecologists, combine them with other data from many sources, and attempt to compile a budget and residence times for nitrogen in terrestrial and oceanic ecosystems (Table 12.3). The residence time of nitrogen is calculated by dividing the amount of nitrogen in a system by the sum of the fluxes, either total

sources or total sinks. For the oceans, the residence time of nitrogen is roughly 2000 years while it is only 500 years in terrestrial systems (Gruber and Galloway, 2008). Both are much shorter than the >10 000 years for phosphorus, leading geochemists to argue that phosphorus, not nitrogen, is the element ultimately limiting biological production on the planet. Such a conclusion seems inconsistent with the many experiments showing limitation by nitrogen or some other element, such as iron. Part of the answer is that all of these elements have a role in controlling growth but on different timescales, ranging from days for elements like nitrogen to geological timescales, which may be the case for phosphorus.

Budgets are also used to examine whether nitrogen inputs and outputs are in balance (Canfield et al., 2010). The fluxes for oceans and terrestrial systems are similar, except for N_2 loss by denitrification and anammox being higher in the oceans than on land. Half of all the denitrification in terrestrial ecosystems actually occurs in freshwaters, but much of the nitrogen is supplied by run off from land (Gruber and Galloway, 2008). Not evident from the table is the larger impact of anthropogenic processes on terrestrial habitats. Table 12.3 also suggests that input and output of nitrogen are roughly in balance for both the oceans and terrestrial systems. However, using different data sets and much higher estimates of N_2 release by denitrification and anammox, marine biogeochemists have argued that outputs substantially exceed the inputs. More data are needed.

Even if nitrogen inputs and outputs are in fact in balance today, they may not have been over geological times, as suggested by the well-documented changes in atmospheric carbon dioxide and biological productivity over millennia. When outputs exceed inputs, as argued for by marine biogeochemists, or when the opposite is true as probably was the case in the past, what processes push the system back into balance? The answer is complicated. According to some biogeochemists, the answer involves the coming and going of glaciers that change the exposure of continental shelves to oxygen, resulting in changes in N_2 production. Iron and phosphorus have roles to play also, and the ultimate answer is a coupling of all of the biogeochemical cycles. This is the stuff of biogeochemistry, but it is built on the microbial processes discussed in this chapter and the rest of the book.

Table 12.3 Summary of N sources and sinks in terrestrial environments and the oceans. "Atmosphere" refers to the deposition of nitrogenous compounds from the atmosphere. Rivers carry dissolved and particulate nitrogen compounds from land (sink) to the oceans (source), so the river flux is not applicable (NA) as a source for land or as a sink for the oceans. Burial is the loss of nitrogen to sediments over geological timescales. Data from Gruber and Galloway (2008).

| | Process | Fluxes (10^{12} g N per year) | |
		Ocean	Terrestrial
Sources			
	Industrial N_2 fixation	0	100
	Natural N_2 fixation	140	110
	Atmosphere	50	25
	River fluxes	80	NA
	Subtotal	270	235
Sinks	N_2 release	240	115
	N_2O release	4	12
	River fluxes	NA	80
	Burial	25	?
	Subtotal	269	207
	Balance (Sources-Sinks)	+1	+28

Summary

1. Nitrogen fixation is carried out by a select but diverse group of prokaryotes. Many of these prokaryotes live symbiotically with eukaryotic microbes and higher plants.

2. Ammonia oxidation, the rate-limiting step of nitrification (the "making" of nitrate), is carried out mainly by Betaproteobacteria and Crenarchaeota. The archaeal group is abundant in the deep oceans and seems to consist mostly of ammonia oxidizers.

3. In contrast to aerobic ammonia oxidation, anaerobic ammonia oxidation (anammox) produces N_2 and is carried out only by bacteria in the phylum Planctomycetes. Anammox can release more N_2 than denitrification does in some environments.

4. Denitrification starts with dissimilatory nitrate reduction, a type of anaerobic respiration, and can produce nitrous oxide or N_2. Dissimilatory nitrate reduction is carried out by a wide variety of bacteria, archaea, and even some eukaryotic microbes. It can produce nitrite or ammonium, in addition to nitrous oxide or nitrogen gas.

5. The greenhouse gas nitrous oxide (N_2O) is produced by nitrification from reduction of nitrite in both soil and marine ecosystems. In both systems it is consumed by denitrifiers.

6. There is some debate about whether the nitrogen cycle is in balance and whether rates of nitrogen fixation are equal to rates of nitrogen loss due to anammox and denitrification. The answer affects global rates of primary production and the carbon cycle.

CHAPTER 13

Introduction to geomicrobiology

Previous chapters have mentioned a few examples of the impact of microbes on the physical world of our planet. Formed by the marriage between geology and microbiology, the field of geomicrobiology is even more focused on those impacts. This chapter will highlight processes important in thinking about how microbes shape the geology of our planet. Some of these processes involve microbe-rock and microbe-mineral interactions—processes between microbes and harder stuff than things most common in the soils and water discussed so far. Many of these processes and interactions are especially important in subsurface environments beneath organic-rich sediments and soils. These subsurface environments are huge and unworldly, home to unique microbes mediating important geochemical reactions, reasons why geomicrobiology has become prominent in the last few years, although the term has been around for several decades. Geomicrobiology is in the newspapers these days because of discoveries in exotic habitats like caves and gold mines (Chivian et al., 2008). But the processes to be discussed in this chapter occur in many habitats in the biosphere.

Geomicrobiologists face the challenge of meshing vast scales of time and space over which microbes interact with geology. We have already seen that environmental conditions of microhabitats have ramifications for global phenomena. The net production or consumption of greenhouse gases such as methane and nitrous oxide, for example, depends on oxygen concentrations in the micron-sized space surrounding bacteria, archaea, and fungi. In geomicrobiology, the range of timescales is especially evident. Microbes important in geomicrobiological processes often grow slowly and their impact

seems minute at any particular time. But these impacts have huge consequences when the process continues for millennia or longer. Oxygen-evolving cyanobacteria first appeared around 2.7 billion years ago, but it took another 300 million years or so before oxygen became prominent in the atmosphere (Kump, 2008). Lots of small things occurring over a long time add up to big consequences.

Cell surface charge, metal sorption, and microbial attachment

Interactions between microbes and metals, minerals, and rocks are important topics in geomicrobiology. An important property governing these interactions is surface charge. The precise value of that charge depends on the microbe, growth conditions, and the environment, but generally it is negative, because exposed carboxyl and phosphate groups at the cell surface are without protons ("deprotonated") at normal pH values. The net negative charge of the cell surface attracts positively charged atoms and compounds, giving rise to a gradient of charge between the cell and the surrounding environment. The classic model for this gradient, referred to as the "electric double layer," consists of an inner layer of positively charged ions (the Stern layer) tightly held next to the negative cell surface, followed by a diffuse layer of counter-ions (the Guoy layer) (Fig. 13.1).

The net charge, the zeta potential, of a cell or any particle can be examined by monitoring its movement in an electric field. Being negatively charged, most cells move towards the positively charged electrode (the cathode),

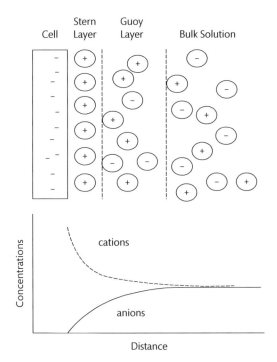

Figure 13.1 A model of a cell surface surrounded by an aqueous solution of various ions. The main feature of the model is the double layer of ions that build up on microbial cell surfaces.

depending on pH, as well as on the microbe and growth conditions. The pH can be manipulated experimentally to determine the value at which the cell does not move either towards the cathode or anode. This pH is defined as the microbe's isoelectric point, meaning that any negative charges of deprotonated carboxyl and phosphoryl groups are neutralized by positive charges of exposed amino groups, leaving the cell with zero net charge. The isoelectric point of most microbes is between pH 2 and 4. When the environment is above this pH, cells are negatively charged. An analogous concept is the point of zero charge (pzc), which is the same as the isoelectric point but often used to describe mineral surfaces.

It is rather easy to measure the isoelectric point of a microbe and to identify the major biochemical groups (carboxyl, phosphoryl, and amino) that contribute to cell charge, but it is much more difficult to understand precisely why a microbe has a particular zeta potential and isoelectric point. There is a correlation between nitrogen and phosphorus content and a cell's electrostatic charge

due to the contribution of phosphoryl groups to anionic surface charge and of amino groups to cationic surface charge (Konhauser, 2007). However, more detailed modeling is difficult, in part due to the variation in pK_a of functional groups. This variation can arise due to seemingly subtle conformational changes in cell surface components.

Metal sorption
Regardless of the precise biochemical underpinning, microbes provide extensive reactive, mostly negatively charged surfaces onto which metals and other positively charged atoms and compounds potentially sorb. The surface area of microbes is larger than that of other particles in aquatic habitats and is substantial even in soils and sediments. Consequently, microbe-metal interactions are important in thinking about the environmental fate of metals, including toxic ones, in addition to understanding the impact of metals on microbes.

Metals can passively sorb onto microbes because of electrostatic attraction. Being positively charged, metals sorb onto negatively charged carboxyl and phosphoryl groups of cell walls, membranes, and extracellular material. The precise identity and number of sorption sites vary with the microbe and growth conditions (Ledin, 2000). In addition to cell surface properties, the amount of sorption also varies with the metal, dissolved metal concentrations, and environmental pH. Several models exist for describing how sorption varies as a function of concentration. The simplest is a first-order dependence of sorption (M_B, the amount of adsorbed metal) with dissolved metal concentration (M_D):

$$M_B = K_d \cdot M_D \qquad (13.1)$$

where K_d is the distribution coefficient for that metal. One problem of many with this simple model is that it implies sorption always increases with increasing concentrations where in fact it must level off at some point. One model including this leveling off is built on the assumption that a surface has a finite number of reactive sites where sorption can occur. Once those sites are filled, sorption stops. These simple assumptions lead to the Langmuir equation:

$$q_e = q_{max} \cdot K \cdot C_e / (1 + K \cdot C_e) \qquad (13.2)$$

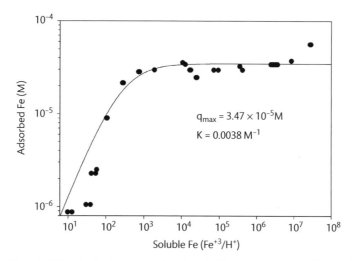

Figure 13.2 Adsorption of iron, Fe(III), onto the bacterium, *Bacillus subtilis* over a pH range of 2–4.5. The curve was found by fitting the data to the Langmuir equation given in the text. Data from Warren and Ferris (1998).

where q_e is the sorption at a particular equilibrium dissolved concentration (C_e), q_{max} is the maximum sorption and K is the Langmuir's equilibrium parameter. As illustrated in Figure 13.2, variation in sorption as a function of concentration is reminiscent of the uptake-concentration relationship modeled by the Michaelis-Menten equation (Chapter 4). Other models for sorption include Freundich and Brunauer–Emmett–Teller (BET) isotherms (Ledin, 2000).

"Passive" describes the mechanism of metal sorption, but the word may be misleading in thinking about metal-microbe interactions. Microbes can modify functional groups at cell surfaces and thus indirectly control sorption to some extent, with several possible rewards. The binding of some metals can help to stabilize cell walls and membranes by neutralizing otherwise destabilizing interactions between anionic wall and membrane components. Sorption to cell surfaces helps in securing some metals, such as iron and copper, needed for biosynthesis, while it may lessen toxic effects of other metals present in high concentrations. Toxic effects may be particularly minimized by sorption to extracellular polymers. Bacteria with capsules (Chapter 2) are able to withstand high metal concentrations better than capsule-less mutants, and the capacity to form capsules is sometimes lost in cultures without heavy metals (Ledin, 2000).

Box 13.1 Heavy metal removal

Because metals such as cadmium, mercury, and lead are toxic to wildlife and humans, there is much interest in removing these dissolved contaminants from waste water before release to natural environments. One approach is to use microbes. The contaminated water is passed over packed beds of microbes onto which the metals sorb. The beds, now loaded with heavy metals, can be discarded or the metals can be desorbed and the microbes re-used. The microbes are usually bacteria, but genetically engineered yeast have also been used (Kuroda and Ueda, 2010, Nisbet and Sleep, 2001). In addition to toxic metals, the same principle can be used to extract valuable metals, such as gold, silver, and platinum.

Iron uptake mediated by siderophores and other metal ligands

In iron-rich environments, some bacteria can become highly encrusted with iron via passive, non-specific sorption onto cell surfaces and extracellular polymers. By contrast, in iron-poor environments, the largest being the open oceans, microbes face the problem of obtaining enough iron for the respiratory chain, photosynthesis, and other redox reactions, such as nitrate reduction and nitrogen fixation (Chapter 12). A further complication is that iron in oxic environments exists as Fe(III) which is insoluble at pHs above 5. Many bacteria and fungi solve this problem by releasing low molecular weight compounds (molecular weight less than 1000 Da) that bind specifically to Fe(III). Metal-binding compounds are called "ligands" or "chelators", and those specific for iron are called "siderophores".

There are over 500 siderophores with a variety of chemical structures, produced mostly by bacteria (Wandersman and Delepelaire, 2004). Siderophores can be divided into three classes based on their iron-binding properties: hydroxamates, catecholates, and alpha-hydroxycarboxylates. Compounds in all three classes have an extremely high affinity for iron; conditional stability constants for Fe(III) can exceed 10^{30} M^{-1}. Once bound to these low molecular weight ligands, Fe(III) is in a soluble form, while it is not when present as inorganic ferric oxides. The high affinity means that siderophores generally outcompete low-affinity ligands for iron but it creates a problem for cells trying to access the iron chelated to a siderophore. Microbes have come up with several solutions to this problem (Fig. 13.3). Much of what is known about iron-microbe interactions comes from studies of pathogenic bacteria (Ratledge and Dover, 2000). One defense of hosts against pathogens is to restrict the supply of iron, which the pathogen tries to circumvent by producing siderophores.

While some siderophores produced by marine bacteria may be associated with cell membranes (Vraspir and Butler, 2009), most siderophores are thought to be released by microbes into the surrounding environment. The advantages of this strategy may outweigh potential disadvantages for siderophore-producing bacteria and fungi in restricted microhabitats of soils, sediments, and suspended particles in aquatic environments. For the plankton in aquatic habits, the advantages are less clear. The siderophore could be used as a carbon source, or the iron in siderophore-iron complex could be stolen by non-siderophore-producing microbes. Siderophore-producing heterotrophic microbes could provide siderophore-bound iron for use by primary producers, leading to higher production of organic

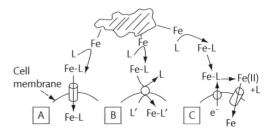

Figure 13.3 Uptake of iron mediated by siderophores (L), which starts with the stripping of iron from an iron-rich source and formation of the iron-siderophore complex (Fe-L). In mechanism A, the iron-siderophore complex (Fe-L) iron is taken up directly by specific transporters. In mechanism B, the iron-siderophore complex is transported into the cell (not shown here), as in mechanism A, but then the iron is exchanged with another siderophore (L'). In mechanism C, the Fe(III) in the Fe-L complex is reduced to Fe(II), which is then taken up. The siderophore does not have a high affinity for Fe(II). Still another mechanism not shown here is the production of weakly complexed iron by photochemical reactions, followed by uptake as described for mechanism C. Based on Hopkinson and Morel (2009) and Stintzi et al. (2000).

material, benefiting the heterotrophs. Eukaryotic algae and the dominant cyanobacteria in the oceans, *Synechococcus* and *Prochlorococcus*, do not produce their own siderophores and instead indirectly rely on siderophore production by prokaryotes and abiotic mechanisms for obtaining iron.

In addition to siderophores, microbes produce ligands with high affinity for other metals. In some environments, chelators help to minimize the toxic effect of high metal concentrations by lowering substantially the concentration of the uncomplexed, "free" form; free metals are toxic while metals bound to other materials often are not. For example, copper is toxic when concentrations of free Cu^{2+} are high but not when copper is chelated by high affinity ligands. *Synechococcus* can reduce free Cu^{2+} concentrations by one thousandfold to tolerable levels for this cyanobacterium. Concentrations of other metals are so low in some environments that microbes have to use chelators to access these metals. The same chelators used to detoxify copper may also aid in copper uptake when concentrations are low. Another example is cobalt, used in vitamin B$_{12}$ and the carbonic anhydrase of some diatoms (Chapter 4). *Synechococcus* and probably other microbes produce ligands for this trace metal (Saito et al.,

2005), contributing to a distribution in the oceans that is reminiscent of iron and other essential nutrients.

Attachment of microbes to surfaces

The association of microbes with surfaces is important in many aspects of microbial ecology, but perhaps especially so in thinking about topics in geomicrobiology, such as interactions between microbes, minerals, and rocks. The weathering of rocks by microbes (discussed below), for example, starts with attachment to the rock surface. The advantages of being able to attach to a solid are numerous, and include surviving turbulent flow and accessing nutrients in the underlying substrata. The classic description of microbial attachment divides the process into two phases: an initial one consisting of reversible adhesion due to surface charges and hydrophobicity, followed by irreversible attachment due to extracellular polysaccharides and other polymers.

Reversible adhesion is often modeled by theories borrowed from colloid sciences, with the most prominent one being the Derjaguin-Landau-Verwey-Overbeek (DLVO) model (van Loosdrecht et al., 1989). In this model, initial adhesion consists of the balance between attractive van der Waals forces and electrostatic repulsion (Fig. 13.4). The model explains how two negatively charged surfaces, the mineral and the bacterium, can come into contact. The answer is that the two surfaces are separated by a double layer of cations over a very small distance determined by the ionic strength of the aqueous micro-environment surrounding the surface and microbe. According to DVLO, the thickness of the double layer decreases with the square of the ionic strength. Divalent cations such as Mg^{2+} are more effective than monovalent cations even if the ionic strength is equal.

In addition to surface charge, hydrophobicity also plays a role in governing the initial stage of adhesion and cell attachment. A simple index of hydrophobicity is the contact angle formed by a water droplet on the surface or on a lawn of bacterial cells. The higher the angle, the more hydrophobic is the surface. Hydrophobicity varies among bacterial taxa and due to growth conditions. Bacteria with hydrophobic surfaces tend to adhere more readily to hydrophobic surfaces, and likewise for hydrophilic bacteria and hydrophilic surfaces (van Loosdrecht et al., 1987). In addition to its contribution to interactions between microbes and solid surfaces, hydrophobic forces play a role in the degradation of hydrocarbons and other hydrophobic compounds that are insoluble in water and tend to sorb to surfaces (Bastiaens et al., 2000).

Models based on easy measures of surface charge and hydrophobicity often are too simple and fail to capture the complex heterogeneity of cell surfaces. Microbes are more than spheres or cylinders with uniform distributions of surface charge and hydrophobic moieties. Still, the simple principles discussed above are useful starting points in devising more realistic but more complicated models of adhesion of microbes and viruses to surfaces.

Biomineralization by microbes

One potential impact of attached bacteria and other microbes is on the formation of solid-phase minerals from dissolved ions, a process called "biomineralization". Chapter 5 used the term "mineralization" in discussing the degradation of organic material to inorganic compounds

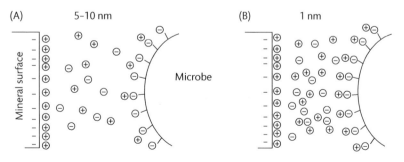

Figure 13.4 Interactions among a mineral surface, ions, and microbial cell, according to the DLVO theory. Panel A depicts a low ionic strength environment while Panel B is for high ionic strength. In both cases, the mineral surface and bacterial cell have net negative charge. Adapted from Konhauser (2007).

such as carbon dioxide, ammonium, and phosphate. The term takes on different meanings in geology and geomicrobiology. It can mean the replacement of organic material with metals and other inorganic material, such as the transformation of a dead organism into a fossil. The term is also used to describe the precipitation of minerals from dissolved constituents. One example is the formation of iron oxide minerals from dissolved Fe^{+2} and colloidal $Fe(III)$ oxides. When microbes are involved in the process, it is called "biomineralization".

A general question facing geomicrobiologists is the extent to which microbes control and affect biomineralization. In cases of "biologically-induced" biomineralization, the microbial role may be quite indirect, affecting mineralization only through the release of metabolic waste products (Dupraz et al., 2009). The microbe may gain no direct advantage from mineral formation, as seems to be the case in the formation of amorphous iron oxide minerals around some bacteria. At the other extreme, "biologically-controlled" biomineralization, microbes regulate the entire process of mineral formation, including nucleation, mineral phase, and location in or around the cell. Two examples were mentioned briefly in Chapter 4; diatoms form the mineral opal ($SiO_2 \cdot nH_2O$) from dissolved silicate to make exquisitely designed cell walls of glass, and coccolithophorids guide the precipitation of calcium carbonate during the construction of their coccolith cell walls.

Carbonate minerals

Whether by biologically-induced or biologically-controlled mineralization, carbonate mineral formation is an important pathway in the carbon cycle. The formation of carbon-rich minerals traps carbon in huge, long-lived pools, in contrast to the fixing of carbon dioxide into organic carbon by primary producers (Fig. 13.5). Carbonate minerals make up the largest pool of carbon in the biosphere, orders of magnitude larger than the carbon in living organisms, detrital organic material including dissolved organic carbon, and atmospheric carbon dioxide. Consequently, the residence time of a carbon atom in carbonate pools is extremely long, on the order of 300 million years, much longer than the 500 to 10 000 year timescale of most organic pool components (Sundquist and Visser, 2004). Artificially promoting carbonate mineral formation and burial is a way to capture carbon dioxide before it is released to the atmosphere. There is much current interest in strategies to enhance carbon sequestration to slow the increase in atmospheric carbon dioxide and to minimize its impact on our climate.

Several microbes and microbial processes indirectly or directly are involved in the formation of calcium carbonate, the most abundant carbonate mineral. These processes affect CO_2 or HCO_3^-, both of which are involved in calcium carbonate formation as described by the equation:

$$Ca^{2+} + 2HCO_3^- \rightarrow CaCO_3 + CO_2 \qquad (13.3).$$

(Note the release of CO_2 by calcium carbonate precipitation. A similar equation explains why manufacturing of concrete contributes to the build-up of atmospheric CO_2 and global warming.) A critical parameter in thinking about carbonate mineral formation is the solubility product (K_{sp}), which is defined as:

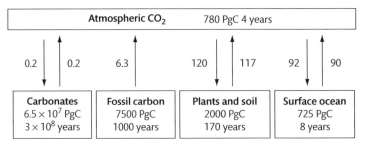

Figure 13.5 Carbon pools and residence times of carbon in carbonates, fossil fuels ("fossil carbon"), terrestrial and oceanic pools. Residence times were calculated by dividing the pool size by the fluxes (average of inputs and outputs). The fossil carbon pool size is the middle of current estimates (5000–10 000 PgC). The surface ocean includes both dissolved inorganic and organic carbon. The residence time for the surface ocean does not include any exchange with the deep ocean. Data from Sundquist and Visser (2004) and Houghton (2007).

$$\left[Ca^{2+}\right]_{eq}\left[CO_3^{2-}\right]_{eq}=K_{sp} \qquad (13.4)$$

where $[Ca^{2+}]_{eq}$ and $[CO_3^{2-}]_{eq}$ indicates the concentrations of Ca^{2+} and CO_3^{2-} at equilibrium with a solid calcium carbonate mineral. The lower the value of K_{sp}, the greater is the tendency for the mineral to form. Solubility constants for common carbonate minerals vary by several orders of magnitude, and even minerals with the same chemical makeup (e.g. aragonite and calcite) can have different constants (Table 13.1). Another, less formal way to express solubility is the maximum concentration which a mineral can be dissolved in water. A sodium salt of carbonate (natron) will stay dissolved even if several kilograms are added to a liter of water, while less than 2 grams of calcium carbonate can be dissolved in a liter.

Two expressions are used to evaluate how close a particular solution is to being saturated with respect to the mineral calcium carbonate. Above saturation, precipitation and mineral formation is possible. Geomicrobiologists use the saturation index (SI) which is

$$SI = \log\left(\left[Ca^{2+}\right]\left[CO_3^{2-}\right]\right)/K_{sp} \qquad (13.5).$$

Experiments have demonstrated that SI has to be greater than about 1, equivalent to $[Ca^{2+}][CO_3^{2-}] = 10$, in order for calcium carbonate to form (Visscher and Stolz, 2005). Chemical oceanographers use a slightly different expression for the degree of saturation (Ω):

$$\begin{aligned}\Omega &=\left[Ca^{2+}\right]\left[CO_3^{2-}\right]/\left[Ca^{2+}\right]_{sat}\left[CO_3^{2-}\right]_{sat} \\ &=\left[Ca^{2+}\right]\left[CO_3^{2-}\right]/K_{sp}\end{aligned} \qquad (13.6)$$

where $[Ca^{2+}]_{sat}$ and $[CO_3^{2-}]_{sat}$ indicate concentrations at which Ca^{2+} and CO_3^{2-} are in equilibrium with $CaCO_3$. The surface ocean is now supersaturated for calcium carbonate, and Ω values for the main forms of calcium carbonate, calcite and aragonite, are about 5.6 and 3.7, respectively (Doney et al., 2009). However, these values have been decreasing over the years due to ocean acidification caused by increasing atmospheric carbon dioxide, which imperils organisms with carbonate shells and walls. Examples of carbonate-containing organisms include coccolithophorids (Chapter 4), other marine protists (foraminifera), and corals. These organisms are involved in biologically-controlled biomineralization.

Several other organisms are involved in biologically-influenced biomineralization of calcium carbonate. Oxygenic and anoxygenic photosynthesizers promote calcium carbonate formation by removing CO_2 and shifting the equilibrium expressed in Equation 13.3 to the right. Of the calcium carbonate photosynthesizers, some of the best studied are cyanobacteria. Some species can form mats made up of layers upon layers of cyanobacteria, other microbes, and calcium carbonate and other inorganic material. Some laminated mats grow to sizes large enough to be called stromatolites; "microbialites" is a more general term to describe mats varying in organization and size. While a few are still living (Fig. 13.6), the heyday of stromatolites was at the beginning of the biosphere when life may have been dominated by these microbial mats (Dupraz et al., 2009). Cyanobacteria in modern-day mats promote calcium carbonate formation

Table 13.1 Some carbonate-containing minerals potentially formed by biomineralization. The solubility constants K_{sp} were calculated, except the measured value for natron (Morse and Mackenzie 1990). The maximum soluble concentration data are from Weast (1987). Above the maximum concentration at 20 °C, the mineral precipitate. See the main text and Konhauser (2007) and Ehrlich and Newman (2009) for information about the microbes potentially involved in forming these minerals.

Mineral	Formula	$-\log(K_{sp})$	Maximum soluble conc (g Liter^{-1})	Microbes
Calcite	$CaCO_3$	8.30	1.4	Cyanobacteria
Aragonite	$CaCO_3$	8.12	1.5	Cyanobacteria
Dolomite	$MgCa(CO_3)_2$	17.1	<1	Sulfate reducers
Magnesite	$MgCO_3$	8.2	10.6	Actinomycetes
Natron	$Na_2CO_3 \cdot 10H_2O$	0.8	7100	Sulfate reducers
Siderite	$FeCO_3$	10.5	6.7	Iron reducers
Rhodochrosite	$MnCO_3$	10.5	6.5	Mn reducers
Strontianite	$SrCO_3$	8.8	1.1	Various

Figure 13.6 Example of modern stromatolites in the Hamelin Pool Marine Nature Reserve, Shark Bay in Western Australia. Each is about 0.5 m in diameter. Photograph by Paul Harrison and used under terms of the GNU Free Documentation License.

by drawing down CO_2 and also indirectly by releasing extracellular polysaccharides which act as nucleation sites for calcium carbonate precipitation.

Other organisms involved in biologically-induced biomineralization of calcium carbonate include sulfate-reducing bacteria and several other anaerobic microbes in anoxic layers of microbial mats. One and a half moles of calcium carbonate should form for every mole of acetate oxidized completely to CO_2 by sulfate reduction (Visscher and Stolz, 2005). Sulfate reducers have been linked to the precipitation of natron ($Na_2CO_3 \cdot 10H_2O$) in the Wadi Natrun of the Libyan Desert (Ehrlich and Newman, 2009). Siderite ($FeCO_3$) formation is thought to result from the activity of sulfate-reducing bacteria in some environments, while Fe(III) reduction is hypothesized to be the main process in others. Some deposits of manganous carbonate have been attributed to microbes, such as *Geobacter metallireducens*, capable of Mn(IV) reduction.

Phosphorus minerals

While carbonate mineral formation has a direct impact on the carbon cycle, phosphorus minerals indirectly affect it by contributing to the control of primary production and other microbial processes. The formation and deposition of calcium phosphate minerals, in particular apatite, is thought to be the largest route by which

phosphorus is removed over geological timescales in marine sediments (Ingall, 2010). Apatite refers to a group of minerals that includes hydroxyapatite, fluorapatite, and chlorapatite, with their respective formulas being $Ca_{10}(PO_4)_6(OH)_2$, $Ca_{10}(PO_4)_6F_2$, and $Ca_{10}(PO_4)_6Cl_2$. Apatite and other calcium phosphate minerals do not form as readily as may be expected from the high concentrations of calcium and phosphate found in sediments. Release of phosphate from decomposing organic material may promote apatite formation, but another mechanism seems necessary.

A recent study provided strong evidence of the role of biologically-induced biomineralization of apatite in the Benguela upwelling system off the coast of Namibia (Goldhammer et al., 2010). These investigators showed that a radioactive form of phosphate ($^{33}PO_4^{2-}$) was incorporated into apatite when sulfide-oxidizing bacteria were present, but not when these bacteria were absent. The reaction was rather quick, with evidence of ^{33}P appearing in apatite within 48 hours. The mechanism involves formation of polyphosphate that either chelates Ca^{2+}, eventually forming apatite, or is hydrolyzed to PO_4^{3-} leading to the precipitation of apatite (Fig. 13.7). Regardless of the precise mechanism, sequestration of phosphate in apatite in anoxic sediments is surprising because usually phosphate is released from sediments under these conditions, at

Box 13.2 Fingerprints of the origin of life

An argument for studying modern microbial mats is to understand ancient ones and to explore questions about the beginning of life during the Precambrian, nearly 4 billion years ago (Gyr). Some of the earliest signs of life have been found in fossil stromatolites such as the 3.5 Gyr Warrawoona Group and the 3.4 Gyr Strelley Pool Chert, both in Australia (Allwood et al., 2006; Konhauser, 2007; Nisbet and Sleep, 2001). These structures consist of 50–200 μm thick wavy bands of chalky carbonate sediments and dark kerogen. Ancient stromatolites have several features reminiscent of modern-day microbial mats. Perhaps most telling, ancient stromatolites have yielded microbe-like fossils ("microfossils") with morphologies similar to those of modern filamentous cyanobacteria or perhaps anoxygenic photosynthetic bacteria such as *Chloroflexus*. Other evidence of microbial life at this time comes from carbon and sulfur isotopes and biomarkers such as 2-methylhopanes indicative of cyanobacteria (Des Marais, 2000). The first stromatolites in the Archean (3.8–2.5 Gyr) were limited to shallow marine evaporitic basins, but they exploded in number and size during the Proterozoic (2.5–0.5 Gyr). During this period, stromatolites reached hundreds of meters in height and extended hundreds of kilometers along shores, similar to fringing reefs and atolls of today. Because of their size and areal coverage, stromatolites and other microbial mats were crucial in the evolution of life and the atmosphere during earth's early days.

least in non-marine systems. The key may be the presence of large sulfide-oxidizing bacteria that promote apatite formation in organic-rich marine sediments but are absent in non-marine anoxic sediments and soils. Polyphosphate from diatoms has also been shown to be important in apatite formation in sediments (Diaz et al., 2008).

Iron mineral formation by non-enzymatic processes

Iron is another element that limits microbial growth in some environments, but in other environments concentrations are so high that microbes have no problem obtaining enough for cellular functions. After all, iron is one of the most abundant elements in the earth's crust (Chapter 2). The role of microbes in iron mineral formation is of interest to geomicrobiologists because of the size and age of some iron-rich sedimentary deposits. Some of these deposits are large enough to attract the attention of mining companies. Among the oldest deposits are the banded iron formations that came into being during the Precambrian about 3.3 to 1.8 billion years ago (Bekker et al., 2010). The name describes 1–2 cm-thick layers that alternate in color due to whether iron or silica (SiO_2 or chert) dominates the layer. The iron-rich minerals include hematite, magnetite, chamosite, and siderite (Table 13.2). A variety of microbes and microbial processes have been implicated in banded iron formation. Banded iron formations have yielded important geological clues about the birth of an oxygen-rich atmosphere during the Precambrian.

One mechanism contributing to banded iron formations is the passive sorption of iron and other minerals onto cell surfaces and extracellular polymers. This mechanism explains mineral formations that are much smaller, much younger, and less spectacular than the banded iron formations. One scenario, similar to the two-step adsorption model of Beveridge and Murray, (1976), is that microbes and their extracellular polymers act as passive nucleation sites for the precipitation of ferric hydroxide ($FeO(OH) \cdot nH_2O$) from dissolved Fe^{2+} in the presence of oxygen. The precipitated ferric hydroxide in turn promotes more precipitation until the entire bacterium is encased in a mineral matrix. This is a good example of biologically-induced biomineralization. The bacterium seems intimately involved in the mineral formation, yet its role is passive and the cell surface, extracellular polymers, and its metabolism are not necessarily designed to promote precipitation of iron minerals.

Other microbes, referred to as iron-depositing bacteria or simply iron bacteria, do synthesize cell surface ligands that promote Fe(II) oxidation, although these bacteria may not necessarily harvest energy from the process, in contrast to the true iron oxidizers discussed

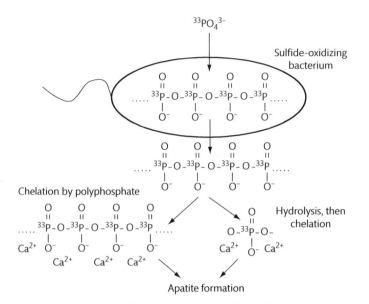

Figure 13.7 Formation of apatite following the uptake of phosphate by sulfide-oxidizing bacteria. Phosphate taken up in phosphorus-rich environments is stored as polyphosphate which is hydrolyzed back to phosphate intracellularly or extracellularly. The released phosphate then chelates Ca^{2+} (Goldhammer et al., 2010). Alternatively, the polyphosphate may be released and directly chelates Ca^{2+} without prior hydrolysis (Ingall, 2010).

Table 13.2 Some iron-rich minerals and possible connections to microbes. "Oxides" includes hydroxides.

Type	Mineral	Composition	Comments
Carbonate	Siderite	$FeCO_3$	Fe(II) produced by iron reducers
Oxides	Amorphous ferric hydroxide	Fe_2O_3	No crystallinity, most easily used by iron reducers
Oxides	Ferrihydrite	$Fe_2O_3 0.5(H_2O)$	Some crystallinity
Oxides	Hematite	Fe_2O_3	Stable form of iron oxide
Oxides	Goethite	$FeO(OH)$	Stable form of iron oxide
Oxides	Magnetite	Fe_2O_4	Produced by magnetotaxic bacteria
Sulfide	Pyrite	FeS_2	Sulfide oxidizers
Sulfide	Greigite	Fe_3S_4	Produced by magnetotaxic bacteria

below. One genus of iron-depositing bacteria is *Leptothrix*, commonly found in iron-rich, low-oxygen freshwater habitats (Fleming et al., 2011, Emerson et al., 2010). Some members of this genus appear to be true chemolithotrophic iron oxidizers while others may be heterotrophs. In either case, this bacterium makes tubular sheaths that become encrusted in iron minerals. Many of these sheaths do not contain living cells, evidence that the bacterium escaped from being entombed in an iron casket. Some organisms are more successful than others in avoiding entrapment in iron (Schadler et al., 2009).

Magnetite and magnetotactic bacteria

The iron mineral magnetite serves an unusual role in some microbes, the magnetotactic bacteria. Geologists are very interested in magnetite because it records changes in the direction and intensity of the earth's magnetic field over geological time. These changes are very useful for understanding plate tectonics and other geological processes. Magnetite is the most magnetic of all minerals on the planet, but only magnetite crystals of the right size are single domain magnets with magnetic properties. The microbial processes producing magnetite can be differentiated by the location of magnetite formation and its role in micro-bial physiology.

Magnetite can be produced as a by-product of Fe(III) reduction coupled to organic carbon oxidation (Chapter 11). The two best-studied dissimilatory Fe(III)-reducing bacteria are *Geobacter metallireducens* and *Shewanella putrefaciens*. The Fe(II) resulting from ferric iron reduction abiotically forms mostly small magnetite crystals outside of the cell, unaligned in chains or other formations (Bazylinski et al., 2007). Most of these crystals are too small to have magnetic properties, and these Fe(III)-reducing bacteria do not respond to magnetic fields. However, even though production is a low fraction of the total, Fe(III)-reducing bacteria still produce 5000 more single domain magnetite per cell than do magnetotactic bacteria.

In contrast to being a by-product, magnetite is an essential and unique feature of the ecophysiology of mag-netotactic bacteria (Bazylinski et al., 2007). These bacteria are from several groups in the Proteobacteria, having in common intracellular membrane-lined structures, called magnetosomes, which carry magnetite or greigite or both. As implied by the name, magnetotactic bacteria swim towards or away from either of the earth's magnetic poles due to the arrangement of the magnetosomes in relation-ship to the earth's magnetic field lines. Even dead magne-totactic bacterial cells become aligned with the magnetic field. The magnetism of magnetite in magnetotactic bac-teria is due to it being chemically pure and having the right size, typically 35–120 nm (Bazylinski et al., 2007). Because of this size and chemical characteristics, magnet-ite crystals in magnetotactic bacteria are permanent, single-domain magnets at ambient temperatures.

The advantage of magnetotactic behavior for microbes is not totally known. It is especially unclear for the few eukaryotes with magnetite. For bacteria, one hypothesis is that magnetotaxis helps the bacterium determine which way is up and down (Fig. 13.8). By cou-pling magnetotaxis with mechanisms for sensing oxygen (aerotaxis), these bacteria potentially find microhabitats with optimal oxygen (many are microaerophilic) or

Box 13.3 Magnetite and life on Mars

Because magnetite is produced by iron-oxidizing bacteria and magnetotactic microbes, its presence has been used to argue for life having once existed on Mars and still persisting in the deep subsurface of our planet (Jimenez-Lopez et al., 2010, Nisbet and Sleep, 2001). In particular, the magnetite and other components of the Martian meteorite ALH84001 have been intensively scrutinized fol-lowing claims that it carried the remains of Martian life (McKay et al., 1996). Since that 1996 report, geomicrobiologists have identified characteristics that distinguish biogenic magnetite from magnet-ite produced by abiotic processes. These character-istics include size, shape, chemical purity, and crystal morphology. Unfortunately, the ALH84001 magnetite fails to pass the test. Since all of the other microbe-like features of the meteorite can now be explained without evoking biology (Jull et al., 1998), life on Mars remains to be demonstrated.

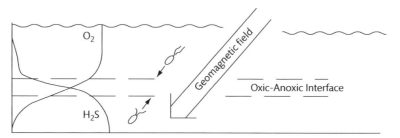

Figure 13.8 One explanation for why microbes have magnetotaxis. The hypothesis is by moving along the geomagnetic field, magnetotactic microbes are better able to find the oxic-anoxic interface, the environment optimal for their growth. This diagram is for the Northern Hemisphere. Based on Bazylinski and Frankel (2004).

hydrogen sulfide concentrations necessary for chemo-lithotrophy. Magnetotaxis reduces the three-dimensional world of these bacteria to one (up and down), increasing the effectiveness of other chemotaxis mechanisms for oxygen and reduced sulfur compounds. One sign of a problem with this hypothesis is that magnetotactic bacteria are found at the equator where magnetotaxis cannot be used to distinguish up from down because the magnetic field lines are horizontal to the plane of the earth's surface. In any case, there must be some advantage to having magnetosomes because cells devote a large amount of iron (3% of cell dry mass) and energy to their synthesis.

Manganese and iron-oxidizing bacteria

Enzyme-mediated oxidation of manganese and iron also can lead to the formation of minerals rich in either one of these two elements or both together. Even though iron-oxidizing bacteria were described back in 1837, much remains unknown about their physiology and ecology (Emerson et al., 2010). Even less is known about manganese-oxidizing bacteria, even whether bacteria that carry out this oxidation are able to harvest energy from it. Iron and manganese-oxidizing bacteria are discussed together in this section because of the similarities in the geochemistry and geomicrobiology of these two elements.

Iron oxidation

A number of prokaryotes are able to harvest energy from the oxidation of ferrous iron (Fe^{2+}), according to the reaction:

$$4Fe^{2+} + O_2 + 4H^+ \rightarrow Fe^{3+} + 2H_2O \qquad (13.7).$$

The energy potentially gained from Equation 13.7 is only 29 kJ mol^{-1}, making it the lowest energy-yielding process of all chemolithoautotrophic metabolisms (Emerson et al., 2010). The energy yield doubles if the ferric iron by-product (Fe^{3+}) precipitates to form ferrihydrite, which occurs spontaneously at near-neutral pH under oxic conditions. Even more energy is to be had ($\Delta G° = -90$ kJ mol^{-1}) under low partial pressures of oxygen. While iron-oxidizing microbes gain more energy when the pH is near neutral, they must compete with abiotic oxidation of Fe^{2+} at high pH values (Fig. 13.9). As mentioned above, another problem facing iron-oxidizing bacteria is the risk of becoming permanently sealed off in an iron oxide case produced by the precipitation of ferric by-products. Iron-encrusted sheaths and stalks are characteristic of both iron-depositing and chemolithotrophic iron-oxidizing bacteria.

Several types of bacteria oxidize iron by several metabolic pathways (Table 13.3). The best-known example is the chemolithotrophic oxidation by *Gallionella*, first described back in the nineteenth century. In contrast to other iron-oxidizers, strains of this betaproteobacterium

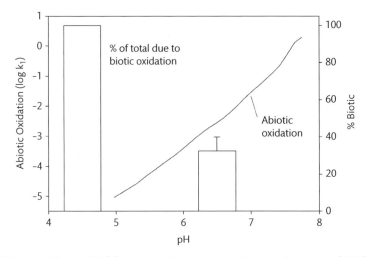

Figure 13.9 Biotic and abiotic oxidation of Fe(II) in oxic environments. Data from Neubauer et al. (2002) and Luther (2010).

are easily cultivated and grown in the lab (Emerson et al., 2010). Being a microaerophilic, it grows best at the oxic-anoxic interfaces where oxygen is present but in low concentrations and Fe^{2+} is available due to fluxes from the anoxic layer. Perhaps to maintain its position at the interface, some *Gallionella* species, such *G. ferruginea*, have a bean-like cell at the top of a stalk. This bacterium as well as others grow in near-neutral pH habitats and are autotrophs. Others use nitrate as an electron acceptor rather than oxygen, according to the equation

$$10Fe^{2+} + 2NO_3^- + 6H_2O \rightarrow 10Fe^{3+} + N_2 + 12OH^- \quad (13.8).$$

Equation 13.8 is important in explaining iron geochemistry in anoxic environments, as it is not clear how Fe^{2+} could be oxidized abiotically in those environments. Other chemolithotrophic iron oxidizers flourish in low pH environments. The acid mine drainage community of iron oxidizers, including *Leptospirillum* and *Ferroplasma*, was discussed in Chapter 10.

These and other acidophilic iron-oxidizing microbes play a key role in generating acidity in run off from abandoned mines. The problem starts with the oxidation of iron sulfur minerals, such as pyrite (FeS_2):

$$FeS_2 + 14Fe^{3+} + 8H_2O \rightarrow 15Fe^{2+} + SO_4^{2-} + 16H^+ \quad (13.9)$$

which produces sulfuric acid. Acid production is minimal in the absence of iron-oxidizing microbes because concentrations of ferric iron (Fe^{3+}) are low relative to pyrite. Production of ferric iron is slow because ferrous iron (Fe^{2+}) is stable when pH is low and iron-oxidizing microbes are absent. Ferric iron is needed because it is a stronger catalyst than the other possibility, oxygen. However, acidophilic iron-oxidizing bacteria oxidize Fe^{2+} back to Fe^{3+}, leading to more pyrite oxidation and faster acid production. The end result is the huge environmental damage due to acid mine run off.

Other iron-oxidizing bacteria are anoxygenic photoautotrophs. These bacteria use Fe^{2+} as a source of electrons for reducing CO_2, in place of H_2O and H_2S used by oxygenic and anoxygenic sulfur-oxidizing photoautotrophs, respectively. Experimental studies indicated that four ferrous atoms are needed to reduce one molecule of CO_2 following the equation (Ehrenreich and Widdel, 1994):

$$4Fe^{2+} + CO_2 + 4H^+ + light \rightarrow CH_2O + 4Fe^{3+} + H_2O \quad (13.10).$$

The organisms carrying out phototrophic iron oxidation are closely related to purple sulfur, purple non-sulfur, and green bacteria which carry out sulfur oxidation and anoxygenic photosynthesis (Chapter 11). Their

Table 13.3. Some types of iron-oxidizing bacteria. The iron oxidizers using nitrate as an electron acceptor are denitrifiers and produce N_2. The photolithotrophic iron oxidizers use the electrons from ferric oxidation to reduce CO_2 and synthesize organic material autotrophically. The chemoorganotrophs oxidize iron but gain energy from the oxidation of organic material rather than from the iron. Based on Canfield et al. (2005) and Emerson et al. (2010).

Metabolism	Electron acceptor	Taxonomic classification	Example
Neutralophilic lithotrophy	O_2	Betaproteobacteria	*Gallionella*
		Zetaproteobacteria	*Mariprofundus ferrooxydans*
	NO_3^-	Betaproteobacteria	*Thiobacillus denitrificans*
		Alphaproteobacteria	FO1 and others*
		Gammaproteobacteria	FO4 and others*
Acidophilic lithotrophy	O_2	Nitrospira	*Leptospirillum*
		Actinobacteria	*Sulfobacillus*
		Crenarchaeota	*Sulfolobus*
		Euryarchaeota	*Ferroplasma acidarmanus*
Chemoorganotrophy	O_2	Betaproteobacteria	*Leptothrix*
		Alphaproteobacteria	*Pedomicrobium*
Photolithotrophy	CO_2	Alphaproteobacteria	*Rhodovulum*
		Chloroflexi	*Chlorobium ferrooxidans*

* Several strains isolated from the Juan de Fuca hydrothermal vent were shown to use oxygen or nitrate as an electron acceptor (Edwards et al. 2003).

distribution is limited today to the few environments with high ferrous iron but low sulfide concentrations and adequate light.

Even so, phototrophic iron oxidizers are of interest to geomicrobiologists exploring the evolution of early life on earth. Iron-oxidizing anoxygenic phototrophs may have been one of the first photosynthetic organisms in the biosphere, predating cyanobacteria. If so and if they were abundant and active enough, the oxidized iron produced by phototrophic iron oxidization may explain banded iron formations in the Precambrian when other data indicate oxygen concentrations were low (Kappler and Straub, 2005, Crowe et al., 2008). The alternative explanation for banded iron formations, oxygen-dependent chemolithotrophic iron oxidation, requires oxygen production by cyanobacteria and mechanisms to prevent oxygen from building up to measurable levels in the atmosphere. Iron- and sulfide-fueled anoxygenic photosynthesis may have been critical in regulating oxygen levels during the Proterozoic (Johnston et al., 2009).

Manganese-oxidizing bacteria

Mn(III) oxyhydroxides and Mn(IV) oxides co-occur with Fe(III) oxides and are part of banded iron formations built during the Precambrian (Konhauser, 2007). These two metals today still commonly precipitate together to form ferromanganese deposits. Famous examples of these deposits are manganese nodules that form on the seafloor, in soils, and in lakes. The bottom of the Pacific Ocean is thought to be covered with 10^{12} tons of nodules, and Oneida Lake, New York has 10^6 tons. Along with being commercially valuable, manganese minerals provide highly reactive surfaces that mediate abiotic transformations of other metals and compounds, such as the oxidation of arsenite, an important process in the spread of arsenic contamination (Ginder-Vogel et al., 2009). Since the abiotic oxidation of Mn(II) is slow, most of the manganese nodules and other mineral formations are thought to be the by-product of manganese-oxidizing bacteria with some help from fungi (Tebo et al., 2005). Like iron oxidizers, manganese-oxidizing bacteria become coated in manganese oxides or have appendages, sheaths, or spore coats where manganese oxides precipitate.

What is odd about manganese-oxidizing bacteria and fungi is that it is not entirely clear why they do it. There is little evidence that microbes gain any energy from manganese oxidation, even though the theoretical gain in energy is 37–76 kJ mol^{-1} (B. Tebo, pers. comm.), depending on environmental conditions and assumptions, about the same or even more than of iron oxidation from which we know bacteria harvest energy. Spores of a *Bacillus* strain can oxidize manganese, evidence that not even active cells are necessary. However, in other cases, manganese oxidation appears to be a specific process mediated by specific enzymes and bacteria (Fig. 13.10). Mutation studies of manganese-oxidizing bacteria and genomic analysis of *Pseudomonas putida* sp. GB-1 have revealed several enzymes potentially involved in manganese oxidation, most notably multicopper oxidase (MCO)-type enzymes (Tebo et al., 2005). This class of enzyme is among the few capable of catalyzing the four electron reduction of O_2 to water, which is necessary if oxygen is the electron acceptor for manganese oxidation.

If bacteria and fungi do not gain any energy directly from the reaction, why do they bother? Several answers have been proposed. The manganese precipitates coating manganese-oxidizing microbes may protect them from reactive oxygen species, such as superoxide or from UV light and the strong oxidants formed by UV-driven photochemistry. The coating may also ward off predators and viruses. Another possible gain is related to energy production, even if ATP is not directly synthesized during Mn(II) oxidation. The Mn(IV) oxides

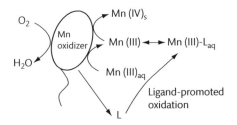

Figure 13.10 Mechanism for Mn(II) oxidation via two one-electron steps. Mn(II) is first oxidized to Mn(III) and then to Mn(IV) oxides. The subscripts "s" and "aq" indicate solid and aqueous (dissolved), respectively. When iron concentrations are low, bacteria produce organic ligands (L) that can promote the oxidation of Mn(II) to Mn(III)-L. Based on Tebo et al. (2010).

Box 13.4 Microbes are not omnipotent

One theme running through this book is the huge metabolic diversity of bacteria, archaea, and protists. Microbes seem to be able to do everything, even gaining energy from reactions that at first seem thermodynamically impossible. But there are some rare cases of microbes seemingly failing to take advantage of an energy-generating reaction. One such case is manganese oxidation. Reduced manganese is also not used as an electron source (reductant) for anoxygenic photosynthesis, nor is ammonium, in contrast to reduced iron. However, it is conceivable that microbial ecologists and microbiologists have not looked in the right places or have not done the proper experiment. Only recently was it demonstrated that some anoxygenic photosynthesizing bacteria use nitrite as a reductant (Schott et al., 2010).

and hydroxides from manganese oxidation may help to oxidize, if only partially, recalcitrant organic material, making it more labile for use by microbes. Some fungi appear to use Mn(IV) oxides to degrade lignin (Chapter 5).

Weathering and mineral dissolution by microbes

While microbes and abiotic reactions are taking ions out of solution to form minerals, other microbes and abiotic reactions are breaking them up and putting ions back into solution. The breaking up, dissolution, of primary minerals and the formation of secondary minerals are called "weathering" by geologists and geomicrobiologists (Uroz et al., 2009). In terms of the carbon cycle, perhaps the most important weathering reaction is the dissolution of rocks by carbon dioxide, with the following reaction being one example:

$$CO_2 + 2H_2O + CaAl_2Si_2O_8 \rightarrow$$
$$Al_2Si_2O_5(OH)_4 + CaCO_3 \qquad (13.11).$$

Hidden in Equation 13.11 is the formation of the weak acid, carbonic acid (H_2CO_3), from CO_2 combining with water. It is carbonic acid and its protons that do the weathering. As illustrated before (Fig. 13.5), the end result is the removal of carbon dioxide from the atmosphere into a geological reservoir, analogous to fixation of atmospheric carbon dioxide by autotrophic organisms into a biological reservoir, the organic material of cells. The huge difference between the two carbon dioxide removal processes is the timescale. While the biotic part of the carbon cycle operates on the day to year timescale, corresponding to the lifespan of photoautotrophic microbes and higher plants, the removal of atmospheric carbon dioxide by geological processes occurs over thousands to millions of years. These geological processes, however, are affected by microbes.

Equation 13.11 is just one of many weathering reactions and processes that have a variety of impacts and roles in biogeochemical cycles. Microbes have a role in most if not all of these processes. Figure 13.11 summarizes some of the mechanisms by which microbes affect mineral dissolution, more colorfully called "eating rocks". Analogous to the distinction between biologically-induced and biologically-influenced biomineralization, microbes have both direct and indirect roles in mineral dissolution and weathering. We have already seen that many microbes produce or consume carbon dioxide, thus indirectly affecting mineral dissolution as indicated in Equation 13.11. At the other extreme, microbes can contribute to mineral dissolution in order to access needed nutrients such as phosphate bound up in apatite (Banfield et al., 1999). Another example is the release of iron and other ions from minerals during iron reduction by bacteria in the genera *Geobacter* and *Shewanella*.

Dissolution by acid and base production
Microbial weathering starts when a fresh rock surface is colonized by bacteria, fungi, and algae. The types of microbes colonizing rock surfaces vary depending on the mineral composition and other environmental properties. DNA fingerprint methods (Chapter 9) have shown differences in community structure of the bacteria colonizing muscovite, plagioclase, K-feldspar, and quartz (Gleeson et al., 2006). These bacteria contribute to dissolution through the production of acid, organic chelators

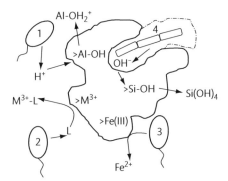

Figure 13.11 Mechanisms by which microbes contribute to mineral dissolution. 1: acidity (protons); 2: ligands (L), including both low molecular weight compounds and extracellular polymers; 3: iron reduction; 4: hydroxides. Similar to what is depicted for the hydroxide mechanism (4), microbes can bore into minerals and rocks and excrete extracellular polymers (the dotted line in mechanism 4) which traps dissolution-promoting compounds. The ">" for the metals, such as >Al-OH, indicates a link with other elements in the solid mineral.

(see below), and in some cases HCN (Frey et al., 2010). In addition to epilithic microbes on the rock surface, other endolithic microbes proliferate in cracks and crevices of rocks to escape the harsh environment of exposed rocks; life on rock surfaces is tough because of low water availability, exposure to full sunlight, and limited availability of nutrients. Fungal hyphae are particularly adept at exploiting narrow channels within rocks and between mineral boundaries (Gadd, 2007). Because of acid production and the release of chelators (see below), these microbes can drastically change rock surfaces, ranging from simple etching and pitting to more extensive disruption of the rock (Konhauser, 2007). Once altered by microbial action, physical forces can more easily erode the rocks and expose new reactive surfaces.

Endolytic algae contribute to the dissolution of sandstones and other silicate-bearing rocks by a different mechanism (Büdel et al., 2004). Photosynthesis by these microbes raises the pH to over 10 in the local microenvironment surrounding the endolytic algal cell. The resulting OH$^-$ ions cause deprotonation of SiOH bonds and loss of soluble ions from the solid rock. Because of their need for light, endolytic algae are active only within millimeters from the rock surface and can grow only on exposed rocks, not covered by soil or vegeta-

tion. These rock environments are found in deserts, such as the Atacama Desert in Chile (Wierzchos et al., 2010) and Antarctica, some of the most extreme environments on the planet. Lichens are also common on rocks where they may contribute to weathering through the production of acidity and extracellular polymers and by creating pathways for water to move into rocks (Banfield et al., 1999).

Dissolution by low and high molecular weight chelators

Acid-producing microbes affect weathering reactions by releasing organic anions as well as protons. These anions can chelate metals, thus increasing the solubility of minerals in the surrounding solution and promoting dissolution from the solid phase. Bidentate (two acid groups) and tridentate (three acid groups) organic acids are more effective than those with only one acidic moiety (Fig. 13.12). A good example is oxalic acid which can occur in concentrations high enough to effect mineral dissolution, but even gluconate with its single acidic functional group enhances dissolution of silicate minerals (Vandevivere et al., 1994). Organic acids can be released by fermenting bacteria (Chapter 11) or by other bacteria partially oxidizing organic material, such as the incomplete oxidation of glucose to gluconate. In either case,

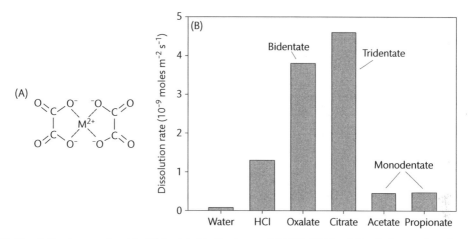

Figure 13.12 Dissolution by organic acids. (A) Oxalate binding to a metal (M^{2+}). (B) Dissolution by inorganic and organic acids of bytownite. Data from Welch and Ullman (1993).

build-up of organic acids occurs only in organic-rich environments.

Siderophores and organic ligands released by microbes to retrieve specific metals can also contribute to mineral dissolution. As with the organic acids, these organic ligands promote solubilization of the metal from the solid phase.

In addition to low molecular weight chelators, microbes can affect mineral dissolution and weathering reactions by the production of extracellular polysaccharides and other polymers. Experiments with isolated polymers have demonstrated direct effects on mineral dissolution (Welch et al., 1999). Polymers such as alginate with acidic groups promote more mineral dissolution than neutral polymers such as starch. Other polymers inhibit dissolution by coating and blocking reactive sites on the mineral. Even if these polymers do not have a direct role, they indirectly affect rates by trapping protons and hydroxides close to the mineral surface, leading to higher local concentrations than would otherwise occur. In dry environments, the water retained by polymers promotes fracturing, hydrolysis, and other chemical reactions that eventually break up rocks.

Geomicrobiology of fossil fuels

While microbes contribute substantially to mineral dissolution and weathering of rocks, they are less successful in degrading one "rock", coal, and its more fluid relative, petroleum. Coal and petroleum deposits exist because of the lack of degradation by microbes. Bacteria and fungi do help indirectly in fossil fuel formation by degrading labile organic compounds, leaving behind refractory organic material that geological processes over time turn into coal and petroleum. Coal comes from higher plant detritus. Substantial amounts of plant litter escaped complete degradation in part because microbes had not solved the problem of breaking down lignin when plants invented this structural polysaccharide during the Carboniferous period 250–350 million years ago. Petroleum originates from undegraded organic material from algae. As mentioned in Chapter 11, another fossil fuel, methane, is produced by methanogenic archaea, in addition to geological processes.

Very few bacteria and fungi can degrade, much less grow on coal, and if growth is observed, it is often because of impurities in coal (Ehrlich and Newman, 2009). Coal can contain substantial amounts of sulfur, occurring as pyrite, elemental sulfur, sulfate, or organic sulfur, depending on the type and origin of the coal. Sulfur-rich coal is less valuable because burning it releases sulfur gases that contribute to air pollution. Likewise, combustion of coal with nitrogenous compounds releases nitrogen dioxide, another atmospheric pollutant. So, removal by microbes of these impurities has practical benefits. Pyrite, for example, can be removed by mechanical methods, but it is also removed by microbial oxidation. Other microbes are used to degrade organic sulfur contaminants in coal. There are similar problems and microbial solutions for sulfur in petroleum.

Degradation of petroleum by microbes has been examined extensively over the years because of many practical concerns and interests, including minimizing the environmental impacts of oil spills (Van Hamme et al., 2003). Petroleum in geological reservoirs is not appreciably degraded because the extreme hydrophobicity of petroleum components severely limits access to microbes and microbial enzymes. Unlike the degradation of plant detritus (Chapter 5), no larger organisms break up petroleum deposits to create more surface area for microbial attack, except for members of *Homo sapiens* who drill and mine these deposits. Even when released via fissures in the earth's crust into natural environments away from subterranean reserves, degradation is slow because the petroleum compounds are still extremely water-insoluble and are toxic to many microbes and other organisms; these lipophilic molecules interfere with membrane structure and function. Other features of petroleum compounds, such as the degree of aromaticity, also contribute to low degradation rates. The following classes of petroleum components are listed from the most easily degraded to the least: linear n-alkanes, branched-chain

alkanes, branched alkenes, low molecular weight n-alkyl aromatics, monoaromatics, cyclic alkanes, polynuclear aromatics, and asphaltenes (Van Hamme et al., 2003).

Several bacteria and even some yeasts are known to degrade petroleum. *Alcanivorax borkumensis*, a marine bacterium, can use linear and branched alkanes, but not aromatic hydrocarbons, sugars, amino acids, fatty acids, and most other common organic carbon compounds (Rojo, 2009). Other bacterial genera known for degrading alkanes include *Thalassolituus*, *Oleiphilus*, and *Oleispira*. The most work has been done on the alkane degradation pathway encoded by the OCT plasmid of *Pseudomonas putida* GPo1 (Rojo, 2009), and in general aerobic pathways (Fig. 13.13) are better understood than anaerobic ones. In addition to having petroleum-degrading enzymes, many petroleum-degrading microbes secrete surfactants to emulsify the oil and thus facilitate degradation of oil components (Rojo, 2009).

The presence of oil-degrading microbes is the reason why bioremediation offers some hope of cleaning up oil spills, even enormous ones like the Deepwater Horizon blowout in the Gulf of Mexico, which started on 20 April 2010. Before 15 July 2010 when it was finally sealed, this deep offshore well spewed out over 4 million barrels of oil over 84 days (Crone and Tolstoy, 2010), making it one of the largest environmental disasters to-date. When this chapter was being written,

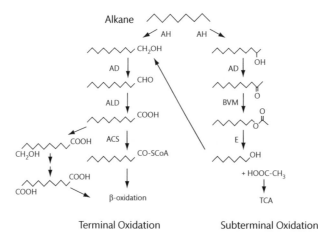

Figure 13.13 Aerobic degradation of alkanes. AH = alkane hydroxylase, AD = alkane dehydrogenase, ALD = aldehyde dehydrogenase, ACS = acyl-CoA synthetase, BVM = Baeyer–Villiger monooxygenase, E = esterase, TCA = tricarboxylic acid cycle. Adapted from Rojo (2009).

much remained unknown about the spill, including its full environmental damage and the microbial response. The available data suggested, however, that degradation by in situ microbes ("intrinsic bioremediation") appears to be occurring. Oxygen concentrations were lower in the oil spill plume than outside it in mid-June 2010, with propane and methane accounting for 70% of oxygen depletion (Valentine et al., 2010). In addition, data from 16S rRNA gene assays conducted between 25 May 2010 and 2 June 2010 indicated that bacteria related to known oil degraders were more abundant in the plume than outside of it (Hazen et al., 2010). But this degradation is not likely to be fast enough to prevent substantial damage to the environment, if previous oil spills are any guide. Oil still fouls coastal habitats decades after spills have ended (Boufadel et al., 2010). For example, oil lies still undegraded below the surface layer of beaches and sediments coated by the Exxon Valdez oil spill in 1989.

While much is known about oil degradation by microbes, much remains to be done and learned. We still cannot adequately predict rates of oil degradation by natural microbial communities. Evaluating the impact of an oil spill is a complex problem, requiring a team of scientists and engineers. As with many environmental problems, key members of the team are microbial ecologists and geomicrobiologists. Arguably, degradation of petroleum contaminating soils and waters is a topic in applied microbial ecology while any degradation in subsurface environments is the purview of geomicrobiology. Regardless, the processes are similar although the microbes and environments may differ.

Even without petroleum pollution, geomicrobiology is an important field. It provides many examples of how microbes rule the world around us, shaping the earth we stand on and the air we breathe. It helps us understand the distant past when life first appeared on the planet and the near future as the world's climate changes.

Summary

1. Geomicrobiology provides many examples of how seemingly small-scale processes mediated by microbes have large impacts on Earth's geology when carried out for thousands and even millions of years.

2. Sorption by dissolved metals to microbial cells and attachment of cells to surfaces are governed by electrostatic and hydrophobic interactions between cells and surfaces.

3. Microbes mediate the precipitation of minerals from dissolved constituents, a process called "mineralization" in geology and geomicrobiology. Biomineralization affects geological formations and has left a footprint of early life on the planet.

4. In "biologically-induced biomineralization", the microbial role may be quite indirect, unlike "biologically-controlled biomineralization" in which microbes actively control every aspect of the mineralization process.

5. Some bacteria are involved in iron reduction in which ferric iron is used as an electron acceptor during organic carbon oxidation while other bacteria oxidize ferrous iron, a form of chemolithotrophy. Analogous processes occur with manganese, except that bacteria do not appear to gain any energy from the oxidation of reduced manganese.

6. Microbes are involved in the weathering and dissolution of minerals and rocks, a complex suite of reactions that are important in removing carbon dioxide from the atmosphere over geological timescales.

7. Bacteria and other microbes are important in bioremediation, one example being the natural degradation of hydrocarbons released by oil spills.

CHAPTER 14

Symbiosis and microbes

One view of the microbial world is of organisms carrying out crucial biogeochemical processes alone, sometimes in competition with one another, always in danger of being eaten, lysed by a virus, or killed off by UV light or desiccation. Life is a four billion year war, it has been said (Majerus et al., 1996). All true, but previous chapters did mention examples of cooperation among microbes and between microbes and larger organisms. We saw the reliance of acidogens on methanogens and the partnership of sulfate reducers with again methanogens in the anaerobic degradation of methane. Chapter 12 pointed out that the nitrogen-fixing bacteria-legume partnership is a crucial component of nitrogen fluxes in soils. All these are examples of symbiotic relationships, the focus of this chapter.

The examples just mentioned were brought up in previous chapters because the symbiosis is important to a particular biogeochemical process. The role of symbiosis in processes is the main reason for this book to explore symbiotic relationships in greater detail. However, there are other reasons. Symbiosis is one type of interaction among microbes and between microbes and larger organisms, all big topics in microbial ecology. Close physical interactions among microbes were essential in the early evolution of eukaryotes, according to the endosymbiotic theory, and continue to be crucial to eukaryotic evolution today. Because symbioses with microbes are so common, the study of ecology of large organisms would be incomplete without considering symbiotic microbes, not to mention all other microbes. Finally, symbioses are fascinating, with many examples of the wonderful, the weird, and the exotic.

This chapter will discuss more examples of symbiotic relationships between microbes and eukaryotes, mainly macroscopic organisms. Chapter 11 already touched on symbioses between prokaryotes.

There are at least two scientific definitions of symbiosis. One is that it is the association between different species in persistent and close contact in which all members receive some benefit (Douglas, 2010). That is the definition slightly favored here. Another definition, the original one proposed by Anton de Bary in 1879, does not require the relationship to be beneficial to all. According to this definition, "symbiosis" covers the entire spectrum of interactions between organisms, ranging from those in which the organisms are indifferent to each other to outright antagonism, as in parasitic relationships (Fig. 14.1). One advantage of de Bary's definition is that "symbiosis" could be used to describe a relationship between two organisms even if the nature of that relationship is unknown. But also according to de Bary's definition, the relationships between humans and malaria-causing protozoa and between potatoes and late blight (this potato pathogen caused the Irish famine in the nineteenth century) would be examples of symbioses. That may sound odd to some readers. The definition is in a state of flux.

In this chapter, we focus on mutualistic relationships where both partners of the relationship receive some benefit. However, in some of the examples discussed, the benefit is not clear, or the association may be occasionally even detrimental to one of the partners.

As suggested by the examples given in Table 14.1, there are symbiotic relationships between all sorts of microbes and eukaryotes. The exception among

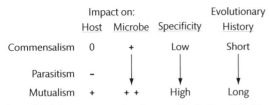

Figure 14.1 Some terms related to symbiosis. One definition of symbiosis would cover all of these interactions whereas another definition is equivalent to mutualism where both partners receive some benefit from the relationship. The plus and negative indicate that the relationship is beneficial or negative to the host or microbe while zero (0) indicates no impact.

microbes is archaea. While some archaea form symbioses with some bacteria, anaerobic methane degradation being one example, there is only one close symbiosis between an archaeon and eukaryote that has been discovered to-date; some anaerobic protists harbor symbiotic methanogens (Fenchel and Finlay, 1995). Also, archaea along with bacteria are prominent residents in sponges (Taylor et al., 2007) and in the gastrointestinal tracts of animals (see below), but these archaea-eukaryote interactions are not as close as seen between bacteria and eukaryotes. Likewise, there is only one example of a pathogenic archaeon, a methanogen involved in

Table 14.1 Examples of symbiotic relationships between eukaryotes and microbes. "Other inverts" refers to invertebrates other than insects. Some eukaryotes harbor archaea, bacteria, and protozoa, called "all" here. Several hosts obtain ammonium ("nitrogen") from diazotrophic symbionts while others have bioluminescent ("Biolumin.") symbionts. "Organic C" is organic carbon while "reduced S" is reduced sulfur compounds like hydrogen sulfide. Based on Douglas (2010) and other sources cited in the main text.

Eukaryote Type	Eukaryote	Symbiont type	Microbe	Benefit to: Host	Benefit to: Symbiont
Microbe	Ciliates	Archaea	Methanogen	Energetics	Hydrogen gas
Microbe	Ciliates	Bacteria	Various	Organic C	Energy
Microbe	Diatom	Bacteria	*Richelia*	Nitrogen	Protection
Microbe	Flagellates	Bacteria	Various	Organic C	Organic C
Microbe	*Paramecium*	Bacteria	*Caedibacter*	Defense	Organic C
Microbe	Fungus	Algae	Various	Organic C	Protection
Plant	*Azolla*	Bacteria	*Anabaena*	Nitrogen	Organic C
Plant	Legumes	Bacteria	*Rhizobium*	Nitrogen	Organic C
Plant	Alder and others	Bacteria	*Frankia*	Nitrogen	Organic C
Plant	Various	Fungi	Fungi	Nutrients	Organic C
Insects	Aphids	Bacteria	*Buchnera*	Organic C	Organic C
Insects	Carpenter ants	Bacteria	*Blochmannia*	Organic C	Organic C
Insects	Cockroaches	Bacteria	Bacteriodetes	Amino acids	Uric acid
Insects	Mealy bugs	Bacteria	*Tremblaya*	Organic C	Organic C
Insects	Sharpshooter	Bacteria	*Homalodisca*	Organic C	Organic C
Insects	Termite	All	Various	Organic C	Organic C
Insects	Tsetse fly	Bacteria	*Wigglesworthia*	Organic C	Organic C
Other Inverts	Corals	Algae	*Symbiodinium*	Organic C	Ammonium
Other Inverts	Deep-sea clam	Bacteria	*Ruthia* and *Vesicomyosocius*	Organic C	Reduced S and oxygen
Other Inverts	Leeches	Bacteria	*Aeromonas* and *Rikenella*	Organic C	Organic C
Other Inverts	Mussels	Bacteria	Proteobacteria	Organic C	Methane
Other Inverts	Nematode	Bacteria	*Xenorhabdus*	Organic C	Organic C
Other Inverts	Oligochaetes	Bacteria	Various	Organic C	Reduced S
Other Inverts	Shipworm	Bacteria	*Teredinibacter*	Organic C	Organic C
Other Inverts	Various	Bacteria	Vibrios	Biolumin.	Organic C and protection?
Vertebrates	Fish	Bacteria	Vibrios	Biolumin.	Organic C
Vertebrates	Ruminants	All	Various	Organic C	Organic C

periodontal disease of humans (Lepp et al., 2004). But even in that case, the microbe arguably has only indirect interactions with a eukaryote. It is unclear why archaea do not form close relationships, negative or positive, with eukaryotes.

Pathogenicity is relevant to the discussion of mutualistic relationships because pathogens as well as parasites can evolve into symbiotic microbes providing benefits to larger organisms. For example, some members of the fungal family Clavicipitaceae are parasites of grasses while others have evolved into symbionts of grasses (Suh et al., 2001). In exchange for organic compounds, these fungal symbionts produce alkaloids that help grasses fend off herbivores. Another example also comes from the Clavicipitaceae. Most species of the genus *Cordiceps* are pathogens of insects, but others are symbionts that provide nitrogen and steroids to their insect hosts. In these cases, phylogenetic analyses and other data indicate an evolution from pathogenicity to symbiosis (Sung et al., 2008), perhaps through a stage in which negative impacts lessen as the host builds up resistance to the invading microbe. To fend off complete eviction, the microbe evolves to provide useful services to the host, reasons for the host to retain it. The end result is a symbiosis. Of course, not all pathogens turn into friends, nor do all mutualistic microbes start off as pathogens. But both pathogens and mutualistic microbes have evolved mechanisms for intimate interactions involved in inhabiting a larger host.

Microbial residents of vertebrates

The rest of this chapter will discuss specific examples of symbiotic relationships between microbes and eukaryotes. We start with vertebrates in part because of our passing interest in one type, *Homo sapiens*, and because some aspects of well-studied microbe-vertebrate interactions can be applied to other, less studied organisms. Humans and other vertebrates are hosts to a huge, diverse community of microbes, as Chapter 1 pointed out. These microbes are only now being examined by the same cultivation-independent approaches used to examine microbial communities in oceans, lakes, and soils (Robinson et al., 2010). Our skin is home to many commensal bacteria and yeasts that live on proteins, such as keratin from skin, secreted oils, lipids, and other compounds. Pyrosequencing studies have revealed that about 19 000 bacterial species reside in human saliva and dental plaque. The largest and most diverse microbial communities inhabiting humans and other vertebrates, however, are found in the gastrointestinal tract.

The gastrointestinal tract, the rumen, of ruminants houses a complex microbial community. The ruminants form a dominant group of herbivores in terrestrial systems. Some examples include deer, bison, and giraffes, as well as several domesticated animals, such as cattle and sheep. There are some aquatic analogs to ruminants. The digestive tracts of herbivorous fish are home to many microbes that help degrade plant material by metabolic pathways seen in ruminants (Clements et al., 2009). Even minke whales have a rumen-like multi-chambered stomach system for digesting herring (Olsen et al., 1994).

The rumen is a large stomach-like pouch consisting of several muscular sacs, containing about 10^{10}–10^{11} bacteria and 10^6 protozoa per gram of rumen fluid as well as fungi and archaea (Hungate, 1975, Russell and Rychlik, 2001). In a metagenomic study of the bovine rumen, 90-95% of all sequences were from bacteria, 2-4% from archaea, and 1-2% from eukaryotes (Brulc et al., 2009). The rumen is the main site where ingested plant material is digested and converted to compounds assimilated by the ruminant. The degradation of plant material in the rumen (Fig. 14.2) is reminiscent of the anaerobic food chain (Chapter 11). Plant polymers, such as cellulose, are hydrolyzed to glucose and other monomers which are then fermented to organic acids, most notably acetate, propionate, and butyrate. The three organic acids are transported across the rumen wall and assimilated by the ruminant. Some of the microbial cells are also broken down, providing protein for the animal.

The symbiotic microbes are essential for converting cellulose, other plant biopolymers, and complex organic material, which the animal alone could not digest, into organic compounds that can be used. Rather than having its own enzymes, the ruminant makes use of hydrolyases and other enzymes synthesized by microbes; rarely are enzymes such as cellulases and xylanases synthesized by eukaryotes. In addition to ruminants, cellulose-degrading microbes are found in many vertebrates and invertebrates eating plant material in aquatic as well as terrestrial environments. These microbes may have been key in the evolution of herbivores from carnivore predecessors (Russell et al., 2009). In terms of the carbon cycle, gut microbes are essential to the success of many

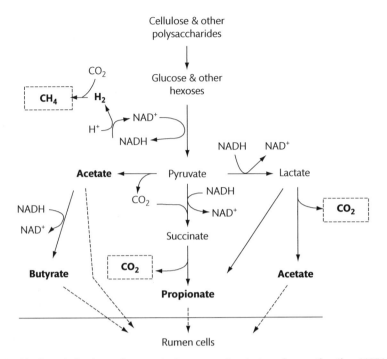

Figure 14.2 Polysaccharide degradation by prokaryotes in the rumen. An electron donor other than NADH is probably involved in methane production. Methane and carbon dioxide are removed by eructations and flatuses from the ruminant. Based on Russell and Rychlik (2001).

detritivores in soils, lakes, and the oceans. These animals couldn't do their main job in the carbon cycle of breaking up organic material were it not for microbes.

Another noteworthy contribution of symbiotic gut microbes is the production of a greenhouse gas, methane (Table 14.2). Methanogens are members of the microbial community in many vertebrate gastrointestinal tracts, including those of humans (Morgavi et al., 2010). Methanogens contribute to the rumen ecosystem by removing hydrogen gas and thus facilitating the production of acetate (Chapter 11). Methane production by ruminant methanogens alone accounts for 17% of the 500–600 Tg methane produced per year (Conrad, 2009). When methanogenesis by termite microbes is added in, symbiotic relationships contribute over 20% of all methane production, more than the methane released from leaky natural gas pipes. Total production by symbiotic methanogens is likely to be higher because some of the production in wetlands and soils now counted as being by free-living archaea is undoubtedly by methanogens in various symbiotic relationships.

Table 14.2. Sources of methane in the biosphere. Some of the methane labeled as being from free-living methanogens is probably from gut-dwelling microbes or other symbiotic methanogens. Data from Conrad (2009).

Process	Source	% of Total
Symbiotic microbes	Ruminants	17
	Termites	3
Subtotal		20
	Wetlands	23
	Rice fields	10
	Landfills	7
Free-living microbes	Plants	6
	Sewage treatment	4
	Gas hydrates	3
	Oceans	3
Subtotal		56
Not by microbes	Fossil fuels	18
	Biomass burning	7
Subtotal		25
	Total	100

While there are many symbiotic interactions between vertebrates and microbes, one type of symbiosis, endo-symbiosis, has not been seen in vertebrates to-date (Douglas, 2010). All mutualistic relationships between microbes and vertebrates are cases of ectosymbioses, meaning the symbiont is on the body surface, in contrast to endosymbiotic relationships in which the microbe is inside a host cell or in host tissue between cells. Many microbes are in mutualistic relationships with vertebrates via their gastrointestinal tracts, but even these microbes are not inside host cells. The examples of endosymbioses discussed below are between microbes and invertebrates and between microbes and plants. Vertebrates may lack endosymbionts because of their highly developed immune system, in contrast to the primitive immune systems of invertebrates and plants. The vertebrate immune system is designed to keep out and destroy microbial intruders, thus erecting a barrier perhaps too high for evolution to surmount. Even bacteria such as *Chlamydia* and *Mycobacterium*, which may infect humans without causing disease-like symptoms, eventually lead to problems in most infected people. Pathogenic microbes have developed the necessary biochemical machinery to enter vertebrate cells, but the eventual end result is death, either of the host cell or of the invading microbe.

Microbial symbioses with insects

As is the case for vertebrates, insects have many types of interactions with microbes, ranging from loose associations and commensualism to mutualism. Unlike vertebrates, the microbe-insect relationships include endosymbiotic ones in which the microbe is nearly an organelle (Douglas, 2010). Many of the symbiotic microbes enable insects to take advantage of imperfect diets, which helps to explain the abundance and diversity of insects. However, other interactions with microbes can be complicated if not detrimental to the insect host. For example, the alphaproteobacterium *Wolbachia*, which infects up to 66% of the 1–10 million insect species, has several complex, often negative effects on insect reproduction (Serbus et al., 2008). Manipulation of host insect reproduction by microbes is common (Engelstädter and Hurst, 2009).

The three examples of microbe-insect symbioses discussed next were chosen because much is known about them and because they involve three different types of microbes: bacteria, fungi, and protozoa.

Microbial symbionts in termites

The microbe-termite system is a good example of how symbiotic microbes enable an insect to flourish in a barren habit it could not survive in without microbial help. Termites depend on symbiotic bacteria, archaea, and protozoa to degrade and transform the ligno-cellulose of wood to organic compounds usable by the termite, as is the case for ruminants and detritus-feeding animals. What stands out about the microbial symbionts of termites is the protozoa. These protists are very abundant in termite hindguts, making up as much as half of the termite's total mass (Brune and Stingl, 2006). Studies using cultivation-independent methods have found three lineages of protozoa in termites: trichomonads, hypermastigids (both in the phylum Parabasalia), and oxymonads in the phylum Loukozoa. Hypermastigids are found only in termite guts, while trichomonads and oxymonads have been found in the guts and body cavities of other animals, including humans. Current models of protozoan roles in termite digestion are based mainly on protozoa isolated from other animals.

Robert Hungate first suggested in the 1940s that protozoa hydrolyze cellulose and produce acetate and hydrogen gas in the anoxic micro-environment of the protozoan hindgut. Later work showed that termite protozoa have hydrogenosomes instead of mitochondria (Brune and Stingl, 2006), essential in cellulose degradation (Fig. 14.3). It is thought that cellulose and other polysaccharides from wood particles are hydrolyzed in food vacuoles yielding simple sugars which are converted to pyruvate by glycolysis. The pyruvate then is metabolized to acetate and hydrogen gas by pyruvate-ferredoxin oxidoreductase and hydrogenase in the hydrogenosome. The hydrogenosome then uses phosphotransacetylase and acetate kinase to synthesize ATP, which is exported back to the protozoan cytoplasm. The hydrogen gas is used by methanogens for methane production (Morgavi et al., 2010).

The model for polysaccharide degradation just described is based on experiments with trichomonad flagellates that are parasites of mammals (Brune and Stingl, 2006). It does not apply to oxymonads, of which there are

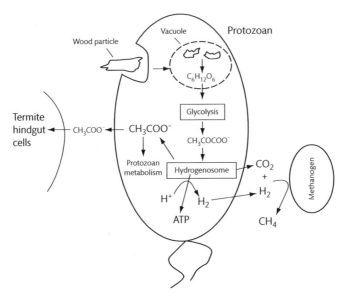

Figure 14.3 Model of cellulose degradation by protozoa in termite hindguts. Cellulase and other enzymes needed to hydrolyze wood to glucose ($C_6H_{12}O_6$) and other sugars are thought to be from protozoa, although some enzymes may be from bacteria. Glucose is then partially oxidized to pyruvate (CH_3COCOO^-) which is then oxidized to acetate (CH_3COO^-), coupled to hydrogen gas production in the hydrogenosome. Acetate taken up by hindgut cells is the main fuel for termite metabolism. Based on Brune and Stingl (2006).

no cultivated representatives. It is known that oxymonads are the dominant protozoa in some termites in spite of not having hydrogenosomes (Ohkuma, 2008). In these termites, lactic acid is a key intermediate, suggesting the need for a model different from the one in Figure 14.3.

In addition to methanogens, the termite gut is home to many bacteria. As is seen time and time again in this book, the termite-gut bacteria sampled by cultivation-independent methods are not very similar (<90% similarity) to cultivated representatives. Some of these bacteria have been placed in Termite Group I, making up potentially a new phylum, Endomicrobia (Geissinger et al., 2009), that appears to be found only in termite guts. In addition to the novelty of these bacteria, what is remarkable is that most of them seem to be associated with protozoa, and only a few are free-living in the termite hindgut or are attached to the gut wall (Strassert et al., 2010). Some of these bacteria are diazotrophic spirochetes and provide much needed nitrogen in the nitrogen-poor wood environment inhabited by termites. The diazotrophs and other bacteria, collectively called epibionts, cover the outer surface of protozoa in dense

rows, one after the other, some providing locomotion for protozoa. These bacteria are symbionts of the protozoa which in turn are symbionts in termites.

Aphids-Buchnera *symbiosis*

Many other insects depend on microbes to flourish in terrestrial habitats in spite of eating an unbalanced diet. The radiation of homopteran insects ("true bugs") into various habitats probably depended on early acquisition of bacterial symbionts in order to feed on the sap provided by vascular plants (Ishikawa, 2003). Today almost all homopterans harbor symbionts. One example of a sap-feeding insect is the aphid. This insect feeds on plant phloem, rich in sugars but poor in essential amino acids. The vital compounds missing from the phloem are provided by endosymbiotic bacteria. Endosymbiotic bacteria, however, have other roles in aphid ecology, such as promoting tolerance of high temperature and conferring resistance to parasites (Tsuchida et al., 2010). Endosymbiont bacteria in the genus *Rickettsiella* affect body color of one aphid species, perhaps lowering predation by ladybird beetles.

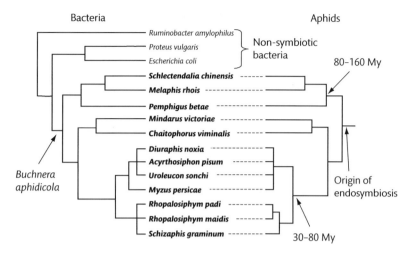

Bacteria   Aphids

Ruminobacter amylophilus
Proteus vulgaris
Escherichia coli — Non-symbiotic bacteria

80–160 My

Schlectendalia chinensis
Melaphis rhois
Pemphigus betae
Mindarus victoriae
Chaitophorus viminalis
Diuraphis noxia
Acyrthosiphon pisum
Uroleucon sonchi
Myzus persicae
Rhopalosiphym padi
Rhopalosiphym maidis
Schizaphis graminum

Buchnera aphidicola

Origin of endosymbiosis

30–80 My

Figure 14.4 Phylogenetic trees of the bacterial symbionts (*Buchnera*) and their aphid hosts. Based on Moran and Baumann (1994) and used with permission of the publisher.

The main endosymbiotic bacteria in aphids belong to the gammaproteobacterial genus, *Buchnera*. The aphid-*Buchnera* symbiosis is thought to be at least 80 million years old, based on analysis of material preserved in amber, while extrapolation from 16S rRNA gene sequences suggest ages of 150–250 million years (Moran et al., 2008, Moran and Baumann, 1994). The estimated age for a particular aphid species and its bacterial symbiont in the *Buchnera* species complex varies from 30 to 160 million years (Fig. 14.4). Aphid and bacterium have evolved together in part because the endosymbiotic bacteria are vertically transmitted to eggs; in vertical transmission, the symbiont is passed from host mother to host offspring during reproduction (Box 14.1).

The genome of *Buchnera* exemplifies three interrelated features of genomes in many bacterial endosymbionts (Douglas, 2010): 1) small genome size; 2) high AT content (the GC content of *Buchnera* is 20–26% (Moya et al., 2008)); and 3) rapid evolution. The genomes of *Buchnera* and *Wigglesworthia* found in tsetse fly are only 0.45–0.66 Mb and 0.7 Mb, respectively, and *Carsonella* in psyllids (jumping plant lice) is even smaller at 0.16 Mb. These are at least twofold smaller than the smallest genome of a free-living bacterium (Chapter 10). Endosymbiotic bacteria can have small genomes because some of the genes essential for an independent lifestyle are no longer necessary and others are redundant when the host takes over a particular function. Any genes not essential for maintaining the symbiosis can tolerate more mutations, leading eventually to the complete loss of the gene. High AT content may be favored because of the greater energetic cost of synthesizing GTP and CTP than ATP and TTP. This high AT content may in turn lead to higher rates of mutation because of a mutational bias towards AT, due to the higher availability of ATP and TTP for DNA synthesis (Douglas, 2010).

Symbiotic relationship between ants and fungi
In addition to insect-bacteria relationships, as exemplified by the aphid-*Buchnera* symbiosis just described,

Box 14.1 Transmission of symbionts

Endosymbionts like *Buchnera* are vertically transmitted to new hosts. The other form of transmission is horizontal in which the symbiont goes through a free-living stage if only briefly. Newly hatched termites, for example, acquire their symbionts by ingesting the symbiont-rich excreta of adult termites (Brune and Stingl, 2006).

insects form various symbiotic relationships with fungi (Gibson and Hunter, 2010). Many of the endosymbiotic fungi in insects are true yeasts in the Saccharomycotina (Fig. 14.5), often more cautiously referred to as "yeast-like cells" (YLC). Insect cells containing endosymbionts are called "mycetocytes" or "bacteriocytes", depending on whether the endosymbiont is a fungus or bacterium. Many yeasts have been found in insect guts but they may not be essential residents there. Yeast-like microbes appear to be obligate in some aphid species, replacing *Buchnera* as the main symbiont. In another type of interaction, the fungus is not inside or attached to the insect, yet similar to other microbe-insect symbioses, the metabolic capacities of the microbe enable the insect to exploit an imperfect diet. One example is between microbes and a group of ants.

Symbiotic fungi and bacteria are essential in providing nutrition for a group of New World myrmecine ants, the Attini, which includes leaf-cutter ants in the genera *Atta* and *Acromyrmex* (Hölldobler and Wilson, 1990). These ants strip bushes or grasses of leaves, and then carry the leaves back to the ant nest on long marches along the forest floor, each ant bearing a leaf fragment much larger

and heavier than its body. Once back in the nest the ants cut the leaves into small bits 1–2 mm in diameter and then they chew these small fragments until they are wet and spongy, sometimes topping them off with a small drop of ant anal fluid containing hydrolytic enzymes. After lining the leaf fragments into a garden, the ants inoculate the new leaf fragments with fungal mycelia from an older part of the garden. In ways not completely understood, nitrogen-fixing bacteria come into the relationship (Pinto-Tomas et al., 2009), providing much needed nitrogen to complement the carbon-rich plant material. The transplanted fungi quickly grow, covering the leaf fragments within a day. The ants then eat the fungi.

The fungi making up the garden are unusual. Analysis of rRNA genes indicated that they are homobasidiomycetes in the order Agaricales (gilled mushrooms) (Hinkle et al., 1994), with each garden consisting of a single fungal strain (Zientz et al., 2005). This lack of diversity helps the fungi avoid the cost of competition, leading to higher fungal yields which benefit the ant. Both the fungus and ant help to maintain the monoculture of the garden. Apparently, the fungus does its share by preventing the growth of fungal strains from other gardens. For its part the ants prevent sexual reproduction by the fungi. Asexual reproduction results in higher fungal yields, but also maintains low diversity within the fungal garden. A negative consequence of low diversity is that fungal gardens are prone to invasion by parasitic fungi, such as the hyphomycete *Escovopsis* (Zientz et al., 2005). Contributing to the defense against parasitic and pathogenic fungi, all of the fungus-gardening ants are covered by actinomycetes (Chapter 9). These bacteria, well known for producing antibiotics, apparently help not only the ant but also the fungi to ward off parasites and pathogens.

Given that one partner is food for the other, the fungi-ant relationship is not an equal one. Still, the fungi do benefit from the arrangement. The fungus as a population gains a steady supply of organic carbon and a controlled growth environment at the expense of some being eaten once in a while by the ants. Outside of ant-tended gardens, fungi would also suffer from predation and viral lysis as well as harsher physical conditions. The ant colony is very dependent on the fungus as it provides the sole food source for most colony members; only some of the workers can supplement their fungal diet

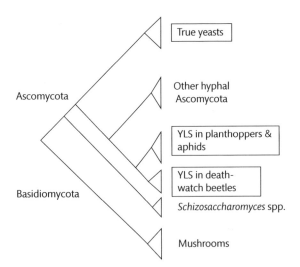

Figure 14.5 Phylogeny of yeast-like symbionts (YLS, in boxes) possibly involved in symbiotic relationships with insects. Many insect symbionts are true yeasts in the Saccharomycotina. Other fungal phyla include Blastocladiomycota, Chytriodiomycota, Glomeromycota, Microsporidia, Neocallimastigomycota, and Zygomycota. Based on Gibson and Hunter (2010).

Box 14.2   Career choice

E.O. Wilson did most of his original field work on ants, and went on to write several award-winning books, including *Sociobiology*, *On Human Nature*, and *The Ants* (with Bert Hölldobler). Near the end of his autobiography, *Naturalist* published in 1994 (Island Press), Wilson mentioned that if he had to do it all over again, he would be a microbial ecologist. He ends with a paean to the microbial world: "The jaguars, ants, and orchids would still occupy distant forests in all their splendor, but now they would be joined by an even stranger and vastly more complex living world virtually without end."

with plant sap (Zientz et al., 2005). The ants apparently have lost some enzymatic capabilities, leaving those to the fungi to carry out (Suen et al., 2011). As with the aphid-*Buchnera* symbiosis, the fungi and leaf-cutting ants appear to have evolved together (Hinkle et al., 1994), indicating tight relationships between the two.

Thanks to the metabolic capacities of the microbes, leaf-cutter ants are the dominant herbivores in tropical savannahs and rainforest and have huge roles in structur-

ing these ecosystems (Hölldobler and Wilson, 1990). The New World leaf-cutter ants are replaced in the Old World tropics by some species of termites (Macrotermitinae) that also cultivate fungal gardens. Fungus-gardening termites do not occur in the Americas.

Symbiotic microbes in marine invertebrates

Ants and termites do not live in the oceans, but some features of microbe-insect symbioses are found in marine organisms. As with termites and ruminants, microbial symbionts expand the metabolic repertoire of many marine animals, allowing them to thrive on otherwise inedible food or an unbalanced diet. Rumen-like microbes in the guts of marine and other aquatic herbivores and detritivores were mentioned above. There is a termite-like marine invertebrate, the shipworm, that depends on cellulytic and nitrogen-fixing bacteria to survive on a wood diet (Distel et al., 1991, Lechene et al., 2007). While most of the well-known symbiotic microbes are essential for nutrition of marine invertebrate hosts (Table 14.3), a few have other roles in host biology, as discussed below for the squid-*Vibrio* symbiosis.

Symbiotic bacteria live in various locations on and in their marine invertebrate hosts (Table 14.3), with some being epibionts or episymbionts that live on the outside

Table 14.3 Examples of marine invertebrates with symbiotic bacteria. "Vents" refers to hydrothermal vents, and "seeps" are cracks in the ocean floor where methane and other gases leak out. Based on Stewart et al. (2005) and other references cited in the text.

Invertebrate	Common name	Location in host	Symbiont metabolism	Habitat
Alvinellidae	worms	surface	chemoautotrophy	vents
Ascidians	sea squirts	cloaca	photoautotrophy	benthos
Bivalvia	shipworms	gills	heterotrophy	wood detritus
Cephalopods	squids	light organ	heterotrophy	water column
Clitellata	oligochaetes	subcuticular	chemoautotrophy	coral sands
Crustacea	shrimp	exoskeleton	chemoautotrophy	vents
Echinodermata	sea urchins	extracellular	chemoautotrophy	anoxic sediments
Mytilidae	mussels	gills	methanotrophy	cold seeps
Polychaete	tubeworm	trophosome	chemoautotrophy	vents
Nemata	nematodes	cuticle	chemoautotrophy	anoxic sediments
Provannidae	snails	gills	chemoautotrophy	vents
Several	detritivores	gut	heterotrophy	sediments
Solemyidae	clams	gills	chemoautotrophy	anoxic sediments

of the animal while others are endosymbionts housed in bacteriocytes within the animal. Most invertebrates have one or the other type of symbiont, with one type of metabolism, but there are interesting exceptions. At least one marine invertebrate, the scaly snail from vents of the Indian Ocean, has both a dense episymbiotic population on its foot and endosymbionts in its esophageal gland (Stewart et al., 2005). A mussel in the genus *Bathymodiolus* has two endosymbionts, one a methanotroph and the other a sulfide-oxidizing chemoautotroph. Episymbionts and endosymbionts differ taxonomically. Most episymbionts are in the Epsilonproteobacteria while endosymbionts are in the Gammaproteobacteria. One argument for these symbionts being highly adapted to the symbiosis lifestyle is our inability, with the exception of symbiotic vibrios, to cultivate them apart from their hosts.

Endosymbionts in Riftia and other sulfide-oxidizing symbionts

In 1977, geologists in the submersible Alvin were hunting in waters near the Galápagos Islands for hydrothermal vents, cracks in the ocean bottom, hypothesized to be the source of anomalously warm waters and heavy metals (Corliss et al., 1979). What they found was much more astonishing: luxurious communities of shrimp, crabs, clams, and plant-like echinoderms, with meter-high tube-like creatures towering over all, waving in the hot water spewing from the vents. This riot of life was in stark contrast to the desert that is most of the ocean floor under deep waters, thousands of meters from the surface. The source of energy and carbon supporting the rich vent community was at first a mystery. Little organic matter reaches the bottom from primary production at the surface, certainly not enough to fuel the dense and diverse communities at vents.

Later work revealed that the entire vent community is based on sulfide oxidation by nonsymbiotic and symbiotic chemoautotrophic bacteria. In contrast to anoxic systems where sulfate reduction supplies the sulfide (Chapter 11), the sulfide at hydrothermal vents comes primarily from the reduction of sulfate by purely geochemical mechanisms (Fig. 14.6). Seawater seeps into the ocean floor where it is superheated up to 350 °C, reducing sulfate

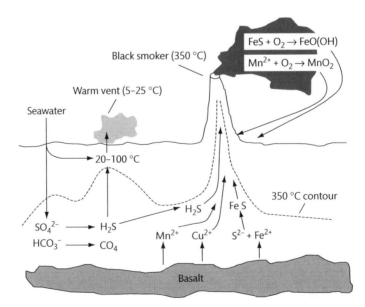

Figure 14.6 Structure of hydrothermal vents. Vents occur at spreading zones between continental plates where the ocean floor is splitting apart, allowing seawater to permeate far enough into the subsurface to be heated geothermally. Compounds in the seawater are reduced abiotically; one example is sulfate (SO_4^{2-})$\rightarrow$ hydrogen sulfide (H_2S). The hydrogen sulfide and reduced metals, such as Fe^{2+} and Mn^{2+}, from basalt are carried back to the surface where they are oxidized either abiotically or by chemolithotrophic reactions, both using oxygen as the electron acceptor. The resulting metal oxides precipitate onto the vent chimney or the ocean floor. Adapted from Madigan et al. (2003).

to sulfide, along with many other compounds. Sulfide, reduced metals, and other compounds in >300 °C acidic seawater then gushes out at vents where the reduced compounds are oxidized by abiotic mechanisms and chemolithotrophy. Sulfide oxidation supports the synthesis of organic carbon by chemoautotrophy, often called just "chemosynthesis". The vent ecosystem is not totally independent of the surface ocean because oxygen, the most common electron acceptor for vent chemolithotrophy, comes from light-dependent photoautotrophy carried out by phytoplankton in the ocean's surface layer. Regardless, chemoautotrophic bacteria form the base of the food chain at vents and support the rich biological community seen from Alvin's portholes.

Initial work focused on the large tube-like creatures, eventually called *Riftia pachyptila*, members of the Annelida phylum, although they were first put into the phyla Pogonophora and Vestimentifera (Hilário et al., 2011). These tubeworms puzzled zoologists because they lack an obvious digestive tract and are too large to live on dissolved organic compounds. Suspecting that *Riftia* relied on chemoautotrophy, zoologists found CO_2 fixa-

tion and sulfide oxidation activity in tubeworm tissue. This enzymatic activity was initially thought to be carried out by tubeworm cells, leading zoologists to conclude that they had found the first "chemoautotrophic animal" (Felbeck, 1981). About the same time, however, microbial ecologists showed that bacteria, not tubeworm cells, were carrying out the measured CO_2 fixation and sulfide oxidation (see Box 14.3). Stable carbon isotope data later made clear that tube worms depend on endosymbiotic bacteria for nutrition.

Tubeworms have a unique structure, the trophosome, designed to support bacterial symbionts (Stewart and Cavanaugh, 2006). This organ consists of blood vessels, coelomic fluid, and bacteriocytes housing the endosymbionts. One gram of trophosome tissue has about 10^9 bacterial cells, taking up 15–35% of the trophosome volume. The morphology of the endosymbiotic bacteria varies with location within the trophosome, due to chemical gradients or growth stages of the bacteria. Tubeworm blood carries sulfide, oxygen (both via hemoglobin), and nitrate to the bacteriocytes where chemolithoautotrophy occurs (Fig. 14.7). Nitrate, which is reduced to ammonium

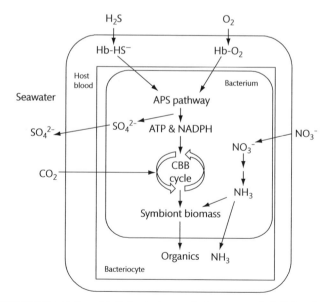

Figure 14.7 Relationships between the endosymbiotic bacterium and a tubeworm. Both H_2S and oxygen are carried to the bacterium via hemoglobin (Hb). Sulfide is oxidized via the adenosine 5'-phosphosulfate (APS) pathway with oxygen as the electron acceptor, yielding ATP and NADPH. These are used by the bacteria to fix CO_2 and synthesize organic compounds via the Calvin-Benson-Bassham (CBB) cycle. Host metabolism is supported by organic compounds leaked from the symbiont or by symbiont biomass directly. Based on Stewart et al. (2005).

by the symbiont, supplies the nitrogen needed by both the host and symbiont, and also serves as an electron acceptor when oxygen concentrations are low. Concentrations of nitrate are high, about 30 μM, in the deep oceanic water surrounding vent communities. As payback for servicing the bacterial symbiont, the tubeworm host gets organic compounds leaked from symbiotic bacteria, supplemented by digesting some of the bacteria from time to time.

After the discovery of endosymbionts in tubeworms, chemoautotrophic symbiotic bacteria were found in many

---

Box 14.3 Bacteria on the brain

Although established zoologists, with input from some equally well-established microbiologists, had the first look at tubeworms, it took a graduate student with a background in microbial ecology to come up with the endosymbiosis hypothesis. Colleen Cavanaugh got her first clue while listening to a zoologist, Meredith Jones, lecturing about mouthless and gutless tubeworms and their strange anatomy in a seminar class at Harvard University. During his presentation, Jones showed a photograph of a dissected tubeworm, noting that he had found numerous sulfur granules within their trophosome tissue. He mentioned that the function of the trophosome was not known. While sitting in the lecture hall, Cavanaugh thought "symbiosis". She thought that the worms fed on symbiotic chemosynthetic bacteria, much like corals living off their symbiotic photosynthetic algae. Cells in the trophosome looked like bacteria she had seen in other scanning electron micrographs. But it took hard work and more definitive evidence to prove this hypothesis. Transmission electron microscopy (TEM) revealed the presence of the two double-layer membranes typical of Gram-negative bacteria, which was confirmed by detection of lipopolysaccharide (Cavanaugh et al., 1981, Nisbet and Sleep, 2001). This evidence along with the enzymatic data strongly supported the endosymbiosis hypothesis, one of the most fascinating chapters in microbial ecology and in all biology.

---

other invertebrates living in sulfide-rich environments. This endosymbiotic relationship is now known to occur in six metazoan phyla and in ciliates (Stewart et al., 2005). One example is the salt marsh clam *Solemya velum*, examined soon after the initial tubeworm studies, although its lack of a gut had long puzzled zoologists (Cavanaugh, 1983). Other gutless invertebrates with sulfide-oxidizing endosymbionts include oligochaete worms. Some of these worms appear to migrate between sediment zones, collecting sulfide in the anoxic, sulfide-rich zone, then swimming to the sulfide-poor but oxygen-rich zone (Dubilier et al., 2006). Other gutless oligochaetes have sulfide-oxidizing symbionts in spite of living in environments without high sulfide concentrations. The sulfide-oxidizing symbionts may depend on other symbiotic bacteria that carry out sulfate reduction and produce sulfide to be used by chemoautolithotrophs.

In contrast to the high diversity of hosts, the symbiotic sulfur-oxidizing bacteria are not very diverse according to 16S rRNA gene sequences (Stewart et al., 2005). These bacteria are limited to a few clades within the Gamma- and Epsilonproteobacteria for endosymbiotic and ectosymbiotic relationships, respectively, as mentioned before. There are some interesting exceptions. Epsilonproteobacterial endosymbionts in the tubeworm *Lamellibrachia* have been reported, but 16S rRNA gene sequences from Alpha-, Beta- and Gammaproteobacteria have been found as well. Several bacteria may inhabit gutless oligochaetes in the genera *Inanidrilus* and *Olavius*, although it is not clear if these bacteria are chemolithotrophic symbionts. Some of these invertebrates may harbor complex endosymbiotic communities, consisting of more than one bacterial type, that vary over time.

For all of these symbiotic relationships, chemoautotrophic bacteria use only reduced sulfur, not other typical chemolithotrophic substrates. Substrate availability and energetic yield may explain why substrates like ammonium and ferrous iron are not used to support symbiotic relationships. Methanotrophs share some similarities with chemolithotrophs and could be considered exceptions to the sulfide-only rule.

Methanotrophic symbiotic bacteria support the metabolism of mussels living at cold seeps with high methane fluxes such as in the Gulf of Mexico (Cavanaugh et al., 1987). Like hydrothermal vents, cold seeps are natural springs from which methane and other fossil fuel components leak out into bottom waters of the oceans from subterranean oil and gas reservoirs. Some of the same type of data was collected for cold seep mussels as had been done for the hydrothermal vent tubeworms. Although these data indicated the presence of symbiotic bacteria in the seep mussels, other data indicated that the bacteria could not be sulfur-oxidizers. ^{13}C content was very low (−74 ‰) and could be explained only if biogenic methane was the source. Then enzymatic and gene assays for methane degradation confirmed that the symbionts were methanotrophs.

Bioluminescent symbionts in the oceans

In the examples of symbioses discussed so far, eukaryotic hosts have symbiotic bacteria and fungi to take advantage of unbalanced diets, wood and plant sap, for example, or energy sources, such as hydrogen sulfide, inaccessible to the eukaryote without help from microbes. The next case study is a different type of symbiosis. Many marine invertebrates and fishes have symbiotic bioluminescent bacteria, for reasons not directly linked to nutrition (Haddock et al., 2010). The anglerfish does rely on symbiotic bacteria for feeding by dangling in front of its mouth a tentacle filled with symbiotic bioluminescent bacteria designed to lure in unsuspecting prey. More commonly, symbiotic bioluminescent bacteria are important components of defenses against predators or for attracting mates. Many of the 43 families of known bioluminescent fish are thought to gain their bio-

luminescence from symbiotic bacteria. Many marine invertebrates are also bioluminescent but without help from symbiotic microbes. The biochemical machinery for bioluminescence may have evolved 40-50 times independently among metazoans and microbes.

Squids in the families Sepiolidae and Loliginidae use bioluminescence provided by symbiotic bacteria, although most species in the 70 genera of bioluminescent squids make their own bioluminescence (Haddock et al., 2010). One of the best known species with symbiotic bioluminescent bacteria is *Euprymna scolopes*, commonly known as the Hawaiian bobtail squid. This cephalopod burrows in the sand of shallow reef flats during the day and then emerges in the early evening to feed. If it weren't for bioluminescence the squid would appear as a dark object backlit by moonlight, an easy target for a predator waiting below in deeper waters. The squid breaks up its dark silhouette by projecting bioluminescent light downward from its light organ (Fig. 14.8). Thanks to a shutter made from a black ink sack and a yellow filter over the light organ, incredibly, the squid can control the intensity and color of the bioluminescence in order to match the background light intensity and color visible at the water depth where the squid swims.

The bioluminescence comes from the bacterium *Vibrio fischeri* colonizing the light organ. This bacterium or closely related strains account for the bioluminescence of several squid species and monocentrid fishes while *Photobacterium leiognathi* and relatives are primarily symbionts for leiognathid, apogonid, and morid fishes

(A)

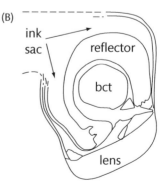

(B)

ink sac

reflector

bct

lens

Figure 14.8 The bobtail squid (A) and its light organ (B) containing symbiotic bacteria, indicated by "bct" for bacteria-containing tissue. The squid is only about 30−mm. The sketch of the squid is from Jones and Nishiguchi (2004) and the light organ diagram is from McFall-Ngai and Montgomery (1990). Used with permission of the authors and publishers.

(Haddock et al., 2010). *Vibrio* and *Photobacterium* are two closely related genera in the gammaproteobacterial family Vibrionaceae. In exchange for providing marine fish and invertebrates with bioluminescence, the bacterial symbionts gain a safe, nutrient-rich home.

The symbiosis starts when a young squid picks up its symbiotic vibrios from the surrounding seawater, a process initiated immediately after hatching (Nyholm and McFall-Ngai, 2004). *V. fischeri* and other vibrios are rare in the free-living bacterial community, present as only a few cells per liter of seawater. In spite of its initial low abundance, *V. fischeri* establishes its dominance in its squid host, by mechanisms that are not completely understood, during the first few hours of the squid's life. Once in the light organ, *V. fischeri* starts to increase in abundance, feasting on the amino acid-rich broth pro-

vided by the squid. The bacterium also induces changes in squid cells that terminate the symbiont collection phase. Symbiont acquisition by newly hatched squid would be even more difficult if adult squids did not release each morning about 95% of their symbiotic bacteria into the surrounding seawater. Along with helping young squid out, the daily release controls population levels in the light organ. Squid also expel by unknown mechanisms vibrio strains not producing enough bioluminescence (Haddock et al., 2010).

Because of the matinal release, vibrio abundance in the light organ starts off low each morning, and increases over the day, reaching sufficient numbers in the early evening when the bobtail squid emerges from the sand and ventures out into open water. Bioluminescence is very low during the day when it is not needed. The bacteria sense it is time to turn on bioluminescence based on its population level, not sunlight or other cues. This detection of population levels, called "quorum sensing", is used by many bacteria in other situations, such as *Pseudomonas aeruginosa* in biofilm formation and *Rhizobium leguminosarum* in root nodulation (Fuqua et al., 1996). The details of the genetic and biochemical machinery vary among different bacteria, but many of the main features are exemplified by *V. fischeri*, the first bacterial quorum sensing system to be described.

V. fischeri takes a census of its population level by two complementary quorum-sensing systems (Lupp and Ruby, 2005). One, the *lux* system, is involved in later stages of the symbiosis and triggers bioluminescence (Fig. 14.9). *V. fischeri* uses the gene *luxI* to produce a signaling compound, N-3-oxo-hexanoyl homoserine, which is an N-acyl-homoserine lactone (AHL). Concentrations of this signaling compound remain low because of diffusion when vibrio cell numbers are low. When population levels reach a threshold of about 10^{10} cells per ml, seen only in the light organ, AHL concentrations can build up to levels high enough for binding of AHL to a sensing protein encoded by *luxR*. Formation of the AHL-LuxR complex causes higher AHL synthesis. For that reason, AHL is called an autoinducer; it induces higher production of itself and thus amplifies the quorum-sensing signal. Most importantly, the AHL-LuxR complex also turns on the *lux* operon, leading to the production of bioluminescence. The now camouflaged squid can venture safely out into open waters.

Box 14.4 Friend or foe?

While *V. fischeri* is beneficial to the bobtail squid, other vibrios, such as *V. parahaemolyticus* and *V. vulnificus*, are pathogenic to larger organisms, including humans. Vibrios have several biochemical features for interacting with eukaryotes both positively and negatively, many of which were revealed by comparing the genome of *V. fischeri* and the cholera-causative pathogen, *V. cholerae* (Ruby et al., 2005). The squid symbiotic vibrio has several genes for Type IV pilus similar to those found in *V. cholerae*. These cell surface-associated structures are used by both bacteria to colonize surfaces, with some being essential for *V. cholerae*'s pathogenicity while others are needed by *V. fischeri* for normal colonization of the light organ. *V. fischeri* also has several genes similar to toxin-producing genes of *V. cholerae*. The comparison of the two vibrio genomes suggests that the shared genes are behind a mutualistic relationship with squids in the case of *V. fischeri* and a pathogenic relationship with humans in the case of *V. cholerae*. The similarities between the two vibrios illustrate again the fine line between mutualism and pathogenicity in microbe-eukaryote interactions.

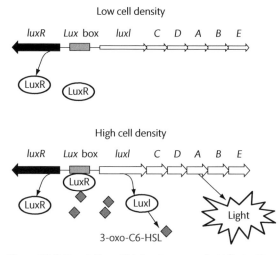

Figure 14.9 Regulation of bioluminescence in *V. fischeri* by quorum sensing. Each box is a gene in the Lux operon with the arrow indicating the direction of transcription and the width indicating its relative strength. When cell abundance is low, the autoinducer (AHL) does not bind to the sensing protein (LuxR), the genes for bioluminescence (*luxCDABE*) are not transcribed, and there is no bioluminescence. When abundance is high, AHL concentrations are high enough so that it binds to LuxR, allowing for transcription and the production of bioluminescence. Quorum sensing is used by several other bacteria in other situations. Based on Miller and Bassler (2001) and Fuqua and Greenberg (2002).

Microbe–plant symbioses

Plants do not have luminescent symbionts, but they do have many of the other relationships with microbes that are seen for animals. As with animals, these relationships range from pathogenic to endosymbiotic. Analogous to the human skin, many microbes are found on exposed surfaces of terrestrial plant leaves and stems, microhabitats collectively referred to as the phyllosphere (Lindow and Brandl, 2003), analogous to the zone around roots, the rhizosphere. Cultivation-dependent approaches have focused on bacteria such as *Pseudomonas syringae*, famous for its role in facilitating the formation of ice crystals in near-freezing weather; less ice forms on leaves when coated with "ice-minus" mutants of *P. syringae* (Hirano and Upper, 2000). As usual, cultivation-independent methods turn up a much more diverse community in the phyllosphere (Yang et al., 2001). Perhaps even more important than microbes in the phyllosphere, however, are the interactions between microbes and plant roots.

Symbiotic relationships between microbes and plants via roots are very common. As much as 90% of all plant species have microbial symbionts (Parniske, 2008), probably a much higher fraction than seen for animals. The symbionts found in or around roots are fungi and diazotrophic bacteria.

Diazotrophic bacteria and plant symbioses

This type of symbiosis is important in agriculture because crop plants need the nitrogen fixed by diazotrophic bacteria, and it is important for many plants and photoautotrophic protists in the many natural environments limited by the supply of nitrogen. Several plants and protists have symbiotic heterotrophic bacteria and cyanobacteria capable of nitrogen fixation. One example is *Frankia*, an actinomycete, which forms symbioses in 194 species in eight dicot families, including woody shrubs and trees that colonize nitrogen-limited land (Benson and Silvester, 1993). Symbiotic *Frankia* are housed in root nodules, large ball-like structures as big as 10 cm in diameter. Another group of diazotrophic bacteria form symbiotic relationships with plants in the Fabaceae family, commonly known as legumes, the third largest family of flowering plants. Among crop plants, legumes include clover, soybeans, and peas while examples of wild legumes include some flowers (lupines and wild indigo) and trees such as black locus (*Robinia pseudoacacia*) and redbud (*Cercis canadensis)*. The best-known legume bacteria are in the alphaproteobacterial genus *Rhizobium*, but bacteria in other genera and even other proteobacterial divisions can form symbioses with legumes (Perret et al., 2000). Collectively all of these bacteria in legume root nodules are called rhizobia.

The root nodule is a ball-like or cylinder-like structure, roughly a millimeter in diameter, depending on the symbiosis, on legume roots that houses the symbiotic rhizobia, similar to the *Frankia* symbiosis. The nodule is the end result of several biochemical exchanges between plant root and rhizobial bacteria. The symbiosis starts when a newly sprouted legume selects its symbiotic bacteria out of the complex microbial community living in soils. Among several possible rhizobial species, only a few strains can form successful symbiotic relationships with any particular legume species. The courtship starts with the constitutive release of flavonoids specific for the legume and its compatible rhizobium (Fig. 14.10); flavonoids are multi-ring compounds with a 2-phenyl-1,4-benzopyrone backbone. The compatible rhizobium responds by releasing the Nod factor, an acylated oligosaccharide. The Nod factor binds to a membrane-associated receptor on the root hair, which in turn triggers other biochemical events in the root. Another component of the legume-rhizobium courtship is the binding of lectins on the root hair surface to specific carbohydrate moieties on the cell surface of compatible rhizobia. These various signaling processes eventually lead to morphological changes both in the root hair and the bacterium. At the end, the transformed bacteria are called "bacteroids".

The plant creates conditions in root nodules to facilitate nitrogen fixation by symbiotic rhizobia. It releases leghemoglobin that binds to oxygen, thus controlling

levels of this nitrogenase-poisoning gas (Chapter 12) and coloring root nodules rust red. The outer cortex of the root nodule is oxic, but oxygen concentrations are much lower (<25 nM) in the center (White et al., 2007). In spite of low free oxygen, rhizobial bacteria can continue to respire and generate the ATP needed for nitrogen fixation and the rest of rhizobial metabolism using the oxygen delivered by leghemoglobin. The plant also releases malate or succinate or both to fuel rhizobial metabolism, and in return, the symbiotic rhizobial bacteria release ammonium to the plant root. There is some evidence that nitrogen is released in the form of alanine rather than ammonium, but this is controversial (White et al., 2007).

The rate of nitrogen fixation by rhizobial symbionts and other root symbiotic bacteria is usually much higher than that by nonsymbiotic soil microbes (Table 14.4). Other studies suggest that nitrogen fixation rates per unit area by *Frankia* and rhizobial symbiotic relationships are about equal overall (Franche et al., 2009); it is harder to estimate the global contribution by both, and clearly more data are needed. Away from roots, nitrogen fixation by heterotrophic diazotrophs is usually limited by the supply of labile organic material whereas substantial

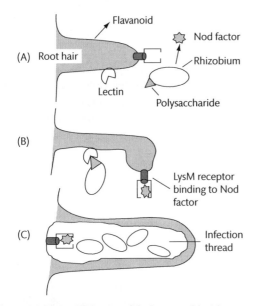

Figure 14.10 Establishment of the legume-rhizobium symbiosis. The interaction begins with the release of a flavonoid specific for the targeted rhizobium, which responds by releasing a Nod factor (A). A secondary factor in establishing the root hair-rhizobium partnership is the binding of a lectin to a polysaccharide on the bacterial cell surface. All signaling events lead to the curling of the root hair and other responses (B). The end result is the formation of the infection thread and proliferation of rhizobial bacteria (C). Based on Parniske and Downie (2003) and Downie (2010).

Table 14.4 N_2 fixation rates by symbiotic and nonsymbiotic prokaryotes in terrestrial systems. "NA" is not applicable. Data from Cleveland et al. (1999) and Evans and Barber (1977).

	N_2 fixation (mmol N m^{-2} d^{-1})		
Environment	Symbiotic	Nonsymbiotic	Symbiont
Tundra	0.112	0.035	Rhizobia
Boreal forest	0.023	0.024	*Frankia*
Temperate forest	2.249	0.133	Rhizobia and *Frankia*
Grasslands	0.062	0.215	Rhizobia
Savannah	0.756	0.251	Rhizobia
Arid shrub land	1.275	<0.01	Rhizobia
Tropical forests	0.365	0.202	Rhizobia
Plants			
Soybeans	148	NA	Rhizobia
Clover	258	NA	Rhizobia
Azolla	613	NA	*Anabaena*
Alnus (alder)	333	NA	*Frankia*
Ceanothus	117	NA	*Frankia*
Coriaria	294	NA	*Frankia*

fixation by cyanobacteria in soil surfaces is prevented by shading by large, higher plants. Other factors affecting nitrogen fixation by both symbiotic and nonsymbiotic bacteria include phosphate, molybdenum, low pH, and iron (Chapter 12).

Fungi-plant symbioses
Symbiotic diazotrophic bacteria supply only nitrogen to their plant hosts and cannot offer any help in securing other necessary nutrients or water potentially limiting plant growth in soils. Plants partially solve this problem by forming symbiotic relationships with fungi. The previously mentioned high fraction (>90%) of plants with microbial symbionts is mainly due to how common fungal symbionts are. They are present in over 85% of all angiosperms (Bonfante and Anca, 2009). Many plants without symbionts are parasites of other plants or are carnivores, while others are aquatic plants (Brundrett, 2009).

As discussed below in more detail, it is thought that the main benefit gained by plants in having symbiotic fungi is help in acquiring nutrients. Fungal symbionts may also help plants fend off pathogenic fungi and bacteria and survive drought (Smith and Read, 2008), but it is possible that the fungi provide no services at all for the plant. It may not be worth the cost to expel the fungi if it is doing the plant no harm. Still, fungi are generally thought to be essential for the success of plants in terrestrial ecosystems. Both fossil and DNA evidence indicate that symbiotic fungi-plant relationships formed 400–460 million years ago when terrestrial plants began to colonize land (Humphreys et al., 2010). The successful plant invasion of land is thought due in part to mycorrhizal fungi in a symbiotic relationship with plant roots. "Mycorrhizal" is derived from Greek for fungus and root.

The various types of mycorrhizal fungus-plant symbioses differ in how the fungus interacts with the plant root, among other features (Table 14.5). Ectomycorrhizal fungi were discovered and examined first because they form large fruiting bodies, such as mushrooms, puffballs, and truffles. These symbiotic fungi are outside root cells, hence the "ecto" prefix, but they can weave their way around root epidermal and cortical cell walls, forming the Hartig net. Another defining characteristic is that these symbiotic fungi form a sheath or mantle around the root tip. A more common type of plant symbiotic fungi, arbuscular mycorrhizae, do penetrate the root cell wall, associating with the plasma membrane, making this relationship an endosymbiotic one. Still another type, ectendomycorrhizae, can penetrate the root cells like arbuscular mycorrhizae, while also forming the Hartig net, like ectomycorrhizal fungi. The fungal-orchid symbiosis is a fourth type not given in Table 14.5. Overall, however, arbuscular mycorrhizae are the most common, accounting for over 85% of all known fungi-plant symbioses among angiosperms (Brundrett, 2009).

Initiation of arbuscular mycorrhizal symbiosis is similar to that of the legume-rhizobium symbiosis (Douglas, 2010, Held et al., 2010), and in fact plant receptors for fungal symbionts may have been hijacked by bacteria to establish the legume-rhizobium symbiosis (Op den Camp et al., 2011). Instead of flavonoids in the case of

Table 14.5 Four types of mycorrhizal fungi. The taxa abbreviations are: Glomero (Glomeromycota), Basidio (Basidiomycota), Asco (Ascomycota), Bryo (Bryophyta), Pterido (Pteridophyta), Gymno (Gymnospermae), and Angio (Angiospermae). Based on Smith and Read (2008) which lists three other types of mycorrhizal fungi.

Characteristic	Arbuscular mycrorrhiza	Ectomycorrhiza	Ectendomycorrhiza	Ericoid
Intracellular symbiosis	Yes	No	Yes	Yes
Mantle around root	No	Yes	Yes or No	No
Hartig net*	Absent	Present	Present	Absent
Aboveground fruiting bodies	No	Yes	No	No
Fungi with septate**	No	Yes	Yes	Yes
Fungal taxa	Glomero	Basidio, Asco	Basidio, Asco	Asco
Plant taxa	Bryo, Pterido, Gymno, Angio	Gymno, Angio	Gymno, Angio	Ericale, Bryo

* A Hartig net, named after the nineteenth century German plant pathologist, Robert Hartig, is a network of fungal hyphae in and around plant roots.
** Septate is the septum dividing fungal cells.

legume-rhizobium symbioses, plant roots initiate the mycorrhizal symbiosis by excreting strigolactones, which in turn trigger release of a "myc factor" by the fungi, analogous to the Nod factor of rhizobia. This signaling between plant and fungi initiates other changes in both organisms, as seen in the legume-rhizobium symbiosis. Arbuscular mycorrhizae earn their name from the formation of an arbuscule, a fungal structure inside of root cells (Fig. 14.11). The term was first coined by Isobel Gallaud in 1905 who thought the structure had tree-like features.

In order to form the arbuscule, the fungus first enters root cells as a fairly wide (10 μm) hypha which bifurcates repeatedly, each time leading to narrower and narrower hyphae, until the process ends in the arbuscule. These smaller-bore hyphae have a higher surface area and can penetrate more easily into and around root cortex cells, both features facilitating uptake of organic compounds supplied by the plant host. Some arbuscular mycorrhizae also form vesicles, which is why these fungi are sometimes called vesicular arbuscular mycorrhizae. Formed from the swelling of a hypha, vesicles are found either inside or between root cortex cells. They store lipids and may serve as propagules for colonizing new habits and starting new symbioses.

Unlike rhizobia, arbuscular mycorrhizal fungi can infect all plants receptive to these fungi (Smith and Read, 2008). This promiscuity has been demonstrated in controlled "pot" experiments in which individual plants grown in pots were successfully inoculated with a variety of arbuscular mycorrhizal fungi. Observations of organisms in the field suggest a more specific relationship between plants and fungi than suggested by the controlled pot experiments (Vandenkoornhuyse et al., 2003). In one study, fungal communities associated with grasses from the same species were more similar to each other

than to the fungus associated with different grass species. The fungi used in pot experiments may not be representative of those found in nature.

Several types of experiments and field observations support the hypothesis that arbuscular mycorrhizal fungi help the plant host acquire nutrients, most notably phosphate (Smith and Read, 2008). The fungi are thought to increase the volume of the soil environment that can be explored for nutrients, in effect becoming an extension of the plant roots. Furthermore, fungi can go where roots cannot because of size; the diameter of a typical fungal hypha is 2–10 μm versus >300 μm for root hairs. Consequently, symbiotic fungi increase the potential area for transporting nutrients by ten to a thousandfold over what can be achieved by roots alone. Symbiotic fungi may also access organic forms of these nutrients not otherwise available to plants. Field observations suggest that symbiotic fungi are most useful to the plant when soil nutrient concentrations are low, such as during late stages of plant community succession. The fungi are less important to the plant when concentrations are high. Indeed, plants can survive without mycorrhizal fungi if the supply of nutrients is adequate.

But arbuscular mycorrhizal fungi cannot survive for long away from their host plant and the organic compounds it supplies (Smith and Read, 2008). The transfer of organic matter from the plant host to symbiotic fungi was demonstrated with ^{14}C tracer experiments. The plant is exposed to $^{14}CO_2$ and the synthesized ^{14}C-organic material is followed into the fungi. These experiments and others with ^{13}C showed that the plant supplies glucose and possibly other hexoses to the fungi, which are quickly converted by the fungi to trehalose and glycogen, a carbon storage compound. As much as 20% of total plant photosynthate may be transferred to arbuscu-

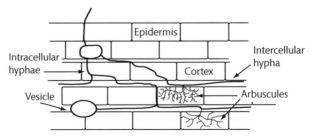

Figure 14.11 Arbuscular mycorrhizal fungus infecting a plant root. Based on Brundrett (2008). See also Parniske (2008).

lar mycorrhizal fungi. Consistent with such a high flux, mycorrhizal fungal biomass is large, 3–20% of root biomass. ^{14}C-labeling experiments have also revealed the possible transfer of organic material and inorganic nutrients along a "common mycelia network" (CMN) linking plants that share a mycorrhizal fungus (Simard et al., 1997). The network may link plants from the same or different species, depending on the plants and the fungus (Beiler et al., 2010). While some interactions mediated by these networks benefit participating plants, others may have adverse effects. There is better evidence of CMN-mediated transfers by ectomycorrhizae than by arbuscular mycorrhizal fungi.

Concluding remarks

The examples of symbioses discussed above illustrate how microbes are essential for the success of larger organisms in the biosphere and for facilitating the contribution by larger organisms to biogeochemical processes. Even when a large organism seems to be the main character and star in the story, microbes are still around, sometimes only behind the scenes, always indispensable in making many things possible. All of life in our world depends on the processes carried out by microbes.

Summary

1. Microbes form close physical relationships with many large organisms, including invertebrates, vertebrates, and higher plants. These relationships range from commensalism to mutualistic symbioses.

2. Some microbe-eukaryotic relationships start as being pathogenic and evolve later into mutualistic relationships. Tight symbiotic relationships often cause morphological changes in both the eukaryotic host and the microbial symbiont.

3. Humans, other vertebrates, and invertebrates benefit from hosting symbiotic microbes in their gastrointestinal tract where microbes aid in digesting otherwise unavailable food. Termites, for example, rely on bacteria, archaea, and especially protozoa to take advantage of cellulose and other complex polysaccharides in wood, which the insect cannot digest alone.

4. Rich communities of metazoans depend on the autotrophic carbon production fueled by oxidation of sulfide from hydrothermal vents, cracks in the deep ocean floor. Tubeworms at vents depend on chemoautotrophic endosymbiotic bacteria, as do other invertebrates in sulfide-rich habitats.

5. Symbiotic bioluminescent bacteria are used by marine fish and invertebrates against predators or for attracting mates. The turning on of bioluminescence by symbiotic vibrios in squid depends on quorum sensing, a census-taking mechanism used by other bacteria in biofilms and pathogenic relationships.

6. Nearly all higher terrestrial plants host symbiotic microbes. Legumes depend on the nitrogen from symbiotic diazotrophic bacteria while many other plants use mycorrhizal fungi to acquire phosphate and other plant nutrients.

References

Adl, S. M. et al. (2005). The new higher level classification of eukaryotes with emphasis on the taxonomy of protists. *Journal of Eukaryotic Microbiology*, **52**, 399–451.

Alexander, M. (1999). *Biodegradation and Bioremediation* Academic Press, San Diego.

Allwood, A. C., Walter, M. R., Kamber, B. S., Marshall, C. P. and Burch, I. W. (2006). Stromatolite reef from the Early Archaean era of Australia. *Nature*, **441**, 714–18.

Andrews, J. A., Harrison, K. G., Matamala, R. and Schlesinger, W. H. (1999). Separation of root respiration from total soil respiration using carbon-13 labeling during free-air carbon dioxide enrichment (FACE). *Soil Science Society of America Journal*, **63**, 1429–35.

Angly, F. et al. (2005). PHACCS, an online tool for estimating the structure and diversity of uncultured viral communities using metagenomic information. *BMC Bioinformatics*, **6**, 41.

Angly, F. E. et al. (2006). The marine viromes of four oceanic regions. *PLoS Biology*, **4**, e368.

Anton, J., Rossello-Mora, R., Rodriguez-Valera, F. and Amann, R. (2000). Extremely halophilic Bacteria in crystallizer ponds from solar salterns. *Applied and Environmental Microbiology*, **66**, 3052–7.

Arístegui, J., Gasol, J. M., Duarte, C. M. and Herndl, G. J. (2009). Microbial oceanography of the dark ocean's pelagic realm. *Limnology and Oceanography*, **54**, 1501–29.

Armbrust, E. V. et al. (2004). The genome of the diatom *Thalassiosira pseudonana*: Ecology, evolution, and metabolism. *Science*, **306**, 79–86.

Auguet, J.-C., Barberan, A. and Casamayor, E. O. (2010). Global ecological patterns in uncultured Archaea. *ISME Journal*, **4**, 182–90.

Azam, F., Fenchel, T., Field, J. G., Gray, J. S., Mayer-Reil, L. A. and Thingstad, T. (1983). The ecological role of water-column microbes in the sea. *Marine Ecology-Progress Series*, **10**, 257–63.

Bååth, E. (1988). Autoradiographic determination of metabolically-active fungal hyphae in forest soil. *Soil Biology and Biochemistry*, **20**, 123–5.

Bååth, E. (1998). Growth rates of bacterial communities in soils at varying pH: A comparison of the thymidine and leucine incorporation techniques. *Microbial Ecology*, **36**, 316–27.

Bååth, E. (2001). Estimation of fungal growth rates in soil using C-14-acetate incorporation into ergosterol. *Soil Biology and Biochemistry*, **33**, 2011–18.

Bailly, J., Fraissinet-Tachet, L., Verner, M.-C., Debaud, J.-C., Lemaire, M., Wesolowski-Louvel, M. and Marmeisse, R. (2007). Soil eukaryotic functional diversity, a metatranscriptomic approach. *ISME Journal*, **1**, 632–42.

Bak, F. and Cypionka, H. (1987). A novel type of energy-metabolism involving fermentation of inorganic sulfur-compounds. *Nature*, **326**, 891–92.

Banfield, J. F., Barker, W. W., Welch, S. A. and Taunton, A. (1999). Biological impact on mineral dissolution: Application of the lichen model to understanding mineral weathering in the rhizosphere. *Proceedings of the National Academy of Sciences of the United States of America*, **96**, 3404–11.

Bapiri, A., Bååth, E. and Rousk, J. (2010). Drying–rewetting cycles affect fungal and bacterial growth differently in an arable soil. *Microbial Ecology*, **60**, 419–28.

Barberan, A. and Casamayor, E. O. (2010). Global phylogenetic community structure and beta-diversity patterns in surface bacterioplankton metacommunities. *Aquatic Microbial Ecology*, **59**, 1–10.

Bardgett, R. D., Freeman, C. and Ostle, N. J. (2008). Microbial contributions to climate change through carbon cycle feedbacks. *ISME Journal*, **2**, 805–14.

Barker, W. W., Welch, S. A. and Banfield, J. F. (1997). 'Geomicrobiology: interactions between microbes and minerals', in Banfield, J. F. and Nealson, K. H., eds. *Geomicrobiology: Interactions between Microbes and Minerals*, pp. 391–428. Mineralogical Society of America, Washington DC.

Barron, A. R., Wurzburger, N., Bellenger, J. P., Wright, S. J., Kraepiel, A. M. L. and Hedin, L. O. (2009). Molybdenum limitation of asymbiotic nitrogen fixation in tropical forest soils. *Nature Geoscience*, **2**, 42–5.

Bartlett, D. H. (2002). Pressure effects on in vivo microbial processes. *Biochimica et Biophysica Acta-Protein Structure and Molecular Enzymology*, **1595**, 367–81.

Barton, L. L. and Fauque, G. D. (2009). Biochemistry, physiology and biotechnology of sulfate-reducing bacteria. *Advances in Applied Microbiology*, **68**, 41–98.

Bastiaens, L., Springael, D., Wattiau, P., Harms, H., Dewachter, R., Verachtert, H. and Diels, L. (2000). Isolation of adherent polycyclic aromatic hydrocarbon (PAH)-degrading bacteria using PAH-sorbing carriers. *Applied and Environmental Microbiology*, **66**, 1834–43.

Bauersachs, T., Speelman, E. N., Hopmans, E. C., Reichart, G.-J., Schouten, S. and Sinninghe Damsté, J. S. (2010). Fossilized glycolipids reveal past oceanic N_2 fixation by heterocystous cyanobacteria. *Proceedings of the National Academy of Sciences of the United States of America*, **107**, 19190–4.

Bazylinski, D. A. and Frankel, R. B. (2004). Magnetosome formation in prokaryotes. *Nature Reviews Microbiology*, **2**, 217–30.

Bazylinski, D. A., Frankel, R. B. and Konhauser, K. O. (2007). Modes of biomineralization of magnetite by microbes. *Geomicrobiology Journal*, **24**, 465–75.

Beal, E. J., House, C. H. and Orphan, V. J. (2009). Manganese- and iron-dependent marine methane oxidation. *Science*, **325**, 184–7.

Beardall, J. and Raven, J. A. (2004). The potential effects of global climate change on microalgal photosynthesis, growth and ecology. *Phycologia*, **43**, 26–40.

Beaumont, H. J. E., Gallie, J., Kost, C., Ferguson, G. C. and Rainey, P. B. (2009). Experimental evolution of bet hedging. *Nature*, **462**, 90–3.

Behrenfeld, M. J. (2010). Abandoning Sverdrup's Critical Depth Hypothesis on phytoplankton blooms. *Ecology*, **91**, 977–89.

Beiler, K. J., Durall, D. M., Simard, S. W., Maxwell, S. A. and Kretzer, A. M. (2010). Architecture of the wood-wide web: *Rhizopogon* spp. genets link multiple Douglas-fir cohorts. *New Phytologist*, **185**, 543–53.

Béjà, O. et al. (2000). Bacterial rhodopsin: Evidence for a new type of phototrophy in the sea. *Science*, **289**, 1902–6.

Bekker, A., Slack, J. F., Planavsky, N., Krapez, B., Hofmann, A., Konhauser, K. O. and Rouxel, O. J. (2010). Iron formation: The sedimentary product of a complex interplay among mantle, tectonic, oceanic, and biospheric processes. *Economic Geology*, **105**, 467–508.

Bell, T., Bonsall, M. B., Buckling, A., Whiteley, A. S., Goodall, T. and Griffiths, R. I. (2010). Protists have divergent effects on bacterial diversity along a productivity gradient. *Biology Letters*, **6**, 639–42.

Beman, J. M. et al. (2011). Global declines in oceanic nitrification rates as a consequence of ocean acidification. *Proceedings of the National Academy of Sciences of the United States of America*, **108**, 208–13.

Beman, J. M., Popp, B. N. and Francis, C. A. (2008). Molecular and biogeochemical evidence for ammonia oxidation by marine Crenarchaeota in the Gulf of California. *ISME Journal*, **2**, 429–41.

Bench, S. R., Hanson, T. E., Williamson, K. E., Ghosh, D., Radosovich, M., Wang, K. and Wommack, K. E. (2007). Metagenomic characterization of Chesapeake Bay virioplankton. *Applied and Environmental Microbiology*, **73**, 7629–41.

Benner, R., Moran, M. A. and Hodson, R. E. (1986). Biogeochemical cycling of lignocellulosic carbon in marine and freshwater ecosystems: relative contributions of procaryotes and eucaryotes. *Limnology and Oceanography*, **31**, 89–100.

Benson, D. R. and Silvester, W. B. (1993). Biology of *Frankia* strains, actinomycete symbionts of actinorhizal plants. *Microbiology and Molecular Biology Reviews*, **57**, 293–319.

Berg, B. and Laskowski, R. (2006). *Litter Decomposition: A guide to carbon and nutrient turnover*, Elsevier, Boston, MA.

Berg, I. A. et al. (2010). Autotrophic carbon fixation in archaea. *Nature Reviews Microbiology*, **8**, 447–60.

Bergh, Ø., Børsheim, K. Y., Bratbak, G. and Heldal, M. (1989). High abundance of viruses found in aquatic environments. *Nature*, **340**, 467–8.

Berglund, J., Muren, U., Bamstedt, U. and Andersson, A. (2007). Efficiency of a phytoplankton-based and a bacteria-based food web in a pelagic marine system. *Limnology and Oceanography*, **52**, 121–31.

Berman, T., Kaplan, B., Chava, S., Viner, Y., Sherr, B. F. and Sherr, E. B. (2001). Metabolically active bacteria in Lake Kinneret. *Aquatic Microbial Ecology*, **23**, 213–24.

Berman-Frank, I., Lundgren, P., Chen, Y. B., Kupper, H., Kolber, Z., Bergman, B. and Falkowski, P. (2001). Segregation of nitrogen fixation and oxygenic photosynthesis in the marine cyanobacterium *Trichodesmium*. *Science*, **294**, 1534–7.

Berner, R. A. (1999). Atmospheric oxygen over Phanerozoic time. *Proceedings of the National Academy of Sciences of the United States of America*, **96**, 10955–7.

Berner, R. A. and Kothavala, Z. (2001). GEOCARB III: A revised model of atmospheric CO_2 over phanerozoic time. *American Journal of Science*, **301**, 182–204.

Bertilsson, S., Berglund, O., Karl, D. M. and Chisholm, S. W. (2003). Elemental composition of marine *Prochlorococcus* and *Synechococcus*: Implications for the ecological stoichiometry of the sea. *Limnology and Oceanography*, **48**, 1721–31.

Beveridge, T. J. and Murray, R. G. E. (1976). Uptake and retention of metals by cell walls of *Bacillus subtilis*. *Journal of Bacteriology*, **127**, 1502–18.

Bhadury, P. and Ward, B. B. (2009). Molecular diversity of marine phytoplankton communities based on key functional genes. *Journal of Phycology*, **45**, 1335–47.

Bianchi, T. S. and Canuel, E. A. (2011). *Chemical Biomarkers in Aquatic Ecosystems*, Princeton University Press, Princeton, NJ.

Biers, E. J., Sun, S. and Howard, E. C. (2009). Prokaryotic genomes and diversity in surface ocean waters: Interrogating the Global Ocean Sampling metagenome. *Applied and Environmental Microbiology*, **75**, 2221–9.

Biersmith, A. and Benner, R. (1998). Carbohydrates in phytoplankton and freshly produced dissolved organic matter. *Marine Chemistry*, **63**, 131–44.

Billings, S. A. (2008). Nitrous oxide in flux. *Nature*, **456**, 888–9.

Birch, H. F. (1958). The effect of soil drying on humus decomposition and nitrogen. *Plant Soil*, **10**, 9–30.

Blattner, F. R. et al. (1997). The complete genome sequence of *Escherichia coli* K-12. *Science*, **277**, 1453–62.

Blom, J. F., Zimmermann, Y. S., Ammann, T. and Pernthaler, J. (2010). Scent of danger: Floc formation by a freshwater bacterium is induced by supernatants from a predator-prey coculture. *Applied and Environmental Microbiology*, **76**, 6156–63.

Boetius, A. et al. (2000). A marine microbial consortium apparently mediating anaerobic oxidation of methane. *Nature*, **407**, 623–26.

Bohannan, B. J. M. and Lenski, R. E. (1999). Effect of prey heterogeneity on the response of a model food chain to resource enrichment. *The American Naturalist*, **153**, 73–82.

Bollmann, J., Elmer, M., Wollecke, J., Raidl, S. and Huttl, R. F. (2010). Defensive strategies of soil fungi to prevent grazing by *Folsomia candida* (Collembola). *Pedobiologia*, **53**, 107–14.

Bond, P. L., Druschel, G. K. and Banfield, J. F. (2000). Comparison of acid mine drainage microbial communities in physically and geochemically distinct ecosystems. *Applied and Environmental Microbiology*, **66**, 4962–71.

Bond-Lamberty, B. and Thomson, A. (2010). A global database of soil respiration data. *Biogeosciences*, **7**, 1915–26.

Bonfante, P. and Anca, I.-A. (2009). Plants, mycorrhizal fungi, and bacteria: A network of interactions. *Annual Review of Microbiology*, **63**, 363–83.

Bongers, L. (1970). Energy generation and utilization in hydrogen bacteria. *Journal of Bacteriology*, **104**, 145–51.

Bonkowski, M. (2004). Protozoa and plant growth: the microbial loop in soil revisited. *New Phytologist*, **162**, 617–31.

Boras, J. A., Sala, M. M., Baltar, F., Arístegui, J., Duarte, C. M. and Vaqué, D. (2010). Effect of viruses and protists on bacteria in eddies of the Canary Current region (subtropical northeast Atlantic). *Limnology and Oceanography*, **55**, 885–98.

Boras, J. A., Sala, M. M., Vázquez-Domínguez, E., Weinbauer, M. G. and Vaqué, D. (2009). Annual changes of bacterial mortality due to viruses and protists in an oligotrophic coastal environment (NW Mediterranean). *Environmental Microbiology*, **11**, 1181–93.

Boschker, H. T. S. and Middelburg, J. J. (2002). Stable isotopes and biomarkers in microbial ecology. *FEMS Microbiology Ecology*, **40**, 85–95.

Boufadel, M. C., Sharifi, Y., Van Aken, B., Wrenn, B. A. and Lee, K. (2010). Nutrient and oxygen concentrations within the sediments of an Alaskan beach polluted with the Exxon Valdez oil spill. *Environmental Science & Technology*, **44**, 7418–24.

Boyd, P. W. et al. (2007). Mesoscale iron enrichment experiments 1993–2005: Synthesis and future directions. *Science*, **315**, 612–17.

Brenner, D. J. and Farmer, J. J. (2005). 'Enterobacteriales', in Brenner, D. J. et al. (eds) *Bergey's Manual of Systematic Bacteriology*, 2nd edn, pp.587–850. Springer, New York, NY.

Brock, T. and Madigan, M. (1991). *Biology of Microorganisms*, Prentice Hall, Englewood Cliffs, NJ.

Brock, T. D. (1966). *Principles of Microbial Ecology*, Prentice-Hall, Englewood Cliffs, NJ.

Brock, T. D. (1967). Bacterial growth rate in the sea: direct analysis by thymidine autoradiography. *Science*, **155**, 81–3.

Brown, M. V. et al. (2009). Microbial community structure in the North Pacific ocean. *ISME Journal*, **3**, 1374–86.

Brulc, J. M. et al. (2009). Gene-centric metagenomics of the fiber-adherent bovine rumen microbiome reveals forage specific glycoside hydrolases. *Proceedings of the National Academy of Sciences of the United States of America*, **106**, 1948–53.

Brundrett, M. (2009). Mycorrhizal associations and other means of nutrition of vascular plants: understanding the global diversity of host plants by resolving conflicting information and developing reliable means of diagnosis. *Plant and Soil*, **320**, 37–77.

Brundrett, M. C. 2008. 'Section 4. Arbuscular mycorrhizas', in Brundrett, M. C. (ed.) *Mycorrhizal Associations: The Web Resource. Version 2.0.* http://mycorrhizas.info/.

Brune, A. and Stingl, U. (2006). 'Prokaryotic symbionts of termite gut flagellates: Phylogenetic and metabolic implications of a tripartite symbiosis', in Overmann, J., ed. *Molecular Basis of Symbiosis*, pp. 39–60. Springer, Berlin.

Brussaard, C. P. D., Kemper, R. S., Kop, A. J., Riegman, R. and Heldal, M. (1996). Virus-like particles in a summer bloom of *Emiliania huxleyi* in the North Sea. *Aquatic Microbial Ecology*, **10**, 105–13.

Buckel, W. (1999). 'Anaerobic energy metabolism', in Lengeler, J. W., Drews, G. and Schlegel, H. G., eds. *Biology of the Prokaryotes*, pp. 278–326. Blackwell Science, Thieme.

Bucklin, A., Steinke, D. and Blanco-Bercial, L. (2011). DNA barcoding of marine metazoa. *Annual Review of Marine Science*, **3**, 471–508.

Büdel, B., Weber, B., Kühl, M., Pfanz, H., Sültemeyer, D. and Wessels, D. (2004). Reshaping of sandstone surfaces by cryptoendolithic cyanobacteria: bioalkalization causes chemical weathering in arid landscapes. *Geobiology*, **2**, 261–8.

Buesing, N. and Gessner, M. O. (2006). Benthic bacterial and fungal productivity and carbon turnover in a freshwater marsh. *Applied and Environmental Microbiology*, **72**, 596–605.

Bui, E. T. N., Bradley, P. J. and Johnson, P. J. (1996). A common evolutionary origin for mitochondria and hydrogenosomes. *Proceedings of the National Academy of Sciences of the United States of America*, **93**, 9651–6.

Bult, C. J. et al. (1996). Complete genome sequence of the methanogenic archaeon, *Methanococcus jannaschii*. *Science*, **273**, 1058–73.

Bushaw, K. L. et al. (1996). Photochemical release of biologically available nitrogen from aquatic dissolved organic matter. *Nature*, **381**, 404–7.

Busse, M. D., Sanchez, F. G., Ratcliff, A. W., Butnor, J. R., Carter, E. A. and Powers, R. F. (2009). Soil carbon sequestration and changes in fungal and bacterial biomass following incorporation of forest residues. *Soil Biology and Biochemistry*, **41**, 220–7.

Cadotte, M. W., Jonathan Davies, T., Regetz, J., Kembel, S. W., Cleland, E. and Oakley, T. H. (2010). Phylogenetic diversity metrics for ecological communities: integrating species richness, abundance and evolutionary history. *Ecology Letters*, **13**, 96–105.

Cameron, V., Vance, D., Archer, C. and House, C. H. (2009). A biomarker based on the stable isotopes of nickel. *Proceedings of the National Academy of Sciences of the United States of America*, **106**, 10944–8.

Campbell, B. J., Waidner, L. A., Cottrell, M. T. and Kirchman, D. L. (2008). Abundant proteorhodopsin genes in the North Atlantic Ocean. *Environmental Microbiology*, **10**, 99–109.

Candy, R. M., Blight, K. R. and Ralph, D. E. (2009). Specific iron oxidation and cell growth rates of bacteria in batch culture. *Hydrometallurgy*, **98**, 148–55.

Canfield, D. E., Glazer, A. N. and Falkowski, P. G. (2010). The evolution and future of earth's nitrogen cycle. *Science*, **330**, 192–6.

Canfield, D. E., Thamdrup, B. and Kristensen, E. (2005). *Aquatic Geomicrobiology*, Elsevier Academic Press, San Diego, CA.

Capone, D. G. (2000). 'The marine microbial nitrogen cycle', in Kirchman, D. L., ed. *Microbial Ecology of the Oceans*, pp. 455–93. Wiley-Liss, New York, NY.

Capone, D. G. and Kiene, R. P. (1988). Comparison of microbial dynamics in marine and freshwater sediments: contrasts in anaerobic carbon catabolism. *Limnology and Oceanography*, **33**, 725–49.

Cardinale, B. J. (2011). Biodiversity improves water quality through niche partitioning. *Nature*, **472**, 86–9.

Carman, K. R. (1990). Radioactive labeling of a natural assemblage of marine sedimentary bacteria and microalgae for trophic studies: an autoradiographic study. *Microbial Ecology*, **19**, 279–90.

Carmichael, W. W. (2001). Health effects of toxin-producing cyanobacteria: "The CyanoHABs." *Human and Ecological Risk Assessment*, **7**, 1393–407.

Caron, D. A. (2000). 'Symbiosis and mixotrophy among pelagic microorganisms', in Kirchman, D. L., ed. *Microbial Ecology of the Ocean*, pp. 495–523. Wiley-Liss, New York, NY.

Caron, D. A., Porter, K. G. and Sanders, R. W. (1990). Carbon, nitrogen, and phosphorus budgets for the mixotrophic phytoflagellate *Poterioochromonas malhamensis* (Chrysophyceae) during bacterial ingestion. *Limnology and Oceanography*, **35**, 433–43.

Carpenter, E. J. and Capone, D. G. (2008). 'Nitrogen fixation in the marine environment', in Capone, D. G., Bronk, D. A., Mulholland, M. A. and Carpenter, E. J., eds. *Nitrogen in the Marine Environment*, 2nd edn, pp. 141–98. Academic Press, Burlington, MA.

Carson, J. K., Gonzalez-Quiñones, V., Murphy, D. V., Hinz, C., Shaw, J. A. and Gleeson, D. B. (2010). Low pore connectivity increases bacterial diversity in soil. *Applied and Environmental Microbiology*, **76**, 3936–42.

Carter, M. D. and Suberkropp, K. (2004). Respiration and annual fungal production associated with decomposing leaf litter in two streams. *Freshwater Biology*, **49**, 1112–22.

Cavanaugh, C. M. (1983). Symbiotic chemoautotrophic bacteria in marine invertebrates from sulphide-rich habitats. *Nature*, **302**, 58–61.

Cavanaugh, C. M., Gardiner, S. L., Jones, M. L., Jannasch, H. W. and Waterbury, J. B. (1981). Prokaryotic cells in the hydrothermal vent tube worm *Riftia pachyptila* Jones - Possible chemoautotrophic symbionts. *Science*, **213**, 340–2.

Cavanaugh, C. M., Levering, P. R., Maki, J. S., Mitchell, R. and Lidstrom, M. E. (1987). Symbiosis of methylotrophic bacteria and deep-sea mussels. *Nature*, **325**, 346–8.

Cebrian, J. (1999). Patterns in the fate of production in plant communities. *The American Naturalist*, **154**, 449–68.

Chapin, F. S., Matson, P. A. and Mooney, H. A. (2002). *Principles of Terrestrial Ecosystem Ecology*, Springer, New York, NY.

Chapuis-Lardy, L., Wrage, N., Metay, A., Chotte, J. L. and Bernoux, M. (2007). Soils, a sink for N_2O? A review. *Global Change Biology*, **13**, 1–17.

Charlson, R. J., Lovelock, J. E., Andreae, M. O. and Warren, S. G. (1987). Oceanic phytoplankton, atmospheric sulfur, cloud albedo and climate. *Nature*, **326**, 655–61.

Chen, B., Liu, H. and Lau, M. (2010). Grazing and growth responses of a marine oligotrichous ciliate fed with two nanoplankton: does food quality matter for micrograzers? *Aquatic Ecology*, **44**, 113–19.

Chen, X. W., Qiu, C. E. and Shao, J. Z. (2006). Evidence for K^+-dependent HCO_3 utilization in the marine diatom *Phaeodactylum tricornutum*. *Plant Physiology*, **141**, 731–6.

Chen, Y. H. and Prinn, R. G. (2006). Estimation of atmospheric methane emissions between 1996 and 2001 using a three-dimensional global chemical transport model. *Journal of Geophysical Research-Atmospheres*, **111**, D10307. doi:029/2005JD6058.

Chisholm, S. W. (1992). 'Phytoplankton size', in Falkowski, P. G. and Woodhead, A. D., eds. *Primary Productivity and Biogeochemical Cycles in the Sea*, pp. 214–37. Plenum Press, New York, NY.

Chisholm, S. W., Olson, R. J., Zettler, E. R., Goericke, R., Waterbury, J. B. and Welschmeyer, N. A. (1988). A novel free-living prochlorophyte abundant in the oceanic euphotic zone. *Nature*, **334**, 340–3.

Chistoserdova, L., Vorholt, J. A. and Lidstrom, M. E. (2005). A genomic view of methane oxidation by aerobic bacteria and anaerobic archaea. *Genome Biology*, **6**, 208.

Chivian, D. et al. (2008). Environmental genomics reveals a single-species ecosystem deep within earth. *Science*, **322**, 275–8.

Choudhary, M., Xie, Z. H., Fu, Y. X. and Kaplan, S. (2007). Genome analyses of three strains of *Rhodobacter sphaeroides*: Evidence of rapid evolution of chromosome II. *Journal of Bacteriology*, **189**, 1914–21.

Church, M. J., Wai, B., Karl, D. M. and DeLong, E. F. (2010). Abundances of crenarchaeal *amoA* genes and transcripts in the Pacific Ocean. *Environmental Microbiology*, **12**, 679–88.

Church, M. J. (2008). 'Resource control of bacterial dynamics in the sea', in Kirchman, D. L., ed. *Microbial Ecology of the Oceans*, 2nd edn, pp. 335–82. John Wiley & Son, New York, NY.

Church, M. J., DeLong, E. F., Ducklow, H. W., Karner, M. B., Preston, C. M. and Karl, D. M. (2003). Abundance and distribution of planktonic Archaea and Bacteria in the waters west of the Antarctic Peninsula. *Limnology and Oceanography*, **48**, 1893–902.

Church, M. J., Hutchins, D. A. and Ducklow, H. W. (2000). Limitation of bacterial growth by dissolved organic matter and iron in the Southern Ocean. *Applied and Environmental Microbiology*, **66**, 455–66.

Claverie, J. M. and Abergel, C. (2009). Mimivirus and its virophage. *Annual Review of Genetics*, **43**, 49–66.

Clements, K. D., Raubenheimer, D. and Choat, J. H. (2009). Nutritional ecology of marine herbivorous fishes: ten years on. *Functional Ecology*, **23**, 79–92.

Cleveland, C. C. and Liptzin, D. (2007). C:N:P stoichiometry in soil: is there a "Redfield ratio" for the microbial biomass? *Biogeochemistry*, **85**, 235–52.

Cleveland, C. C. et al. (1999). Global patterns of terrestrial biological nitrogen (N_2) fixation in natural ecosystems. *Global Biogeochemical Cycles*, **13**, 623–45.

Cole, J. J., Findlay, S. and Pace, M. L. (1988). Bacterial production in fresh and saltwater ecosystems: a cross-system overview. *Marine Ecology-Progress Series*, **43**, 1–10.

Cole, J. J., Pace, M. L., Carpenter, S. R. and Kitchell, J. F. (2000). Persistence of net heterotrophy in lakes during nutrient addition and food web manipulations. *Limnology and Oceanography*, **45**, 1718–30.

Coleman, D. C. (1994). The microbial loop concept as used in terrestrial soil ecology studies *Microbial Ecology*, **28**, 245–50.

Coleman, D. C. (2008). From peds to paradoxes: Linkages between soil biota and their influences on ecological processes. *Soil Biology and Biochemistry*, **40**, 271–89.

Coleman, D. C. and Wall, D. H. (2007). 'Fauna: The engine for microbial activity and transport', in Paul, E. A., ed. *Soil Microbiology and Biochemistry*, 3rd edn, pp. 163–532. Academic Press, San Diego, CA.

Coleman, M. L. and Chisholm, S. W. (2007). Code and context: *Prochlorococcus* as a model for cross-scale biology. *Trends in Microbiology*, **15**, 398.

Conrad, R. (2009). The global methane cycle: recent advances in understanding the microbial processes involved. *Environmental Microbiology Reports*, **1**, 285–92.

Conti, L. and Scardi, M. (2010). Fisheries yield and primary productivity in large marine ecosystems. *Marine Ecology-Progress Series*, **410**, 233–44.

Corliss, J. B. et al. (1979). Submarine thermal springs on the Galápagos Rift. *Science*, **203**, 1073–83.

Costa, E., Pérez, J. and Kreft, J.-U. (2006). Why is metabolic labour divided in nitrification? *Trends in Microbiology*, **14**, 213–19.

Cotner, J. B., Ammerman, J. W., Peele, E. R. and Bentzen, E. (1997). Phosphorus-limited bacterioplankton growth in the Sargasso Sea. *Aquatic Microbial Ecology*, **13**, 141–9.

Cotner, J. B., Hall, E. K., Scott, T. and Heldal, M. (2010). Freshwater bacteria are stoichiometrically flexible with a nutrient composition similar to seston. *Frontiers in Microbiology*, **1**, 132. doi: 10.3389/fmicb.2010.00132.

Cottrell, M. T., Malmstrom, R. R., Hill, V., Parker, A. E. and Kirchman, D. L. (2006a). The metabolic balance between autotrophy and heterotrophy in the western Arctic Ocean. *Deep Sea Research*, **53**, 1831–44.

Cottrell, M. T., Mannino, A. and Kirchman, D. L. (2006b). Aerobic anoxygenic phototrophic bacteria in the Mid-Atlantic Bight and the North Pacific Gyre. *Applied and Environmental Microbiology*, **72**, 557-64.

Cottrell, M. T., Wood, D. N., Yu, L. Y. and Kirchman, D. L. (2000). Selected chitinase genes in cultured and uncultured marine bacteria in the alpha- and gamma-subclasses of the proteobacteria. *Applied and Environmental Microbiology*, **66**, 1195-201.

Craine, J. M., Fierer, N. and McLauchlan, K. K. (2010). Widespread coupling between the rate and temperature sensitivity of organic matter decay. *Nature Geoscience*, **3**, 854-7.

Crone, T. J. and Tolstoy, M. (2010). Magnitude of the 2010 Gulf of Mexico oil leak. *Science*, **330**, 634.

Cross, W. F., Benstead, J. P., Frost, P. C. and Thomas, S. A. (2005). Ecological stoichiometry in freshwater benthic systems: recent progress and perspectives. *Freshwater Biology*, **50**, 1895-912.

Crowe, S. A. et al. (2008). Photoferrotrophs thrive in an Archean Ocean analogue. *Proceedings of the National Academy of Sciences of the United States of America*, **105**, 15938-43.

Cullen, D. and Kersten, P. J. (2004). 'Enzymology and molecular biology of lignin degradation', in Brambl, R. and Marzluf, G. A., eds. *The Mycota III. Biochemistry and Molecular Biology*, pp. 249-73. Springer-Verlag, Berlin.

Culley, A. I., Lang, A. S. and Suttle, C. A. (2006). Metagenomic analysis of coastal RNA virus communities. *Science*, **312**, 1795-8.

Czaja, A. D. (2010). Early Earth: Microbes and the rise of oxygen. *Nature Geoscience*, **3**, 522-3.

Daniel, R. (2005). The metagenomics of soil. *Nature Reviews Microbiology*, **3**, 470-8.

Danovaro, R. et al. (2008). Viriobenthos in freshwater and marine sediments: a review. *Freshwater Biology*, **53**, 1186-213.

Daszak, P., Cunningham, A. A. and Hyatt, A. D. (2000). Emerging infectious diseases of wildlife—Threats to biodiversity and human health. *Science*, **287**, 443-9.

Daubin, V., Moran, N. A. and Ochman, H. (2003). Phylogenetics and the cohesion of bacterial genomes. *Science*, **301**, 829-32.

Davidson, E. A. and Janssens, I. A. (2006). Temperature sensitivity of soil carbon decomposition and feedbacks to climate change. *Nature*, **440**, 165-73.

Davidson, E. A., Janssens, I. A. and Luo, Y. Q. (2006). On the variability of respiration in terrestrial ecosystems: moving beyond Q(10). *Global Change Biology*, **12**, 154-64.

Davies, J. (2009). Everything depends on everything else. *Clinical Microbiology and Infection*, **15**, 1-4.

Dawes, C. J. (1981). *Marine Botany*, Wiley, New York, NY.

De Corte, D., Yokokawa, T., Varela, M. M., Agogue, H. and Herndl, G. J. (2009). Spatial distribution of Bacteria and Archaea and *amoA* gene copy numbers throughout the water column of the Eastern Mediterranean Sea. *ISME Journal*, **3**, 147-58.

De La Torre, J. R., Christianson, L. M., Béjà, O., Suzuki, M. T., Karl, D. M., Heidelberg, J. and DeLong, E. F. (2003). Proteorhodopsin genes are distributed among divergent marine bacterial taxa. *Proceedings of the National Academy of Sciences of the United States of America*, **100**, 12830-5.

Deines, P., Matz, C. and Jürgens, K. (2009). Toxicity of violacein-producing bacteria fed to bacterivorous freshwater plankton. *Limnology and Oceanography*, **54**, 1343-52.

del Giorgio, P. A., Cole, J. J., Caraco, N. F. and Peters, R. H. (1999). Linking planktonic biomass and metabolism to net gas fluxes in northern temperate lakes. *Ecology*, **80**, 1422-31.

del Giorgio, P. A. and Gasol, J. M. (2008). 'Physiological structure and single-cell activity in marine bacterioplankton', in Kirchman, D. L., ed. *Microbial Ecology of the Ocean*, 2nd ed, pp. 243-98., Wiley, New York, NY.

del Giorgio, P. A. and Cole, J. J. (1998). Bacterial growth efficiency in natural aquatic systems. *Annual Review of Ecology and Systematics*, **29**, 503-541.

del Giorgio, P. A., Gasol, J. M., Vaqué, D., Mura, P., Agustí, S. and Duarte, C. M. (1996). Bacterioplankton community structure: Protists control net production and the proportion of active bacteria in a coastal marine community. *Limnology and Oceanography*, **41**, 1169-79.

Delwiche, C. F. (1999). Tracing the thread of plastid diversity through the tapestry of life. *The American Naturalist*, **154**, S164-77.

Demoling, F., Figueroa, D. and Bååth, E. (2007). Comparison of factors limiting bacterial growth in different soils. *Soil Biology and Biochemistry*, **39**, 2485-95.

Denef, V. J., et al. (2010). Proteogenomic basis for ecological divergence of closely related bacteria in natural acidophilic microbial communities. *Proceedings of the National Academy of Sciences of the United States of America*, **107**, 2383-90.

Derelle, E. et al. (2006). Genome analysis of the smallest free-living eukaryote *Ostreococcus tauri* unveils many unique features. *Proceedings of the National Academy of Sciences of the United States of America*, **103**, 11647-52.

Des Marais, D. J. (2000). When did photosynthesis emerge on earth? *Science*, **289**, 1703-5.

Diaz, J., Ingall, E., Benitez-Nelson, C., Paterson, D., De Jonge, M. D., McNulty, I. and Brandes, J. A. (2008). Marine polyphosphate: A key player in geologic phosphorus sequestration. *Science*, **320**, 652–5.

Distel, D. L., DeLong, E. F. and Waterbury, J. B. (1991). Phylogenetic characterization and in situ localization of the bacterial symbiont of shipworms (Teredinidae: bivalvia) by using 16S rRNA sequence analysis and oligodeoxynucleotide probe hybridization. *Applied and Environmental Microbiology*, **57**, 2376–82.

Dittmar, T. and Paeng, J. (2009). A heat-induced molecular signature in marine dissolved organic matter. *Nature Geoscience*, **2**, 175–9.

Doney, S. C., Fabry, V. J., Feely, R. A. and Kleypas, J. A. (2009). Ocean acidification: The other CO_2 problem. *Annual Review of Marine Science*, **1**, 169–92.

Donoghue, P. C. J. and Antcliffe, J. B. (2010). Early life: Origins of multicellularity. *Nature*, **466**, 41–2.

Dorrepaal, E., Toet, S., Van Logtestijn, R. S. P., Swart, E., Van De Weg, M. J., Callaghan, T. V. and Aerts, R. (2009). Carbon respiration from subsurface peat accelerated by climate warming in the subarctic. *Nature*, **460**, 616–19.

Douglas, A. E. (2010). *The Symbiotic Habit*, Princeton University Press, Princeton, NJ.

Downie, J. A. (2010). The roles of extracellular proteins, polysaccharides and signals in the interactions of rhizobia with legume roots. *FEMS Microbiology Reviews*, **34**, 150–70.

Drake, H. L., Gößner, A. S. and Daniel, S. L. (2008). Old acetogens, new light. *Annals of The New York Academy of Sciences*, **1125**, 100–28.

Dubilier, N., Blazejak, A. and Rühland, C. (2006). 'Symbioses between bacteria and gutless marine oligochaetes', in Overmann, J., ed. *Molecular Basis of Symbiosis*, pp. 251–75. Springer, Berlin.

Ducklow, H. W., Purdie, D. A., Williams, P. J. L. and Davies, J. M. (1986). Bacterioplankton: a sink for carbon in a coastal marine plankton community. *Science*, **232**, 865–7.

Dupraz, C., Reid, R. P., Braissant, O., Decho, A. W., Norman, R. S. and Visscher, P. T. (2009). Processes of carbonate precipitation in modern microbial mats. *Earth-Science Reviews*, **96**, 141–62.

Dusenbery, D. B. (1997). Minimum size limit for useful locomotion by free-swimming microbes. *Proceedings of the National Academy of Science of the United States of America*, **94**, 10949–54.

Edwards, K. J., Rogers, D. R., Wirsen, C. O. and McCollom, T. M. (2003). Isolation and characterization of novel psychrophilic, neutrophilic, Fe-oxidizing, chemolithoautotrophic Alpha- and Gammaproteobacteria from the deep sea. *Applied and Environmental Microbiology*, **69**, 2906–13.

Edwards, R. A. and Rohwer, F. (2005). Viral metagenomics. *Nature Reviews Microbiology*, **3**, 504–10.

Ehrenreich, A. and Widdel, F. (1994). Anaerobic oxidation of ferrous iron by purple bacteria, a new type of phototrophic metabolism. *Applied and Environmental Microbiology*, **60**, 4517–26.

Ehrlich, H. L. and Newman, D. K. (2009). *Geomicrobiology*, CRC Press, Boca Raton, FLA.

Eickhorst, T. and Tippkötter, R. (2008). Improved detection of soil microorganisms using fluorescence in situ hybridization (FISH) and catalyzed reporter deposition (CARD-FISH). *Soil Biology and Biochemistry*, **40**, 1883–91.

Ekelund, F. and Rønn, R. (1994). Notes on protozoa in agricultural soil with emphasis on heterotrophic flagellates and naked amoebae and their ecology. *FEMS Microbiology Ecology*, **15**, 321–53.

Emerson, D., Fleming, E. J. and McBeth, J. M. (2010). Iron-oxidizing bacteria: An environmental and genomic perspective. *Annual Review of Microbiology*, **64**, 561–83.

Engelstädter, J. and Hurst, G. D. D. (2009). The ecology and evolution of microbes that manipulate host reproduction. *Annual Review of Ecology, Evolution, and Systematics*, **40**, 127–49.

Eppley, R. W. (1972). Temperature and phytoplankton growth in the sea. *Fish Bulletin*, **70**, 1063–85.

Erguder, T. H., Boon, N., Wittebolle, L., Marzorati, M. and Verstraete, W. (2009). Environmental factors shaping the ecological niches of ammonia-oxidizing archaea. *FEMS Microbiology Ecology*, **33**, 855–69.

Etheridge, D. M., Steele, L. P., Langenfelds, R. L., Francey, R. J., Barnola, J. M. and Morgan, V. I. (1996). Natural and anthropogenic changes in atmospheric CO_2 over the last 1000 years from air in Antarctic ice and firn. *Journal of Geophysical Research*, **101**, 4115–28.

Ettwig, K. F. et al. (2010). Nitrite-driven anaerobic methane oxidation by oxygenic bacteria. *Nature*, **464**, 543–8.

Evans, C., Pearce, I. and Brussaard, C. P. D. (2009). Viral-mediated lysis of microbes and carbon release in the sub-Antarctic and Polar Frontal zones of the Australian Southern Ocean. *Environmental Microbiology*, **11**, 2924–34.

Evans, H. J. and Barber, L. E. (1977). Biological nitrogen fixation for food and fiber production. *Science*, **197**, 332–9.

Falkowski, P. G. and Raven, J. A. (2007). *Aquatic Photosynthesis*, Princeton University Press, Princeton, NJ.

Falloon, P. D. and Smith, P. (2000). Modelling refractory soil organic matter. *Biology and Fertility of Soils*, **30**, 388–98.

Fang, C. M., Smith, P., Moncrieff, J. B. and Smith, J. U. (2005). Similar response of labile and resistant soil organic matter pools to changes in temperature. *Nature*, **433**, 57–9.

Fang, J., Zhang, L. and Bazylinski, D. A. (2010). Deep-sea piezosphere and piezophiles: geomicrobiology and biogeochemistry. *Trends in Microbiology*, **18**, 413–22.

Farmer, S. and Jones, C. W. (1976). The energetics of *Escherichia coli* during aerobic growth in continuous culture. *European Journal of Biochemistry*, **67**, 115–22.

Faruque, S. M., Biswas, K., Udden, S. M. N., Ahmad, Q. S., Sack, D. A., Nair, G. B. and Mekalanos, J. J. (2006). Transmissibility of cholera: In vivo-formed biofilms and their relationship to infectivity and persistence in the environment. *Proceedings of the National Academy of Sciences of the United States of America*, **103**, 6350–5.

Felbeck, H. (1981). Chemoautotrophic potential of the hydrothermal vent tube worm, *Riftia pachyptila* Jones (Vestimentifera). *Science*, **213**, 336–8.

Fenchel, T. (1980). Suspension feeding in ciliated protozoa: functional response and particle-size selection *Microbial Ecology*, **6**, 1–11.

Fenchel, T. (1987). *Ecology of Protozoa*, Science Tech Publishers, Madison, WI.

Fenchel, T. (2005). Cosmopolitan microbes and their "cryptic" species. *Aquatic Microbial Ecology*, **41**, 49–54.

Fenchel, T. and Blackburn, T. H. (1979). *Bacteria and Mineral Cycling*, Academic Press, London.

Fenchel, T. and Finlay, B. J. (1991). Endosymbiotic methanogenic bacteria in anaerobic ciliates: significance for the growth efficiency of the host. *Journal of Protozoology*, **38**, 18–22.

Fenchel, T. and Finlay, B. J. (1995). *Ecology and Evolution in Anoxic Worlds*, Oxford University Press, Oxford.

Field, C. B., Behrenfeld, M. J., Randerson, J. T. and Falkowski, P. (1998). Primary production of the biosphere: integrating terrestrial and oceanic components. *Science*, **281**, 237–40.

Fierer, N., Bradford, M. A. and Jackson, R. B. (2007a). Toward an ecological classification of soil bacteria. *Ecology*, **88**, 1354–64.

Fierer, N. et al. (2007b). Metagenomic and small-subunit rRNA analyses reveal the genetic diversity of bacteria, archaea, fungi, and viruses in soil. *Applied and Environmental Microbiology*, **73**, 7059–66.

Fierer, N. and Jackson, R. B. (2006). The diversity and biogeography of soil bacterial communities. *Proceedings of the National Academy of Science of the United States of America*, **103**, 626–31.

Fierer, N., Strickland, M. S., Liptzin, D., Bradford, M. A. and Cleveland, C. C. (2009). Global patterns in belowground communities. *Ecology Letters*, **12**, 1238–49.

Findlay, S. et al. (2002). A cross-system comparison of bacterial and fungal biomass in detritus pools of headwater streams. *Microbial Ecology*, **43**, 55–66.

First, M. R. and Hollibaugh, J. T. (2008). Protistan bacterivory and benthic microbial biomass in an intertidal creek mudflat. *Marine Ecology-Progress Series*, **361**, 59–68.

Fleischmann, R. D. et al. (1995). Whole-genome random sequencing and assembly of *Haemophilus influenzae* Rd. *Science*, **269**, 496–512.

Fleming, E. J., Langdon, A. E., Martinez-Garcia, M., Stepanauskas, R., Poulton, N. J., Masland, E. D. P. and Emerson, D. (2011). What's new Is old: Resolving the identity of *Leptothrix ochracea* using single cell genomics, pyrosequencing and FISH. *PLoS ONE*, **6**, e17769. doi:10.1371/journal.pone.0017769.

Flemming, H.-C. and Wingender, J. (2010). The biofilm matrix. *Nature Reviews Microbiology*, **8**, 623–33.

Foissner, W. (1987). Soil protozoa: fundamental problems, ecological significance, adaptions in ciliates and testaceans, bioindicators, and guide to the literature. *Progress in Protozoology*, **2**, 69–212.

Foissner, W. (1999). Protist diversity: estimates of the near-imponderable. *Protist*, **150**, 363–8.

Forster, J., Famili, I., Fu, P., Palsson, B. O. and Nielsen, J. (2003). Genome-scale reconstruction of the *Saccharomyces cerevisiae* metabolic network. *Genome Research*, **13**, 244–53.

Forster, P. et al. (2007). 'Changes in atmospheric constituents and in radiative forcing' in Solomon, S. et al. (eds), *Climate Change 2007: The Physical Science Basis. Contribution of Working Group I to the Fourth Assessment Report of the Intergovernmental Panel on Climate Change*, pp. 130–234. Cambridge University Press, Cambridge.

Fouilland, E. and Mostajir, B. (2010). Revisited phytoplanktonic carbon dependency of heterotrophic bacteria in freshwaters, transitional, coastal and oceanic waters. *FEMS Microbiology Ecology*, **73**, 419–29.

Franche, C., Lindström, K. and Elmerich, C. (2009). Nitrogen-fixing bacteria associated with leguminous and non-leguminous plants. *Plant and Soil*, **321**, 35–59.

Fraser, C. M. et al. (1995). The minimal gene complement of *Mycoplasma genitalium*. *Science*, **270**, 397–403.

Frey, B. et al. (2010). Weathering-associated bacteria from the Damma Glacier Forefield: Physiological capabilities and impact on granite dissolution. *Applied and Environmental Microbiology*, **76**, 4788–96.

Frey, S. D., Elliott, E. T. and Paustian, K. (1999). Bacterial and fungal abundance and biomass in conventional and no-tillage agroecosystems along two climatic gradients. *Soil Biology and Biochemistry*, **31**, 573–85.

Fuchs, G. (2011). Alternative pathways of carbon dioxide fixation: Insights into the early evolution of life? *Annual Review of Microbiology*, **65**, 631–658.

Fuhrman, J. A. (2000). 'Impact of viruses on bacterial processes', in Kirchman, D. L., ed. *Microbial Ecology of the Oceans*, pp. 327–50. Wiley-Liss, New York, NY.

Fuhrman, J. A. and Azam, F. (1980). Bacterioplankton secondary production estimates for coastal waters of British Columbia, Antarctica, and California. *Applied and Environmental Microbiology*, **39**, 1085–95.

Fuhrman, J. A., Steele, J. A., Hewson, I., Schwalbach, M. S., Brown, M. V., Green, J. L. and Brown, J. H. (2008). A latitudinal diversity gradient in planktonic marine bacteria. *Proceedings of the National Academy of Sciences of the United States of America*, **105**, 7774–8.

Fuqua, C. and Greenberg, E. P. (2002). Listening in on bacteria: acyl-homoserine lactone signalling. *Nature Reviews Molecular Cell Biology*, **3**, 685–95.

Fuqua, C., Winans, S. C. and Greenberg, E. P. (1996). Census and consensus in bacterial ecosystems: The LuxR-LuxI family of quorum-sensing transcriptional regulators. *Annual Review of Microbiology*, **50**, 727–51.

Gadd, G. M. (2007). Geomycology: biogeochemical transformations of rocks, minerals, metals and radionuclides by fungi, bioweathering and bioremediation. *Mycological Research*, **111**, 3–49.

Galagan, J. E. et al. (2003). The genome sequence of the filamentous fungus *Neurospora crassa*. *Nature*, **422**, 859–68.

Galloway, J. N. et al. (2008). Transformation of the nitrogen cycle: Recent trends, questions, and potential solutions. *Science*, **320**, 889–92.

Gao, H. et al. (2010). Aerobic denitrification in permeable Wadden Sea sediments. *ISME Journal*, **4**, 417–26.

Gasol, J. M., del Giorgio, P. A., Massana, R. and Duarte, C. M. (1995). Active versus inactive bacteria: Size-dependence in a coastal marine plankton community. *Marine Ecology-Progress Series*, **128**, 91–7.

Gause, G. F. (1964). *The Struggle for Existence*. Hafner, New York, NY.

Geider, R. J. (1987). Light and temperature dependence of the carbon to chlorophyll *a* ratio in microalgae and cyanobacteria: implications for physiology and growth of phytoplankton *New Phytologist*, **106**, 1–34.

Geider, R. J. and La Roche, J. (2002). Redfield revisited: variability of C: N: P in marine microalgae and its biochemical basis. *European Journal of Phycology*, **37**, 1–17.

Geissinger, O., Herlemann, D. P. R., Morschel, E., Maier, U. G. and Brune, A. (2009). The ultramicrobacterium "*Elusimicrobium minutum*" gen. nov., sp. nov., the first cultivated representative of the Termite Group 1 phylum. *Applied and Environmental Microbiology*, **75**, 2831–40.

Gerbersdorf, S. U., Jancke, T., Westrich, B. and Paterson, D. M. (2008). Microbial stabilization of riverine sediments by extracellular polymeric substances. *Geobiology*, **6**, 57–69.

Ghabrial, S. A. (1998). Origin, adaptation and evolutionary pathways of fungal viruses. *Virus Genes*, **16**, 119–31.

Ghosh, D., Roy, K., Williamson, K. E., White, D. C., Wommack, K. E., Sublette, K. L. and Radosevich, M. (2008). Prevalence of lysogeny among soil bacteria and presence of 16S rRNA and trzN genes in viral-community DNA. *Applied and Environmental Microbiology*, **74**, 495–502.

Gibson, C. M. and Hunter, M. S. (2010). Extraordinarily widespread and fantastically complex: comparative biology of endosymbiotic bacterial and fungal mutualists of insects. *Ecology Letters*, **13**, 223–34.

Gilbert, J. A. et al. (2009). The seasonal structure of microbial communities in the Western English Channel. *Environmental Microbiology*, **11**, 3132–9.

Ginder-Vogel, M., Landrot, G., Fischel, J. S. and Sparks, D. L. (2009). Quantification of rapid environmental redox processes with quick-scanning x-ray absorption spectroscopy (Q-XAS). *Proceedings of the National Academy of Sciences of the United States of America*, **106**, 16124–8.

Giovannoni, S. J. et al. (2005). Genome streamlining in a cosmopolitan oceanic bacterium. *Science*, **309**, 1242–5.

Glass, J. B., Wolfe-Simon, F., Elser, J. J. and Anbar, A. D. (2010). Molybdenum-nitrogen co-limitation in freshwater and coastal heterocystous cyanobacteria. *Limnology and Oceanography*, **55**, 667–76.

Gleeson, D., Kennedy, N., Clipson, N., Melville, K., Gadd, G. and McDermott, F. (2006). Characterization of bacterial community structure on a weathered pegmatitic granite. *Microbial Ecology*, **51**, 526–34.

Goldberg, I., Rock, J. S., Ben-Bassat, A. and Mateles, R. I. (1976). Bacterial yields on methanol, methylamine, formaldehyde, and formate. *Biotechnology and Bioengineering*, **18**, 1657–68.

Goldhammer, T., Bruchert, V., Ferdelman, T. G. and Zabel, M. (2010). Microbial sequestration of phosphorus in anoxic upwelling sediments. *Nature Geoscience*, **3**, 557–61.

Goldman, J. G., Caron, D. A. and Dennett, M. R. (1987). Nutrient cycling in a microflagellate food chain: IV. Phytoplankton-microflagellate interactions. *Marine Ecology-Progress Series*, **38**, 75–87.

Gómez-Consarnau, L. et al. (2010). Proteorhodopsin phototrophy promotes survival of marine bacteria during starvation. *PLoS Biol*, **8**, e1000358.

Gómez-Consarnau, L. et al. (2007). Light stimulates growth of proteorhodopsin-containing marine Flavobacteria. *Nature*, **445**, 210–13.

González, J. M., Kiene, R. P. and Moran, M. A. (1999). Transformation of sulfur compounds by an abundant lineage of marine bacteria in the alpha-subclass of the class Proteobacteria. *Applied and Environmental Microbiology*, **65**, 3810–19.

Goreau, T. J., Kaplan, W. A., Wofsy, S. C., McElroy, M. B., Valois, F. W. and Watson, S. W. (1980). Production of NO_2^- and N_2O by nitrifying bacteria at reduced concentrations of oxygen. *Applied and Environmental Microbiology*, **40**, 526–32.

Gregory, T. R. (2010). Animal Genome Size Database. <http://www.genomesize.com>

Gruber, N. and Galloway, J. N. (2008). An Earth-system perspective of the global nitrogen cycle. *Nature*, **451**, 293–6.

Gulis, V., Suberkropp, K. and Rosemond, A. D. (2008). Comparison of fungal activities on wood and leaf litter in unaltered and nutrient-enriched headwater streams. *Applied and Environmental Microbiology*, **74**, 1094–101.

Gundersen, K., Heldal, M., Norland, S., Purdie, D. A. and Knap, A. H. (2002). Elemental C, N, and P cell content of individual bacteria collected at the Bermuda Atlantic Time-Series Study (BATS) site. *Limnology and Oceanography*, **47**, 1525–30.

Haddock, S. H. D., Moline, M. A. and Case, J. F. (2010). Bioluminescence in the sea. *Annual Review of Marine Science*, **2**, 443–93.

Hagström, Å., Pommier, T., Rohwer, F., Simu, K., Stolte, W., Svensson, D. and Zweifel, U. L. (2002). Use of 16S ribosomal DNA for delineation of marine bacterioplankton species. *Applied and Environmental Microbiology*, **68**, 3628–33.

Hallam, S. J., Mincer, T. J., Schleper, C., Preston, C. M., Roberts, K., Richardson, P. M. and DeLong, E. F. (2006). Pathways of carbon assimilation and ammonia oxidation suggested by environmental genomic analyses of marine crenarchaeota. *PLoS Biology*, **4**, 520–36.

Hamberger, A., Horn, M. A., Dumont, M. G., Murrell, J. C. and Drake, H. L. (2008). Anaerobic consumers of monosaccharides in a moderately acidic fen. *Applied and Environmental Microbiology*, **74**, 3112–20.

Hansell, D. A., Carlson, C. A., Repeta, D. J. and Schlitzer, R. (2009). Dissolved organic matter in the ocean: A controversy stimulates new insights. *Oceanography*, **22**, 202–11.

Hansen, J., Sato, M., Ruedy, R., Lo, K., Lea, D. W. and Medina-Elizade, M. (2006). Global temperature change. *Proceedings of the National Academy of Sciences of the United States of America*, **103**, 14288–93.

Hansen, P. J. (1991). Quantitative importance and trophic role of heterotrophic dinoflagellates in a coastal pelagial food web. *Marine Ecology-Progress Series*, **73**, 253–61.

Hansman, R. L., Griffin, S., Watson, J. T., Druffel, E. R. M., Ingalls, A. E., Pearson, A. and Aluwihare, L. I. (2009). The radiocarbon signature of microorganisms in the mesopelagic ocean. *Proceedings of the National Academy of Sciences of the United States of America*, **106**, 6513–18.

Hanson, T. E., Alber, B. E. and Tabita, F. R. (2012). Phototrophic CO_2 fixation: recent insights into ancient metabolisms. In Burnap, R. L. and Vermaas, W., eds. *Functional Genomics and Evolution of Photosynthetic Systems*, pp. 225–251 Springer, Dordrecht, The Netherlands.

Haring, M., Vestergaard, G., Rachel, R., Chen, L., Garrett, R. A. and Prangishvili, D. (2005). Virology: Independent virus development outside a host. *Nature*, **436**, 1101–2.

Harte, J. (1985). *Consider a Spherical Cow: A Course in Environmental Problem Solving*. William Kaufmann, Los Altos, CA.

Hartel, P. G. (1998). 'The soil habitat', in Sylvia, D. M. et al. (eds), *Principles and Applications of Soil Microbiology*, pp. 21–43. Prentice Hall, Upper Saddle River, NJ.

Hayatsu, M., Tago, K. and Saito, M. (2008). Various players in the nitrogen cycle: Diversity and functions of the microorganisms involved in nitrification and denitrification. *Soil Science and Plant Nutrition*, **54**, 33–45.

Hazen, T. C. et al. (2010). Deep-sea oil plume enriches indigenous oil-degrading bacteria. *Science*, **330**, 204–8.

Hedges, J. I., Baldock, J. A., Gelinas, Y., Lee, C., Peterson, M. L. and Wakeham, S. G. (2002). The biochemical and elemental compositions of marine plankton: A NMR perspective. *Marine Chemistry*, **78**, 47–63.

Held, M., Hossain, M. S., Yokota, K., Bonfante, P., Stougaard, J. and Szczyglowski, K. (2010). Common and not so common symbiotic entry. *Trends in Plant Science*, **15**, 540–5.

Heldal, M. and Bratbak, G. (1991). Production and decay of viruses in aquatic environments. *Marine Ecology-Progress Series*, **72**, 205–12.

Herron, P. M., Stark, J. M., Holt, C., Hooker, T. and Cardon, Z. G. (2009). Microbial growth efficiencies across a soil moisture gradient assessed using C-13-acetic acid vapor and N-15-ammonia gas. *Soil Biology and Biochemistry*, **41**, 1262–9.

Hilário, A. et al. (2011). New perspectives on the ecology and evolution of siboglinid tubeworms. *PLoS ONE*, **6**, e16309.

Hill, S. (1976). Apparent ATP requirement for nitrogen-fixation in growing *Klebsiella pneumonia. Journal of General Microbiology*, **95**, 297–312.

Hinkle, G., Wetterer, J., Schultz, T. and Sogin, M. (1994). Phylogeny of the attine ant fungi based on analysis of small subunit ribosomal RNA gene sequences. *Science*, **266**, 1695–7.

Hirano, S. S. and Upper, C. D. (2000). Bacteria in the leaf ecosystem with emphasis on *Pseudomonas syringae* - a pathogen, ice nucleus, and epiphyte. *Microbiology and Molecular Biology Reviews*, **64**, 624–53.

Hjort, K., Goldberg, A. V., Tsaousis, A. D., Hirt, R. P. and Embley, T. M. (2010). Diversity and reductive evolution of mitochondria among microbial eukaryotes. *Philosophical Transactions of the Royal Society B-Biological Sciences*, **365**, 713–27.

Hoffmaster, A. R. et al. (2004). Identification of anthrax toxin genes in a *Bacillus cereus* associated with an illness resembling inhalation anthrax. *Proceedings of the National Academy of Sciences of the United States of America*, **101**, 8449–54.

Högberg, P. and Read, D. J. (2006). Towards a more plant physiological perspective on soil ecology. *Trends in Ecology & Evolution*, **21**, 548–54.

Hoiczyk, E. and Hansel, A. (2000). Cyanobacterial cell walls: News from an unusual prokaryotic envelope. *Journal of Bacteriology*, **182**, 1191–99.

Hölldobler, B. and Wilson, E. O. (1990). *The Ants*, Harvard University Press, Cambridge, MA.

Hopkinson, B. M. and Morel, F. M. M. (2009). The role of siderophores in iron acquisition by photosynthetic marine microorganisms. *Biometals*, **22**, 659–69.

Horvath, P. and Barrangou, R. (2010). CRISPR/Cas, the immune system of bacteria and archaea. *Science*, **327**, 167–70.

Houghton, R. A. (2007). Balancing the global carbon budget. *Annual Review of Earth and Planetary Sciences*, **35**, 313–47.

House, C. H., Runnegar, B. and Fitz-Gibbon, S. T. (2003). Geobiological analysis using whole genome-based tree building applied to the Bacteria, Archaea, and Eukarya. *Geobiology*, **1**, 15–26.

Howard, D. M. and Howard, P. J. A. (1993). Relationships between CO_2 evolution, moisture-content and temperature for a range of soil types *Soil Biology and Biochemistry*, **25**, 1537–46.

Howe, A. T., Bass, D., Vickerman, K., Chao, E. E. and Cavalier-Smith, T. (2009). Phylogeny, taxonomy, and astounding genetic diversity of Glissomonadida ord. nov., the dominant gliding zooflagellates in soil (Protozoa: Cercozoa). *Protist*, **160**, 159–89.

Humbert, S., Tarnawski, S., Fromin, N., Mallet, M. P., Aragno, M. and Zopfi, J. (2010). Molecular detection of anammox bacteria in terrestrial ecosystems: distribution and diversity. *ISME Journal*, **4**, 450–54.

Humphreys, C. P., Franks, P. J., Rees, M., Bidartondo, M. I., Leake, J. R. and Beerling, D. J. (2010). Mutualistic mycorrhiza-like symbiosis in the most ancient group of land plants. *Nature Communications*, **1**, 103 doi: 10.1038/ncomms1105.

Hungate, R. E. (1966). *The Rumen and its Microbes*, Academic Press, New York.

Hungate, R. E. (1975). Rumen microbial ecosystem. *Annual Review of Ecology and Systematics*, **6**, 39–66.

Hunt, H. W. et al. (1987). The detrital food web in a shortgrass prairie. *Biology and Fertility of Soils*, **3**, 57–68.

Hutchinson, G. E. (1961). The paradox of the plankton. *The American Naturalist*, **95**, 137–45.

Ianora, A. et al. (2004). Aldehyde suppression of copepod recruitment in blooms of a ubiquitous planktonic diatom. *Nature*, **429**, 403–7.

Ingall, E. D. (2010). Phosphorus burial. *Nature Geoscience*, **3**, 521–2.

Ingraham, J. L. (2010). *March of the Microbes: Sighting the Unseen*. Belknap Press of Harvard University Press, Cambridge, MA.

Inselsbacher, E. et al. (2010). Short-term competition between crop plants and soil microbes for inorganic N fertilizer. *Soil Biology and Biochemistry*, **42**, 360–72.

Iovieno, P. and Bååth, E. (2008). Effect of drying and rewetting on bacterial growth rates in soil. *FEMS Microbiology Ecology*, **65**, 400–7.

Ishikawa, H. (2003). 'Insect symbiosis: An introduction', in Bourtzis, K. and Miller, T. A., eds. *Insect Symbiosis*, pp. 1–21. CRC Press, Boca Raton, FLA.

Itoh, T., Suzuki, K.-I. and Nakase, T. (1998). *Thermocladium modestius* gen. nov., sp. nov., a new genus of rod-shaped, extremely thermophilic crenarchaeote. *Int J Syst Bacteriol*, **48**, 879–87.

Itoh, T., Suzuki, K.-I., Sanchez, P. C. and Nakase, T. (1999). *Caldivirga maquilingensis* gen. nov., sp. nov., a new genus of rod-shaped crenarchaeote isolated from a hot spring in the Philippines. *International Journal of Systems Bacteriology*, **49**, 1157–63.

Jannasch, H. W., Eimhjell K, Wirsen, C. O. and Farmanfa, A. (1971). Microbial degradation of organic matter in deep sea. *Science*, **171**, 672–5.

Jannasch, H. W. and Jones, G. E. (1959). Bacterial populations in sea water as determined by different methods of enumeration. *Limnology and Oceanography*, **4**, 128–39.

Janssen, P. H. (2006). Identifying the dominant soil bacterial taxa in libraries of 16S rRNA and 16S rRNA genes. *Applied and Environmental Microbiology*, **72**, 1719–28.

Jeong, H. J. et al. (2007). Feeding by the *Pfiesteria*-like heterotrophic dinoflagellate *Luciella masanensis*. *Journal of Eukaryotic Microbiology*, **54**, 231–41.

Jeong, H. J. et al. (2005). *Stoeckeria algicida* n. gen., n. sp (Dinophyceae) from the coastal waters off Southern Korea: Morphology and small subunit ribosomal DNA gene sequence. *Journal of Eukaryotic Microbiology*, **52**, 382–90.

Jetten, M. S. M. (2001). New pathways for ammonia conversion in soil and aquatic systems. *Plant and Soil*, **230**, 9–19.

Jetten, M. S. M., Van Niftrik, L., Strous, M., Kartal, B., Keltjens, J. T. and Op Den Camp, H. J. M. (2009). Biochemistry and molecular biology of anammox bacteria. *Critical Reviews in Biochemistry and Molecular Biology*, **44**, 65–84.

Jia, Z. J. and Conrad, R. (2009). Bacteria rather than Archaea dominate microbial ammonia oxidation in an agricultural soil. *Environmental Microbiology*, **11**, 1658–71.

Jiang, S. C. and Paul, J. H. (1998). Gene transfer by transduction in the marine environment. *Applied and Environmental Microbiology*, **64**, 2780-7.

Jimenez-Lopez, C., Romanek, C. S. and Bazylinski, D. A. (2010). Magnetite as a prokaryotic biomarker: A review. *Journal of Geophysical Research-Biogeosciences*, **115**, G00g03. doi 10.1029/2009jg001152.

Joergensen, R. G. and Wichern, F. (2008). Quantitative assessment of the fungal contribution to microbial tissue in soil. *Soil Biology and Biochemistry*, **40**, 2977-91.

Johnson, M. D., Oldach, D., Delwiche, C. F. and Stoecker, D. K. (2007). Retention of transcriptionally active cryptophyte nuclei by the ciliate *Myrionecta rubra*. *Nature*, **445**, 426-8.

Johnson, P. T. J., Stanton, D. E., Preu, E. R., Forshay, K. J. and Carpenter, S. R. (2006). Dining on disease: how interactions between infection and environment affect predation risk. *Ecology*, **87**, 1973-80.

Johnston, D. T., Wolfe-Simon, F., Pearson, A. and Knoll, A. H. (2009). Anoxygenic photosynthesis modulated Proterozoic oxygen and sustained Earth's middle age. *Proceedings of the National Academy of Sciences of the United States of America*, **106**, 16925-9.

Jones, B. W. and Nishiguchi, M. K. (2004). Counterillumination in the Hawaiian bobtail squid, *Euprymna scolopes* Berry (Mollusca: Cephalopoda). *Marine Biology*, **144**, 1151-5.

Jones, D. L., Kielland, K., Sinclair, F. L., Dahlgren, R. A., Newsham, K. K., Farrar, J. F. and Murphy, D. V. (2009a). Soil organic nitrogen mineralization across a global latitudinal gradient. *Global Biogeochem. Cycles*, **23**, doi:10.1029/2008GB003250

Jones, R. T., Robeson, M. S., Lauber, C. L., Hamady, M., Knight, R. and Fierer, N. (2009b). A comprehensive survey of soil acidobacterial diversity using pyrosequencing and clone library analyses. *ISME Journal*, **3**, 442-53.

Jørgensen, B. B. (1982). Ecology of the bacteria of the sulphur cycle with special reference to anoxic-oxic interface environments *Philosophical Transactions of the Royal Society of London. B, Biological Sciences*, **298**, 543-61.

Jørgensen, B. B. (2000). 'Bacteria and marine biogeochemistry', in Schulz, H. D. and Zabel, M., eds. *Marine Geochemistry*, pp. 173-203. Springer, Berlin.

Jürgens, K. and Massana, R. (2008). 'Protistan grazing on marine bacterioplankton', in Kirchman, D. L., ed. *Microbial Ecology of the Oceans*, 2nd ed,pp. 383-441. John Wiley & Sons, Hoboken, NJ.

Jull, A. J. T., Courtney, C., Jeffrey, D. A. and Beck, J. W. (1998). Isotopic evidence for a terrestrial source of organic compounds found in Martian meteorites Allan Hills 84001 and Elephant Moraine 79001. *Science*, **279**, 366-9.

Junier, P., Kim, O.-S., Molina, V., Limburg, P., Junier, T., Imhoff, J. F. and Witzel, K.-P. (2008). Comparative in silico analysis of PCR primers suited for diagnostics and cloning of ammonia monooxygenase genes from ammonia-oxidizing bacteria. *FEMS Microbiology Ecology*, **64**, 141-52.

Kamp, A., De Beer, D., Nitsch, J. L., Lavik, G. and Stief, P. (2011). Diatoms respire nitrate to survive dark and anoxic conditions. *Proceedings of the National Academy of Sciences of the United States of America*, **108**, 5649-54.

Kamp, A., Stief, P. and Schulz-Vogt, H. N. (2006). Anaerobic sulfide oxidation with nitrate by a freshwater *Beggiatoa* enrichment culture. *Applied and Environmental Microbiology*, **72**, 4755-60.

Kappler, A. and Straub, K. L. (2005). Geomicrobiological cycling of iron. *Reviews in Mineralogy and Geochemistry*, **59**, 85-108.

Kapuscinski, R. B. and Mitchell, R. (1983). Sunlight-induced mortality of viruses and *Escherichia coli* in coastal seawater. *Environmental Science & Technology*, **17**, 1-6.

Karlsson, F., Ussery, D., Nielsen, J. and Nookaew, I. (2011). A closer look at *Bacteroides*: Phylogenetic relationship and genomic implications of a life in the human gut. *Microbial Ecology*, **61**, 473-85.

Karner, M. B., DeLong, E. F. and Karl, D. M. (2001). Archaeal dominance in the mesopelagic zone of the Pacific Ocean. *Nature*, **409**, 507-10.

Kashefi, K. and Lovley, D. R. (2003). Extending the upper temperature limit for life. *Science*, **301**, 934.

Keller, J. K. and Bridgham, S. D. (2007). Pathways of anaerobic carbon cycling across an ombrotrophic-minerotrophic peatland gradient. *Limnology and Oceanography*, **52**, 96-107.

Kelly, D. P. (1999). Thermodynamic aspects of energy conservation by chemolithotrophic sulfur bacteria in relation to the sulfur oxidation pathways. *Archives of Microbiology*, **171**, 219-29.

Kemp, P. F. (1990). The fate of benthic bacterial production. *Reviews in Aquatic Sciences*, **2**, 109-24.

Kemp, P. F., Lee, S. and Laroche, J. (1993). Estimating the growth rate of slowly growing marine bacteria from RNA content. *Applied and Environmental Microbiology*, **59**, 2594-601.

Kerr, B., West, J. and Bohannan, B. J. M. (2008). 'Bacteriophages: models for exploring basic principles of ecology', in Abedon, S. T., ed. *Bacteriophage Ecology: Population Growth, Evolution, and Impact of Bacterial Viruses*, pp. 31-63. Cambridge University Press, Cambridge.

Khoruts, A., Dicksved, J., Jansson, J. K. and Sadowsky, M. J. (2010). Changes in the composition of the human fecal microbiome after bacteriotherapy for recurrent *Clostridium difficile*-associated diarrhea. *Journal of Clinical Gastroenterology*, 10.1097/MCG.0b013e3181c87e02.

Kiene, R. P. and Linn, L. J. (2000). Distribution and turnover of dissolved DMSP and its relationship with bacterial production and dimethylsulfide in the Gulf of Mexico. *Limnology and Oceanography*, **45**, 849–61.

Kimura, M., Jia, Z.-J., Nakayama, N. and Asakawa, S. (2008). Ecology of viruses in soils: Past, present and future perspectives. *Soil Science & Plant Nutrition*, **54**, 1–32.

King, A. J. et al. (2010). Molecular insight into lignocellulose digestion by a marine isopod in the absence of gut microbes. *Proceedings of the National Academy of Sciences of the United States of America*, **107**, 5345–50.

King, G. M. (2011). Enhancing soil carbon storage for carbon remediation: potential contributions and constraints by microbes. *Trends in Microbiology*, **19**, 75–84.

King, J. D. and White, D. C. (1977). Muramic acid as a measure of microbial biomass in estuarine and marine samples. *Applied and Environmental Microbiology*, **33**, 777–83.

Kirchman, D. L. (2000). 'Uptake and regeneration of inorganic nutrients by marine heterotrophic bacteria', in Kirchman, D. L., ed. *Microbial Ecology of the Oceans*, pp. 261–88. Wiley-Liss, New York, NY.

Kirchman, D. L. (2002a). The ecology of *Cytophaga-Flavobacteria* in aquatic environments. *FEMS Microbiol. Ecology*, **39**, 91–100.

Kirchman, D. L. (2002b). 'Inorganic nutrient use by marine bacteria', in Bitton, G., ed. *Encyclopedia of Environmental Microbiology*, pp. 1697–709. Wiley, New York, NY.

Kirchman, D. L. (2003). 'The contribution of monomers and other low molecular weight compounds to the flux of DOM in aquatic ecosystems', in Findlay, S. and Sinsabaugh, R. L., eds. *Aquatic Ecosystems -Dissolved Organic Matter*, pp. 217–41. Academic Press, New York.

Kirchman, D. L., K'nees, E. and Hodson, R. E. (1985). Leucine incorporation and its potential as a measure of protein synthesis by bacteria in natural aquatic systems. *Applied and Environmental Microbiology*, **49**, 599–607.

Kirchman, D. L., Meon, B., Cottrell, M. T., Hutchins, D. A., Weeks, D. and Bruland, K. W. (2000). Carbon versus iron limitation of bacterial growth in the California upwelling regime. *Limnology and Oceanography*, **45**, 1681–8.

Kirchman, D. L., Morán, X. A. G. and Ducklow, H. (2009). Microbial growth in the polar oceans- role of temperature and potential impact of climate change. *Nature Reviews Microbiology*, **7**, 451–9.

Kirschbaum, M. U. F. (2006). The temperature dependence of organic-matter decomposition—still a topic of debate. *Soil Biology and Biochemistry*, **38**, 2510–18.

Kivelson, D. and Tarjus, G. (2001). H_2O below 277 K: A novel picture. *Journal of Physical Chemistry B*, **105**, 6620–7.

Kleber, M. and Johnson, M. G. (2010). Advances in understanding the molecular structure of soil organic matter: Implications for interactions in the environment. *Advances in Agronomy*, **106**, 77–142.

Kleidon, A. (2004). Beyond Gaia: Thermodynamics of life and earth system functioning. *Climatic Change*, **66**, 271–319.

Klein, D. A. and Paschke, M. W. (2004). Filamentous fungi: The indeterminate lifestyle and microbial ecology. *Microbial Ecology*, **47**, 224–35.

Klein, D. A., Paschke, M. W. and Heskett, T. L. (2006). Comparative fungal responses in managed plant communities infested by spotted (*Centaurea maculosa* Lam.) and diffuse (*C. diffusa* Lam.) knapweed. *Applied Soil Ecology*, **32**, 89–97.

Kleinsteuber, S., Muller, F. D., Chatzinotas, A., Wendt-Potthoff, K. and Harms, H. (2008). Diversity and in situ quantification of Acidobacteria subdivision 1 in an acidic mining lake. *FEMS Microbiology Ecology*, **63**, 107–17.

Klenk, H.-P. et al. (1997). The complete genome sequence of the hyperthermophilic, sulphate-reducing archaeon *Archaeoglobus fulgidus*. *Nature*, **390**, 364–70.

Klironomos, J. N. and Hart, M. M. (2001). Food-web dynamics: Animal nitrogen swap for plant carbon. *Nature*, **410**, 651–2.

Knittel, K. and Boetius, A. (2009). Anaerobic oxidation of methane: Progress with an unknown process. *Annual Review of Microbiology*, **63**, 311–34.

Knorr, W., Prentice, I. C., House, J. I. and Holland, E. A. (2005). Long-term sensitivity of soil carbon turnover to warming. *Nature*, **433**, 298–301.

Kolber, Z. S., Van Dover, C. L., Niederman, R. A. and Falkowski, P. G. (2000). Bacterial photosynthesis in surface waters of the open ocean. *Nature*, **407**, 177–9.

Konhauser, K. O. (2007). *Introduction to Geomicrobiology*, Blackwell Publishing, Malden, MA.

Könneke, M., Bernhard, A. E., De La Torre, J. R., Walker, C. B., Waterbury, J. B. and Stahl, D. A. (2005). Isolation of an autotrophic ammonia-oxidizing marine archaeon. *Nature*, **437**, 543–6.

Konstantinidis, K. T. and Tiedje, J. M. (2004). Trends between gene content and genome size in prokaryotic species with larger genomes. *Proceedings of the National Academy of Sciences of the United States of America*, **101**, 3160–5.

Konstantinidis, K. T. and Tiedje, J. M. (2007). Prokaryotic taxonomy and phylogeny in the genomic era: advancements and challenges ahead. *Current Opinion in Microbiology*, **10**, 504–9.

Koonin, E. V. and Wolf, Y. I. (2008). Genomics of bacteria and archaea: the emerging dynamic view of the prokaryotic world. *Nucleic Acids Research*, **36**, 6688–719.

Koop-Jakobsen, K. and Giblin, A. E. (2010). The effect of increased nitrate loading on nitrate reduction via denitrification and DNRA in salt marsh sediments. *Limnology and Oceanography*, **55**, 789–802.

Kortzinger, A., Hedges, J. I. and Quay, P. D. (2001). Redfield ratios revisited: Removing the biasing effect of anthropogenic CO_2. *Limnology and Oceanography*, **46**, 964-70.

Kristensen, D. M., Mushegian, A. R., Dolja, V. V. and Koonin, E. V. (2010). New dimensions of the virus world discovered through metagenomics. *Trends in Microbiology*, **18**, 11-19.

Kump, L. R. (2008). The rise of atmospheric oxygen. *Nature*, **451**, 277-8.

Kuo, C.-H. and Ochman, H. (2009). Inferring clocks when lacking rocks: the variable rates of molecular evolution in bacteria. *Biology Direct*, **4**, 35.

Kuroda, K. and Ueda, M. (2010). Engineering of microorganisms towards recovery of rare metal ions. *Applied Microbiology and Biotechnology*, **87**, 53-60.

Lafferty, K. D., Porter, J. W. and Ford, S. E. (2004). Are diseases increasing in the ocean? *Annual Review of Ecology, Evolution, and Systematics*, **35**, 31-54.

Lami, R. et al.(2007). High abundances of aerobic anoxygenic photosynthetic bacteria in the South Pacific Ocean. *Applied and Environmental Microbiology*, **73**, 4198-205.

Landry, M. R. and Hassett, R. P. (1982). Estimating the grazing impact of marine micro-zooplankton. *Marine Biology*, **67**, 283-8.

Lane, N., Allen, J. F. and Martin, W. (2010). How did LUCA make a living? Chemiosmosis in the origin of life. *Bioessays*, **32**, 271-80.

Lane, T. W. and Morel, F. M. M. (2000). A biological function for cadmium in marine diatoms. *Proceedings of the National Academy of Sciences of the United States of America*, **97**, 4627-31.

Langdon, C. (1993). The significance of respiration in production measurements based on oxygen. *ICES Mairne Science Symposium*, **197**, 69-78.

Lauber, C. L., Hamady, M., Knight, R. and Fierer, N. (2009). Pyrosequencing-based assessment of soil pH as a predictor of soil bacterial community structure at the continental scale. *Applied and Environmental Microbiology*, **75**, 5111-20.

Laughlin, R. J. and Stevens, R. J. (2002). Evidence for fungal dominance of denitrification and codenitrification in a grassland soil. *Soil Science Society of America Journal*, **66**, 1540-8.

Laughlin, R. J., Stevens, R. J., Muller, C. and Watson, C. J. (2008). Evidence that fungi can oxidize NH_4^+ to NO_3^- in a grassland soil. *European Journal of Soil Science*, **59**, 285-91.

Lauro, F. M., Chastain, R. A., Blankenship, L. E., Yayanos, A. A. and Bartlett, D. H. (2007). The unique 16S rRNA genes of piezophiles reflect both phylogeny and adaptation. *Applied and Environmental Microbiology*, **73**, 838-45.

Lechene, C. P., Luyten, Y., McMahon, G. and Distel, D. L. (2007). Quantitative imaging of nitrogen fixation by individual bacteria within animal cells. *Science*, **317**, 1563-6.

Ledin, M. (2000). Accumulation of metals by microorganisms - processes and importance for soil systems. *Earth-Science Reviews*, **51**, 1-31.

Lee, C. C., Hu, Y. and Ribbe, M. W. (2010). Vanadium nitrogenase reduces CO. *Science*, **329**, 642.

Lee, Z. M., Bussema, C. and Schmidt, T. M. (2009). rrnDB: documenting the number of rRNA and tRNA genes in bacteria and archaea. *Nucleic Acids Research*, **37**, D489-493.

Lehtovirta, L. E., Prosser, J. I. and Nicol, G. W. (2009). Soil pH regulates the abundance and diversity of Group 1.1c Crenarchaeota. *FEMS Microbiology Ecology*, **70**, 35-44.

Leininger, S. et al. (2006). Archaea predominate among ammonia-oxidizing prokaryotes in soils. *Nature*, **442**, 806-9.

Lenski, R. E. (2011). Evolution in action: a 50,000-generation salute to Charles Darwin. *Microbe*, **6**, 30-3.

Lepp, P. W., Brinig, M. M., Ouverney, C. C., Palm, K., Armitage, G. C. and Relman, D. A. (2004). Methanogenic Archaea and human periodontal disease. *Proceedings of the National Academy of Sciences of the United States of America*, **101**, 6176-81.

Lesen, A., Juhl, A. and Anderson, R. (2010). Heterotrophic microplankton in the lower Hudson River Estuary: potential importance of naked, planktonic amebas for bacterivory and carbon flux. *Aquatic Microbial Ecology*, **61**, 45-56.

Levine, S. N. and Schindler, D. W. (1999). Influence of nitrogen to phosphorus supply ratios and physicochemical conditions on cyanobacteria and phytoplankton species composition in the Experimental Lakes Area, Canada. *Canadian Journal of Fisheries and Aquatic Sciences*, **56**, 451-66.

Li, J., Yuan, H. and Yang, J. (2009). Bacteria and lignin degradation. *Frontiers of Biology in China*, **4**, 29-38.

Li, W. K. W. and Dickie, P. M. (1987). Temperature characteristics of photosynthetic and heterotrophic activities: seasonal variations in temperate microbial plankton. *Applied and Environmental Microbiology*, **53**, 2282-95.

Li, W. K. W., Subba-Rao, D., V, Harrison, W. G., Smith, J. C., Cullen, J. J., Irwin, B. and Platt, T. (1983). Autotrophic picoplankton in the tropical ocean. *Science*, **219**, 292-5.

Lin, S., Zhang, H., Zhuang, Y., Tran, B. and Gill, J. (2010). Spliced leader-based metatranscriptomic analyses lead to recognition of hidden genomic features in dinoflagellates. *Proceedings of the National Academy of Sciences of the United States of America*, **107**, 20033-8.

Lindell, D., Jaffe, J. D., Johnson, Z. I., Church, G. M. and Chisholm, S. W. (2005). Photosynthesis genes in marine viruses yield proteins during host infection. *Nature*, **438**, 86-9.

Lindow, S. E. and Brandl, M. T. (2003). Microbiology of the phyllosphere. *Applied and Environmental Microbiology*, **69**, 1875-83.

Liu, J., McBride, M. J. and Subramaniam, S. (2007). Cell surface filaments of the gliding bacterium *Flavobacterium johnsoniae* revealed by cryo-electron tomography. *Journal of Bacteriology*, **189**, 7503-6.

Liu, K. K. and Kaplan, I. R. (1984). Denitrification rates and availability of organic-matter in marine environments *Earth and Planetary Science Letters*, **68**, 88-100.

Logan, B. E. (1999). *Environmental Transport Processes*, Wiley-Interscience, New York, NY.

Long, R. A. and Azam, F. (2001). Microscale patchiness of bacterioplankton assemblage richness in seawater. *Aquatic Microbial Ecology*, **26**, 103-13.

López, D., Fischbach, M. A., Chu, F., Losick, R. and Kolter, R. (2009). Structurally diverse natural products that cause potassium leakage trigger multicellularity in *Bacillus subtilis*. *Proceedings of the National Academy of Sciences of the United States of America*, **106**, 280-5.

Lopez-Garcia, P., Rodriguez-Valera, F., Pedrós-Alió, C. and Moreira, D. (2001). Unexpected diversity of small eukaryotes in deep-sea Antarctic plankton. *Nature*, **409**, 603-7.

Lovley, D. R. (2003). Cleaning up with genomics. *Nature Reviews Microbiology*, **1**, 35-44.

Lovley, D. R. and Klug, M. J. (1983). Sulfate reducers can outcompete methanogens at freshwater sulfate concentrations. *Applied and Environmental Microbiology*, **45**, 187-92.

Lozupone, C. A. and Knight, R. (2007). Global patterns in bacterial diversity. *Proceedings of the National Academy of Sciences of the United States of America*, **104**, 11436-40.

Lupp, C. and Ruby, E. G. (2005). *Vibrio fischeri* uses two quorum-sensing systems for the regulation of early and late colonization factors. *Journal of Bacteriology*, **187**, 3620-9.

Luther, G. (2010). The role of one- and two-electron transfer reactions in forming thermodynamically unstable intermediates as barriers in multi-electron redox reactions. *Aquatic Geochemistry*, **16**, 395-420.

MacArthur, R. H. and Wilson, E. O. (1967). *The Theory of Island Biogeography*, Princeton University Press, Princeton, NJ.

Madigan, M. and Ormerod, J. (2004). 'Taxonomy, physiology and ecology of Heliobacteria', in Blankenship, R., Madigan, M. and Bauer, C., eds. *Anoxygenic Photosynthetic Bacteria*, pp. 17-30. Kluwer Academic Publishers, Dordrecht.

Madigan, M. T., Martinko, J. M. and Parker, J. (2003). *Brock Biology of Microorganisms*, Prentice Hall, Upper Saddle River, NJ.

Madsen, E. L. (2008). *Environmental Microbiology: from Genomes to Biogeochemistry,*, Blackwell Publishers, Oxford.

Magurran, A. E. (2004). *Measuring Biological Diversity*. Blackwell Science, Malden, MA.

Mahecha, M. D. et al. (2010). Global convergence in the temperature sensitivity of respiration at ecosystem level. *Science*, **329**, 838-40.

Majerus, M. E. N., Amos, W. and Hurst, G. (1996). *Evolution: the Four Billion Year War*. Longman, Harlow.

Mandelstam, J., McQuillen, K. and Dawes, I. (1982). *Biochemistry of Bacterial Growth*. John Wiley & Sons, New York.

Maranger, R., Bird, D. F. and Juniper, S. K. (1994). Viral and bacterial dynamics in Arctic sea ice during the spring algal bloom near Resolute, N.W.T., Canada. *Marine Ecology-Progress Series*, **111**, 121-7.

Martel, C. M. (2009). Conceptual bases for prey biorecognition and feeding selectivity in the microplanktonic marine phagotroph *Oxyrrhis marina*. *Microbial Ecology*, **57**, 589-97.

Martens-Habbena, W., Berube, P. M., Urakawa, H., De La Torre, J. R. and Stahl, D. A. (2009). Ammonia oxidation kinetics determine niche separation of nitrifying Archaea and Bacteria. *Nature*, **461**, 976-9.

Martin-Creuzburg, D. and Elert, E. V. (2009). Ecological significance of sterols in aquatic food webs. In Kainz, M., Brett, M. T. and Arts, M. T., eds. *Lipids in Aquatic Ecosystems*, pp. 43-64. Springer, New York.

Martiny, J. B. H. et al. (2006). Microbial biogeography: putting microorganisms on the map. *Nature Reviews Microbiology*, **4**, 102-12.

Mayer, F. (1999). 'Cellular and subcellular organization of prokaryotes', in Lengeler, J. W., Drews, G. and Schlegel, H. G., eds. *Biology of the Prokaryotes*, pp. 20-46. Blackwell Science, Thieme.

McCarren, J. and Brahamsha, B. (2005). Transposon mutagenesis in a marine *Synechococcus* strain: Isolation of swimming motility mutants. *Journal of Bacteriology*, **187**, 4457-62.

McDaniel, L. D., Young, E., Delaney, J., Ruhnau, F., Ritchie, K. B. and Paul, J. H. (2010). High frequency of horizontal gene transfer in the oceans. *Science*, **330**, 50.

McFall-Ngai, M. and Montgomery, M. K. (1990). The anatomy and morphology of the adult bacterial light organ of *Euprymna scolopes* Berry (Cephalopoda:Sepiolidae). *Biological Bulletin*, **179**, 332-9.

McKay, D. S. et al. (1996). Search for past life on Mars: Possible relic biogenic activity in Martian meteorite ALH84001. *Science*, **273**, 924-30.

McLaughlin, D. J., Hibbett, D. S., Lutzoni, F., Spatafora, J. W. and Vilgalys, R. (2009). The search for the fungal tree of life. *Trends in Microbiology*, **17**, 488-97.

McManus, G. B. and Katz, L. A. (2009). Molecular and morphological methods for identifying plankton: what makes a successful marriage? *Journal of Plankton Research*, **31**, 1119-29.

McNamara, C., Perry, T., Bearce, K., Hernandez-Duque, G. and Mitchell, R. (2006). Epilithic and endolithic bacterial communities in limestone from a Maya archaeological site. *Microbial Ecology*, **51**, 51-64.

Methe, B. A. et al. (2005). The psychrophilic lifestyle as revealed by the genome sequence of *Colwellia psychrerythraea* 34H through genomic and proteomic analyses. *Proceedings of the National Academy of Sciences of the United States of America*, **102**, 10913-18.

Miller, M. B. and Bassler, B. L. (2001). Quorum sensing in bacteria. *Annual Review of Microbiology*, **55**, 165-99.

Miller, S. R., Strong, A. L., Jones, K. L. and Ungerer, M. C. (2009). Bar-coded pyrosequencing reveals shared bacterial community properties along the temperature gradients of two alkaline hot springs in Yellowstone National Park. *Applied and Environmental Microbiology*, **75**, 4565-72.

Miralto, A. et al. (1999). The insidious effect of diatoms on copepod reproduction. *Nature*, **402**, 173-6.

Mitchell, J. G. and Kogure, K. (2006). Bacterial motility: links to the environment and a driving force for microbial physics. *FEMS Microbiology Ecology*, **55**, 3-16.

Montagnes, D. J. S. et al. (2008). Selective feeding behaviour of key free-living protists: avenues for continued study. *Aquatic Microbial Ecology*, **53**, 83-98.

Moore, J. C., McCann, K. and De Ruiter, P. C. (2005). Modeling trophic pathways, nutrient cycling, and dynamic stability in soils. *Pedobiologia*, **49**, 499-510.

Moran, M. A. (2008). 'Genomics and metagenomics of marine prokaryotes', in Kirchman, D. L., ed. *Microbial Ecology of the Oceans*, 2nd ed, pp. 91-129. John Wiley & Sons, Hoboken, NJ.

Moran, M. A. and Miller, W. L. (2007). Resourceful heterotrophs make the most of light in the coastal ocean. *Nature Reviews Microbiology*, **5**, 792-800.

Moran, N. and Baumann, P. (1994). Phylogenetics of cytoplasmically inherited microorganisms of arthropods. *Trends in Ecology & Evolution*, **9**, 15-20.

Moran, N. A., McCutcheon, J. P. and Nakabachi, A. (2008). Genomics and evolution of heritable bacterial symbionts. *Annual Review of Genetics*, **42**, 165-90.

Morgavi, D. P., Forano, E., Martin, C. and Newbold, C. J. (2010). Microbial ecosystem and methanogenesis in ruminants. *Animal*, **4**, 1024-36.

Moriarty, D. J. W. (1977). Improved method using muramic acid to estimate biomass of bacteria in sediments. *Oecologia*, **26**, 317-23.

Morris, R. M., Nunn, B. L., Frazar, C., Goodlett, D. R., Ting, Y. S. and Rocap, G. (2010). Comparative metaproteomics reveals ocean-scale shifts in microbial nutrient utilization and energy transduction. *ISME Journal*, **4**, 673-85.

Morse, J. W. and Mackenzie, F. T. (1990). *Geochemistry of Sedimentary Carbonates*, Elsevier, Amsterdam.

Moya, A., Pereto, J., Gil, R. and Latorre, A. (2008). Learning how to live together: genomic insights into prokaryote-animal symbioses. *Nature Reviews Genetics*, **9**, 218-29.

Mukohata, Y., Ihara, K., Tamura, T. and Sugiyama, Y. (1999). Halobacterial rhodopsins. *Journal of Biochemistry*, **125**, 649-57.

Mulholland, M. R. and Lomas, M. W. (2008). 'Nitrogen uptake and assimilation', in Capone, D. G., Bronk, D. A., Mulholland, M. R. and Carpenter, E. J., eds. *Nitrogen in the Marine Environment*, 2nd edn, pp. 303-84. Elsevier, New York, NY.

Muyzer, G. and Stams, A. J. M. (2008). The ecology and biotechnology of sulphate-reducing bacteria. *Nature Reviews Microbiology*, **6**, 441-54.

Nagata, T. (2000). 'Production mechanisms of dissolved organic matter', in Kirchman, D. L., ed. *Microbial Ecology of the Oceans*, pp. 121-52. Wiley-Liss, New York.

Nagle, D. G. and Paul, V. J. (1999). Production of secondary metabolites by filamentous tropical marine cyanobacteria: Ecological functions of the compounds. *Journal of Phycology*, **35**, 1412-21.

Nealson, K. H. and Saffarini, D. (1994). Iron and manganese in anaerobic respiration: environmental significance, physiology, and regulation. *Annual Review of Microbiology*, **48**, 311-43.

Neidhardt, F. C., Ingraham, J. L. and Schaechter, M. (1990). *Physiology of the Bacterial Cell: A Molecular Approach*, Sinauer Associates, Sunderland, MA.

Neubauer, S. C., Emerson, D. and Megonigal, J. P. (2002). Life at the energetic edge: Kinetics of circumneutral iron oxidation by lithotrophic iron-oxidizing bacteria isolated from the wetland-plant rhizosphere. *Applied and Environmental Microbiology*, **68**, 3988-95.

Newell, S. Y. and Fallon, R. D. (1991). Toward a method for measuring instantaneous fungal growth in field samples. *Ecology*, **72**, 1547-59.

Newton, R. J., Jones, S. E., Eiler, A., McMahon, K. D. and Bertilsson, S. (2011). A guide to the natural history of freshwater lake bacteria. *Microbiology and Molecular Biology Reviews*, **75**, 14-49.

Nisbet, E. G. and Sleep, N. H. (2001). The habitat and nature of early life. *Nature*, **409**, 1083-91.

Norton, J. M. and Firestone, M. K. (1991). Metabolic status of bacteria and fungi in the rhizosphere of Ponderosa pine seedlings. *Applied and Environmental Microbiology*, **57**, 1161-7.

Nyholm, S. V. and McFall-Ngai, M. (2004). The winnowing: establishing the squid-vibrio symbiosis. *Nature Reviews Microbiology*, **2**, 632-42.

Oberbauer, S. F. et al. (2007). Tundra CO_2 fluxes in response to experimental warming across latitudinal and moisture gradients. *Ecological Monographs*, **77**, 221–38.

Obernosterer, I., Kawasaki, N. and Benner, R. (2003). P-limitation of respiration in the Sargasso Sea and uncoupling of bacteria from P-regeneration in size-fractionation experiments. *Aquatic Microbial Ecology*, **32**, 229–37.

Ochsenreiter, T., Selezi, D., Quaiser, A., Bonch-Osmolovskaya, L. and Schleper, C. (2003). Diversity and abundance of Crenarchaeota in terrestrial habitats studied by 16S RNA surveys and real time PCR. *Environmental Microbiology*, **5**, 787–97.

Ogawa, H., Amagai, Y., Koike, I., Kaiser, K. and Benner, R. (2001). Production of refractory dissolved organic matter by bacteria. *Science*, **292**, 917–20.

Ohkuma, M. (2008). Symbioses of flagellates and prokaryotes in the gut of lower termites. *Trends in Microbiology*, **16**, 345–52.

Oliver, J., D. (2010). Recent findings on the viable but nonculturable state in pathogenic bacteria. *FEMS Microbiology Ecology*, **34**, 415–25.

Olsen, G. J. and Woese, C. R. (1993). Ribosomal RNA: a key to phylogeny. *The FASEB Journal*, **7**, 113–23.

Olsen, M. A., Aagnes, T. H. and Mathiesen, S. D. (1994). Digestion of herring by indigenous bacteria in the minke whale forestomach. *Applied and Environmental Microbiology*, **60**, 4445–55.

Op Den Camp, R. et al. (2011). LysM-type mycorrhizal receptor recruited for rhizobium symbiosis in nonlegume *Parasponia*. *Science*, **331**, 909–12.

Opdyke, M. R., Ostrom, N. E. and Ostrom, P. H. (2009). Evidence for the predominance of denitrification as a source of N_2O in temperate agricultural soils based on isotopologue measurements. *Global Biogeochemial Cycles*, **23**, doi: 10.1029/2009gb003523.

Orchard, V. A. and Cook, F. J. (1983). Relationship between soil respiration and soil moisture. *Soil Biology and Biochemistry*, **15**, 447–53.

Oren, A. (1999). Bioenergetic aspects of halophilism. *Microbiology and Molecular Biology Reviews*, **63**, 334–48.

Orphan, V. J., House, C. H., Hinrichs, K. U., McKeegan, K. D. and DeLong, E. F. (2001). Methane-consuming archaea revealed by directly coupled isotopic and phylogenetic analysis. *Science*, **293**, 484–7.

Ostfeld, R. S., Keesing, F. and Eviner, V. T. (eds.) 2008. *Infectious Disease Ecology: Effects of Ecosystems on Disease and of Disease on Ecosystems*, Princeton University Press. Princeton, New Jersey.

Pace, M. L., Cole, J. J., Carpenter, S. R. and Kitchell, J. F. (1999). Trophic cascades revealed in diverse ecosystems. *Trends in Ecology & Evolution*, **14**, 483–8.

Pace, M. L. and Prairie, Y. T. (2005). 'Respiration in lakes', in del Giorgio, P. A. and Williams, P. J. L., eds. *Respiration in Aquatic Ecosystems*, pp. 103–21.Oxford University Press, New York, NY.

Pace, N. R. (2006). Time for a change. *Nature*, **441**, 289–289.

Paerl, H. W. and Huisman, J. (2009). Climate change: a catalyst for global expansion of harmful cyanobacterial blooms. *Environmental Microbiology Reports*, **1**, 27–37.

Pagaling, E., Wang, H., Venables, M., Wallace, A., Grant, W. D., Cowan, D. A., Jones, B. E., Ma, Y., Ventosa, A. and Heaphy, S. (2009). Microbial biogeography of six salt lakes in Inner Mongolia, China, and a salt lake in Argentina. *Applied and Environmental Microbiology*, **75**, 5750–60.

Pakulski, J. D., Kase, J. P., Meador, J. A. and Jeffrey, W. H. (2008). Effect of stratospheric ozone depletion and enhanced ultraviolet radiation on marine bacteria at Palmer Station, Antarctica in the early austral spring. *Photochemistry and Photobiology*, **84**, 215–21.

Panikov, N. S., Flanagan, P. W., Oechel, W. C., Mastepanov, M. A. and Christensen, T. R. (2006). Microbial activity in soils frozen to below -39 °C. *Soil Biology and Biochemistry*, **38**, 785–94.

Parkes, R. J., Gibson, G. R., Mueller-Harvey, I., Buckingham, W. J. and Herbert, R. A. (1989). Determination of the substrates for sulfate-reducing bacteria within marine and estuarine sediments with different rates of sulfate reduction. *Journal of General Microbiology*, **135**, 175–87.

Parniske, M. (2008). Arbuscular mycorrhiza: the mother of plant root endosymbioses. *Nature Reviews Microbiology*, **6**, 763–75.

Parniske, M. and Downie, J. A. (2003). Plant biology: Locks, keys and symbioses. *Nature*, **425**, 569–70.

Pascal, P.-Y., et al. (2009). Seasonal variation in consumption of benthic bacteria by meio- and macrofauna in an intertidal mudflat. *Limnology and Oceanography*, **54**, 1048–59.

Pascoal, C. and Cassio, F. (2004). Contribution of fungi and bacteria to leaf litter decomposition in a polluted river. *Applied and Environmental Microbiology*, **70**, 5266–73.

Paul, J. H. (2008). Prophages in marine bacteria: dangerous molecular time bombs or the key to survival in the seas? *ISME Journal*, **2**, 579–89.

Payne, J. L. et al. (2009). Two-phase increase in the maximum size of life over 3.5 billion years reflects biological innovation and environmental opportunity. *Proceedings of the National Academy of Sciences of the United States of America*, **106**, 24–7.

Pearce, D. A., Cockell, C. S., Lindstrom, E. S. and Tranvik, L. J. (2007). First evidence for a bipolar distribution of dominant freshwater lake bacterioplankton. *Antarctic Science*, **19**, 245–52.

Pellicer, J., Fay, M. F. and Leitch, I. J. (2010). The largest eukaryotic genome of them all? *Botanical Journal of the Linnean Society*, **164**, 10–15.

Pereyra, L. P., Hiibel, S. R., Riquelme, M. V. P., Reardon, K. F. and Pruden, A. (2010). Detection and quantification of functional genes of cellulose-degrading, fermentative, and sulfate-reducing bacteria and methanogenic archaea. *Applied and Environmental Microbiology*, **76**, 2192–202.

Perret, X., Staehelin, C. and Broughton, W. J. (2000). Molecular basis of symbiotic promiscuity. *Microbiology and Molecular Biology Reviews*, **64**, 180–201.

Philippot, L., Andersson, S. G. E., Battin, T. J., Prosser, J. I., Schimel, J. P., Whitman, W. B. and Hallin, S. (2010). The ecological coherence of high bacterial taxonomic ranks. *Nature Reviews Microbiology*, **8**, 523–9.

Pietikåinen, J., Pettersson, M. and Bååth, E. (2005). Comparison of temperature effects on soil respiration and bacterial and fungal growth rates. *FEMS Microbiology Ecology*, **52**, 49–58.

Pinto-Tomas, A. A. et al. (2009). Symbiotic nitrogen fixation in the fungus gardens of leaf-cutter ants. *Science*, **326**, 1120–3.

Pitois, S., Jackson, M. H. and Wood, B. J. B. (2000). Problems associated with the presence of cyanobacteria in recreational and drinking waters. *International Journal of Environmental Health Research*, **10**, 203–18.

Pomeroy, L. R. (1974). The ocean food web - a changing paradigm. *Bioscience*, **24**, 499–504.

Poorvin, L., Rinta-Kanto, J. M., Hutchins, D. A. and Wilhelm, S. W. (2004). Viral release of iron and its bioavailability to marine plankton. *Limnology and Oceanography*, **49**, 1734–41.

Poretsky, R. S., Sun, S., Mou, X. and Moran, M. A. (2010). Transporter genes expressed by coastal bacterioplankton in response to dissolved organic carbon. *Environmental Microbiology*, **12**, 616–27.

Poretsky, R. S., Hewson, I., Sun, S., Allen, A. E., Zehr, J. P. and Moran, M. A. (2009). Comparative day/night metatranscriptomic analysis of microbial communities in the North Pacific subtropical gyre. *Environmental Microbiology*, **11**, 1358–75.

Prangishvili, D., Forterre, P. and Garrett, R. A. (2006). Viruses of the Archaea: a unifying view. *Nature Reviews Microbiology*, **4**, 837–48.

Prentice, I. C. et al. (2001). 'The carbon cycle and atmospheric CO_2', in Houghton, J. T., ed. *Climate Change 2001: the Scientific Basis: Contribution of Working Group I to the Third Assessment Report of the Intergovernmental Panel on Climate Change*, pp. 183–237. Cambridge University Press, Cambridge.

Proctor, L. M. and Fuhrman, J. A. (1990). Viral mortality of marine bacteria and cyanobacteria. *Nature*, **343**, 60–2.

Pruzzo, C., Vezzulli, L. and Colwell, R. R. (2008). Global impact of *Vibrio cholerae* interactions with chitin. *Environmental Microbiology*, **10**, 1400–10.

Purcell, E. M. (1977). Life at low Reynolds number. *American Journal of Physics*, **45**, 3–11.

Raich, J. W. and Mora, G. (2005). Estimating root plus rhizosphere contributions to soil respiration in annual croplands. *Soil Science Society of America Journal*, **69**, 634–9.

Raich, J. W. and Schlesinger, W. H. (1992). The global carbon dioxide flux in soil respiration and its relationship to vegetation and climate. *Tellus Series B-Chemical and Physical Meteorology*, **44**, 81–99.

Rajapaksha, R. M. C. P., Tobor-Kaplon, M. A. and Bååth, E. (2004). Metal toxicity affects fungal and bacterial activities in soil differently. *Applied and Environmental Microbiology*, **70**, 2966–73.

Randlett, D. L., Zak, D. R., Pregitzer, K. S. and Curtis, P. S. (1996). Elevated atmospheric carbon dioxide and leaf litter chemistry: influences on microbial respiration and net nitrogen mineralization. *Soil Science Society of America Journal*, **60**, 1571–7.

Rasmussen, B., Fletcher, I. R., Brocks, J. J. and Kilburn, M. R. (2008). Reassessing the first appearance of eukaryotes and cyanobacteria. *Nature*, **455**, 1101–4.

Ratledge, C. and Dover, L. G. (2000). Iron metabolism in pathogenic bacteria. *Annual Review of Microbiology*, **54**, 881–941.

Ravishankara, A. R., Daniel, J. S. and Portmann, R. W. (2009). Nitrous oxide (N_2O): The dominant ozone-depleting substance emitted in the 21st century. *Science*, **326**, 123–5.

Redfield, A. C. (1958). The biological control of chemical factors in the environment. *American Scientist*, **46**, 205–21.

Reyes-Prieto, A., Weber, A. P. M. and Bhattacharya, D. (2007). The origin and establishment of the plastid in algae and plants. *Annual Review of Genetics*, **41**, 147–68.

Riemann, L., Holmfeldt, K. and Titelman, J. (2009). Importance of viral lysis and dissolved DNA for bacterioplankton activity in a P-limited estuary, northern Baltic Sea. *Microbial Ecology*, **57**, 286–94.

Rigby, M. et al. (2008). Renewed growth of atmospheric methane. *Geophysical Research Letters*, **35**, L22805.

Rinke, C., Schmitz-Esser, S., Loy, A., Horn, M., Wagner, M. and Bright, M. (2009). High genetic similarity between two geographically distinct strains of the sulfur-oxidizing symbiont "*Candidatus* Thiobios zoothamnicoli." *FEMS Microbiology Ecology*, **67**, 229–41.

Rinnan, R. and Bååth, E. (2009). Differential utilization of carbon substrates by bacteria and fungi in tundra soil. *Applied and Environmental Microbiology*, **75**, 3611–20.

Risgaard-Petersen, N. et al. (2006). Evidence for complete denitrification in a benthic foraminifer. *Nature*, **443**, 93–6.

Robinson, C. J., Bohannan, B. J. M. and Young, V. B. (2010). From structure to function: the ecology of host-associated microbial communities. *Microbiology and Molecular Biology Reviews*, **74**, 453–76.

Rocap, G. et al. (2003). Genome divergence in two *Prochlorococcus* ecotypes reflects oceanic niche differentiation. *Nature*, **424**, 1042–7.

Roden, E. E. et al. (2010). Extracellular electron transfer through microbial reduction of solid-phase humic substances. *Nature Geoscience*, **3**, 417–21.

Roden, E. E. and Wetzel, R. G. (1996). Organic carbon oxidation and suppression of methane production by microbial Fe(III) oxide reduction in vegetated and unvegetated freshwater wetland sediments. *Limnology and Oceanography*, **41**, 1733–48.

Roesler, C. S., Culbertson, C. W., Etheridge, S. M., Goericke, R., Kiene, R. P., Miller, L. G. and Oremland, R. S. (2002). Distribution, production, and ecophysiology of *Picocystis* strain ML in Mono Lake, California. *Limnology and Oceanography*, **47**, 440–52.

Rogers, S. W., Moorman, T. B. and Ong, S. K. (2007). Fluorescent in situ hybridization and micro-autoradiography applied to ecophysiology in soil. *Soil Science Society of America Journal*, **71**, 620–31.

Rohr, J. R. and Raffel, T. R. (2010). Linking global climate and temperature variability to widespread amphibian declines putatively caused by disease. *Proceedings of the National Academy of Sciences of the United States of America*, **107**, 8269–74.

Rohrlack, T., Christoffersen, K., Dittmann, E., Nogueira, I., Vasconcelos, V. and Borner, T. (2005). Ingestion of microcystins by *Daphnia*: Intestinal uptake and toxic effects. *Limnology and Oceanography*, **50**, 440–8.

Rojo, F. (2009). Degradation of alkanes by bacteria. *Environmental Microbiology*, **11**, 2477–90.

Roossinck, M. J. (2011). The good viruses: viral mutualistic symbioses. *Nature Reviews Microbiology*, **9**, 99–108.

Rousk, J. and Bååth, E. (2007). Fungal and bacterial growth in soil with plant materials of different C/N ratios. *FEMS Microbiology Ecology*, **62**, 258–67.

Rousk, J. and Bååth, E. (2007). Fungal biomass production and turnover in soil estimated using the acetate-in-ergosterol technique. *Soil Biology and Biochemistry*, **39**, 2173–7.

Rousk, J., Brookes, P. C. and Bååth, E. (2009). Contrasting soil pH effects on fungal and bacterial growth suggest functional redundancy in carbon mineralization. *Applied and Environmental Microbiology*, **75**, 1589–96.

Rousk, J., Demoling, L. A., Bahr, A. and Bååth, E. (2008). Examining the fungal and bacterial niche overlap using selective inhibitors in soil. *FEMS Microbiology Ecology*, **63**, 350–8.

Rousk, J. and Nadkarni, N. M. (2009). Growth measurements of saprotrophic fungi and bacteria reveal differences between canopy and forest floor soils. *Soil Biology and Biochemistry*, **41**, 862–5.

Rowe, J. M., Fabre, M.-F., Gobena, D., Wilson, W. H. and Wilhelm, S. W. (2011). Application of the major capsid protein as a marker of the phylogenetic diversity of *Emiliania huxleyi* viruses. *FEMS Microbiology Ecology*, **76**, 373–80.

Ruby, E. G. et al. (2005). Complete genome sequence of *Vibrio fischeri*: A symbiotic bacterium with pathogenic congeners. *Proceedings of the National Academy of Sciences of the United States of America*, **102**, 3004–9.

Ruess, L. and Chamberlain, P. M. (2010). The fat that matters: Soil food web analysis using fatty acids and their carbon stable isotope signature. *Soil Biology and Biochemistry*, **42**, 1898–910.

Russell, J. A., Moreau, C. S., Goldman-Huertas, B., Fujiwara, M., Lohman, D. J. and Pierce, N. E. (2009). Bacterial gut symbionts are tightly linked with the evolution of herbivory in ants. *Proceedings of the National Academy of Sciences of the United States of America*, **106**, 21236–41. doi:10.1073/pnas.0907926106.

Russell, J. B. and Rychlik, J. L. (2001). Factors that alter rumen microbial ecology. *Science*, **292**, 1119–22.

Saito, M. A., Goepfert, T. J. and Ritt, J. T. (2008). Some thoughts on the concept of colimitation: Three definitions and the importance of bioavailability. *Limnology and Oceanography*, **53**, 276–90.

Saito, M. A., Rocap, G. and Moffett, J. W. (2005). Production of cobalt binding ligands in a *Synechococcus* feature at the Costa Rica upwelling dome. *Limnology and Oceanography*, **50**, 279–90.

Sampson, D. A., Janssens, I. A., Yuste, J. C. and Ceulemans, R. (2007). Basal rates of soil respiration are correlated with photosynthesis in a mixed temperate forest. *Global Change Biology*, **13**, 2008–17.

Santos, S. R. and Ochman, H. (2004). Identification and phylogenetic sorting of bacterial lineages with universally conserved genes and proteins. *Environmental Microbiology*, **6**, 754–759.

Sañudo-Wilhelmy, S. A. et al. (2001). Phosphorus limitation of nitrogen fixation by *Trichodesmium* in the central Atlantic Ocean. *Nature*, **411**, 66–9.

Sarmiento, J. L. and Gruber, N. (2006). *Ocean biogeochemical dynamics*, Princeton University Press, Princeton, NJ.

Sarmiento, J. L. and Toggweiler, J. R. (1984). A new model for the role of oceans in determining atmospheric CO_2. *Nature*, **308**, 621–4.

Sattley, W. M. et al. (2008). The genome of *Heliobacterium modesticaldum*, a phototrophic representative of the Firmicutes containing the simplest photosynthetic apparatus. *Journal of Bacteriology*, **190**, 4687–96.

Schadler, S., Burkhardt, C., Hegler, F., Straub, K. L., Miot, J., Benzerara, K. and Kappler, A. (2009). Formation of cell-iron-mineral aggregates by phototrophic and nitrate-reducing anaerobic Fe(II)-oxidizing bacteria. *Geomicrobiology Journal*, **26**, 93–103.

Schadt, C. W., Martin, A. P., Lipson, D. A. and Schmidt, S. K. (2003). Seasonal dynamics of previously unknown fungal lineages in tundra soils. *Science*, **301**, 1359–61.

Schleper, C. (2010). Ammonia oxidation: different niches for bacteria and archaea? *ISME Journal*, **4**, 1092–4.

Schleper, C., Jurgens, G. and Jonuscheit, M. (2005). Genomic studies of uncultivated archaea. *Nature Reviews Microbiology*, **3**, 479–88.

Schlesinger, W. H. (1997). *Biogeochemistry. An Analysis of Global Change*. Academic Press, San Diego.

Schott, J., Griffin, B. M. and Schink, B. (2010). Anaerobic phototrophic nitrite oxidation by *Thiocapsa* sp strain KS1 and *Rhodopseudomonas* sp strain LQ17. *Microbiology*, **156**, 2428–37.

Schoustra, S. E., Bataillon, T., Gifford, D. R. and Kassen, R. (2009). The properties of adaptive walks in evolving populations of fungus. *PLoS Biology*, **7**, 10.1371/journal.pbio.1000250

Schulz, H. N. and Jorgensen, B. B. (2001). Big bacteria. *Annual Review of Microbiology*, **55**, 105–37.

Serbus, L. R., Casper-Lindley, C., Landmann, F. and Sullivan, W. (2008). The genetics and cell biology of *Wolbachia*-host interactions. *Annual Review of Genetics*, **42**, 683–707.

Seymour, J. R., Simo, R., Ahmed, T. and Stocker, R. (2010). Chemoattraction to dimethylsulfoniopropionate throughout the marine microbial food web. *Science*, **329**, 342–5.

Shamir, I., Zahavy, E. and Steinberger, Y. (2009). Bacterial viability assessment by flow cytometry analysis in soil. *Frontiers of Biology in China*, **4**, 424–35.

Sharma, A. K., Zhaxybayeva, O., Papke, R. T. and Doolittle, W. F. (2008). Actinorhodopsins: proteorhodopsin-like gene sequences found predominantly in non-marine environments. *Environmental Microbiology*, **10**, 1039–56.

Sherr, B. F. and Sherr, E. B. (1991). Proportional distribution of total numbers, biovolume, and bacterivory among size classes of 2–20 μm nonpigmented marine flagellates. *Marine Microbial Food Webs*, **5**, 227–37.

Sherr, B. F., Sherr, E. B. and McDaniel, J. (1992). Effect of protistan grazing on the frequency of dividing cells in bacterioplankton assemblages. *Applied and Environmental Microbiology*, **58**, 2381–5.

Sherr, E. B. and Sherr, B. F. (2000). 'Marine microbes: An overview', in Kirchman, D. L., ed. *Microbial Ecology of the Oceans*, pp. 13–46. Wiley-Liss, New York.

Sherr, E. B. and Sherr, B. F. (2009). Capacity of herbivorous protists to control initiation and development of mass phytoplankton blooms. *Aquatic Microbial Ecology*, **57**, 253–62.

Simard, S. W., Perry, D. A., Jones, M. D., Myrold, D. D., Durall, D. M. and Molina, R. (1997). Net transfer of carbon between ectomycorrhizal tree species in the field. *Nature*, **388**, 579–82.

Six, J., Frey, S. D., Thiet, R. K. and Batten, K. M. (2006). Bacterial and fungal contributions to carbon sequestration in agroecosystems. *Soil Science Society of America Journal*, **70**, 555–69.

Smith, H. O., Hutchison, C. A., Pfannkoch, C. and Venter, J. C. (2003). Generating a synthetic genome by whole genome assembly: φX174 bacteriophage from synthetic oligonucleotides. *Proceedings of the National Academy of Sciences of the United States of America*, **100**, 15440–5.

Smith, S. E. and Read, D. J. (2008). *Mycorrhizal Symbiosis*, Academic Press, Amsterdam.

Smith, V. H. (2007). Microbial diversity-productivity relationships in aquatic ecosystems. *FEMS Microbiology Ecology*, **62**, 181–6.

Sogin, M. L. et al. (2006). Microbial diversity in the deep sea and the underexplored "rare biosphere." *Proceedings of the National Academy of Sciences of the United States of America*, **103**, 12115–20.

Sonntag, B., Summerer, M. and Sommaruga, R. (2007). Sources of mycosporine-like amino acids in planktonic *Chlorella*-bearing ciliates (Ciliophora). *Freshwater Biology*, **52**, 1476–85.

Sørensen, J., Jørgensen, B. B. and Revsbech, N. P. (1979). A comparison of oxygen, nitrate, and sulfate respiration in coastal marine sediments. *Microbial Ecology*, **5**, 105–15.

Spudich, J. L., Yang, C. S., Jung, K. H. and Spudich, E. N. (2000). Retinylidene proteins: Structures and functions from archaea to humans. *Annual Review of Cell and Developmental Biology*, **16**, 365–92.

Srinivasiah, S., Bhavsar, J., Thapar, K., Liles, M., Schoenfeld, T. and Wommack, K. E. (2008). Phages across the biosphere: contrasts of viruses in soil and aquatic environments. *Research in Microbiology*, **159**, 349–57.

Stackebrandt, E. and Goebel, B. M. (1994). A place for DNA-DNA reassociation and 16S ribosomal-RNA sequence-analysis in the present species definition in bacteriology. *International Journal of Systems Bacteriology*, **44**, 846–9.

Stal, L. J. (2009). Is the distribution of nitrogen-fixing cyanobacteria in the oceans related to temperature? *Environmental Microbiology*, **11**, 1632–45.

Stams, A. J. M. and Plugge, C. M. (2009). Electron transfer in syntrophic communities of anaerobic bacteria and archaea. *Nature Reviews Microbiology*, **7**, 568–77.

Staroscik, A. M. and Smith, D. C. (2004). Seasonal patterns in bacterioplankton abundance and production in Narragansett Bay, Rhode Island, USA. *Aquatic Microbial Ecology*, **35**, 275–82.

Stein, L. Y. and Yung, Y. L. (2003). Production, isotopic composition, and atmospheric fate of biologically produced nitrous oxide. *Annual Review of Earth and Planetary Sciences*, **31**, 329–56.

Steinbeiss, S., Gleixner, G. and Antonietti, M. (2009). Effect of biochar amendment on soil carbon balance and soil microbial activity. *Soil Biology and Biochemistry*, **41**, 1301–10.

Stepanauskas, R. and Sieracki, M. E. (2007). Matching phylogeny and metabolism in the uncultured marine bacteria, one cell at a time. *Proceedings of the National Academy of Sciences of the United States of America*, **104**, 9052–7.

Stephens, B. B. et al. (2007). Weak northern and strong tropical land carbon uptake from vertical profiles of atmospheric CO_2. *Science*, **316**, 1732–5.

Sterner, R. W. and Elser, J. J. (2002). *Ecological Stoichiometry: the Biology of Elements from Molecules to the Biosphere*, Princeton University Press, Princeton, NJ.

Stevenson, F. J. (1994). *Humus Chemistry: Genesis, Composition, Reactions*. Wiley, New York.

Steward, G. F., Wikner, J., Smith, D. C., Cochlan, W. P. and Azam, F. (1992). Estimation of virus production in the sea: I. method development. *Marine Microbial Food Webs*, **6**, 57–78.

Stewart, F. J. and Cavanaugh, C. M. (2006). 'Symbiosis of thioautotrophic bacteria with *Riftia pachyptila*', in Overmann, J., ed. *Molecular Basis of Symbiosis*, pp. 197–225. Springer, Heidelberg.

Stewart, F. J., Newton, I. L. G. and Cavanaugh, C. M. (2005). Chemosynthetic endosymbioses: adaptations to oxic-anoxic interfaces. *Trends in Microbiology*, **13**, 439–48.

Stewart, P. S. and Franklin, M. J. (2008). Physiological heterogeneity in biofilms. *Nature Reviews Microbiology*, **6**, 199–210.

Stintzi, A., Barnes, C., Xu, J. and Raymond, K. N. (2000). Microbial iron transport via a siderophore shuttle: A membrane ion transport paradigm. *Proceedings of the National Academy of Sciences of the United States of America*, **97**, 10691–6.

Stoica, E. and Herndl, G. J. (2007). Contribution of Crenarchaeota and Euryarchaeota to the prokaryotic plankton in the coastal northwestern Black Sea. *Journal of Plankton Research*, **29**, 699–706.

Stolper, D. A., Revsbech, N. P. and Canfield, D. E. (2010). Aerobic growth at nanomolar oxygen concentrations. *Proceedings of the National Academy of Sciences of the United States of America*, **107**, 18755–60.

Straile, D. (1997). Gross growth efficiencies of protozoan and metazoan zooplankton and their dependence on food concentration, predator-prey weight ratio, and taxonomic group. *Limnology and Oceanography*, **42**, 1375–85.

Strassert, J. F. H., Desai, M. S., Radek, R. and Brune, A. (2010). Identification and localization of the multiple bacterial symbionts of the termite gut flagellate *Joenia annectens*. *Microbiology*, **156**, 2068–79.

Straza, T. R. A., Cottrell, M. T., Ducklow, H. W. and Kirchman, D. L. (2009). Geographic and phylogenetic variation in bacterial biovolume as revealed by protein and nucleic acid staining. *Applied and Environmental Microbiology*, **75**, 4028–34.

Strom, S. L. (2000). 'Bacterivory: interactions between bacteria and their grazers', in Kirchman, D. L., ed. *Microbial Ecology of the Oceans*, pp. 351–86. Wiley-Liss, New York, NY.

Strom, S. L., Macri, E. L. and Olson, M. B. (2007). Microzooplankton grazing in the coastal Gulf of Alaska: Variations in top-down control of phytoplankton. *Limnology and Oceanography*, **52**, 1480–94.

Strous, M. and Jetten, M. S. M. (2004). Anaerobic oxidation of methane and ammonium. *Annual Review of Microbiology*, **58**, 99–117.

Strous, M. et al. (2006). Deciphering the evolution and metabolism of an anammox bacterium from a community genome. *Nature*, **440**, 790–4.

Stumm, W. and Morgan, J. J. (1996). *Aquatic Chemistry: Chemical Equilibria and Rates in Natural Waters*, Wiley, New York, NY.

Suen, G. et al. (2011). The genome sequence of the leaf-cutter ant *Atta cephalotes* reveals insights into its obligate symbiotic lifestyle. *PLoS Genetics*, **7**, e1002007. doi: 10.1371/journal.pgen.1002007.

Suh, S.-O., Noda, H. and Blackwell, M. (2001). Insect symbiosis: derivation of yeast-like endosymbionts within an entomopathogenic filamentous lineage. *Molecular Biology and Evolution*, **18**, 995–1000.

Sunda, W., Kieber, D. J., Kiene, R. P. and Huntsman, S. (2002). An antioxidant function for DMSP and DMS in marine algae. *Nature*, **418**, 317–20.

Sundquist, E. T. and Visser, K. (2004). 'The geological history of the carbon cycle', in Holland, H. D. and Turekian, K. K., eds. *Biogeochemistry*, pp. 425–72. Elsevier Pergamon, Amsterdam.

Sung, G. H., Poinar, G. O. and Spatafora, J. W. (2008). The oldest fossil evidence of animal parasitism by fungi supports a

Cretaceous diversification of fungal-arthropod symbioses. *Molecular Phylogenetics and Evolution*, **49**, 495–502.

Sutherland, T. F., Grant, J. and Amos, C. L. (1998). The effect of carbohydrate production by the diatom *Nitzschia curvilineata* on the erodibility of sediment. *Limnology and Oceanography*, **43**, 65–72.

Suttle, C. A. (2000). 'Ecological, evolutionary, and geochemical consequences of viral infection of cyanobacteria and eukaryotic algae', in Hurst, C. J., ed. *Viral Ecology*, pp. 247–96. Academic Press, San Diego, CA.

Suttle, C. A. (2005). Viruses in the sea. *Nature*, **437**, 356–61.

Sverdrup, H. U. (1953). On conditions for the vernal blooming of phytoplankton. *Journal du Conseil International pour l'Exploration de la Mer*, **18**, 287–95.

Szollosi-Janze, M. (2001). Pesticides and war: the case of Fritz Haber. *European Review*, **9**, 97–108.

Tabita, F. R., Hanson, T. E., Li, H. Y., Satagopan, S., Singh, J. and Chan, S. (2007). Function, structure, and evolution of the RubisCO-like proteins and their RubisCO homologs. *Microbiology and Molecular Biology Reviews*, **71**, 576–99.

Tabita, F. R., Satagopan, S., Hanson, T. E., Kreel, N. E. and Scott, S. S. (2008). Distinct form I, II, III, and IV Rubisco proteins from the three kingdoms of life provide clues about Rubisco evolution and structure/function relationships. *Journal of Experimental Botany*, **59**, 1515–24.

Tamames, J., Abellan, J., Pignatelli, M., Camacho, A. and Moya, A. (2010). Environmental distribution of prokaryotic taxa. *BMC Microbiology*, **10**, 85.

Tarao, M., Jezbera, J. and Hahn, M. W. (2009). Involvement of cell surface structures in size-independent grazing resistance of freshwater Actinobacteria. *Applied and Environmental Microbiology*, **75**, 4720–6.

Taylor, M. W., Radax, R., Steger, D. and Wagner, M. (2007). Sponge-associated microorganisms: Evolution, ecology, and biotechnological potential. *Microbiology and Molecular Biology Reviews*, **71**, 295–347.

Tebo, B. M., Geszvain, K. and Lee, S.-W. (2010). 'The molecular geomicrobiology of bacterial manganese(II) oxidation', in Barton, L., Mandl, M. and Loy, A., eds. *Geomicrobiology: Molecular and Environmental Perspectives*, pp. 285–308, Springer, New York, NY.

Tebo, B. M., Johnson, H. A., McCarthy, J. K. and Templeton, A. S. (2005). Geomicrobiology of manganese(II) oxidation. *Trends in Microbiology*, **13**, 421–8.

Thamdrup, B. and Dalsgaard, T. (2008). 'Nitrogen cycling in sediments', in Kirchman, D. L., ed. *Microbial Ecology of the Oceans*, 2nd ed, pp. 527–93. John Wiley & Sons, New York, NY.

Thauer, R. K., Kaster, A. K., Seedorf, H., Buckel, W. and Hedderich, R. (2008). Methanogenic archaea: ecologically relevant differences in energy conservation. *Nature Reviews Microbiology*, **6**, 579–91.

Thingstad, T. F. (2000). Elements of a theory for the mechanisms controlling abundance, diversity, and biogeochemical role of lytic bacterial viruses in aquatic systems. *Limnology and Oceanography*, **45**, 1320–8.

Thomsen, U., Thamdrup, B., Stahl, D. A. and Canfield, D. E. (2004). Pathways of organic carbon oxidation in a deep lacustrine sediment, Lake Michigan. *Limnology and Oceanography*, **49**, 2046–57.

Thyrhaug, R., Larsen, A., Thingstad, T. F. and Bratbak, G. (2003). Stable coexistence in marine algal host-virus systems. *Marine Ecology-Progress Series*, **254**, 27–35.

Toberman, H., Freeman, C., Evans, C., Fenner, N. and Artz, R. R. E. (2008). Summer drought decreases soil fungal diversity and associated phenol oxidase activity in upland Calluna heathland soil. *FEMS Microbiology Ecology*, **66**, 426–36.

Tomaru, Y., Takao, Y., Suzuki, H., Nagumo, T. and Nagasaki, K. (2009). Isolation and characterization of a single-stranded RNA virus infecting the bloom-forming diatom *Chaetoceros socialis*. *Applied and Environmental Microbiology*, **75**, 2375–81.

Torrella, F. and Morita, R. Y. (1979). Evidence by electron micrographs for a high incidence of bacteriophage particles in the waters of Yaquina Bay, Oregon - Ecological and taxonomical implications. *Applied and Environmental Microbiology*, **37**, 774–8.

Tranvik, L. J. et al. (2009). Lakes and reservoirs as regulators of carbon cycling and climate. *Limnology and Oceanography*, **54**, 2298–314.

Tripp, H. J. et al. (2010). Metabolic streamlining in an open-ocean nitrogen-fixing cyanobacterium. *Nature*, **464**, 90–4.

Trumbore, S. (2009). Radiocarbon and soil carbon dynamics. *Annual Review of Earth and Planetary Sciences*, **37**, 47–66.

Tsuchida, T. et al. (2010). Symbiotic bacterium modifies aphid body color. *Science*, **330**, 1102–4.

Tyson, G. W. et al. (2004). Community structure and metabolism through reconstruction of microbial genomes from the environment. *Nature*, **428**, 37–43.

Urich, T., Lanzén, A., Qi, J., Huson, D. H., Schleper, C. and Schuster, S. C. (2008). Simultaneous assessment of soil microbial community structure and function through analysis of the meta-transcriptome. *PLoS ONE*, **3**, e2527.

Uroz, S., Calvaruso, C., Turpault, M.-P. and Frey-Klett, P. (2009). Mineral weathering by bacteria: ecology, actors and mechanisms. *Trends in Microbiology*, **17**, 378–87.

Vadstein, O. (1998). Evaluation of competitive ability of two heterotrophic planktonic bacteria under phosphorus limitation. *Aquatic Microbial Ecology*, **14**, 119–27.

Vahatalo, A. V. and Wetzel, R. G. (2004). Photochemical and microbial decomposition of chromophoric dissolved organic matter during long (months-years) exposures. *Marine Chemistry*, **89**, 313–26.

Valentine, D. L. (2007). Adaptations to energy stress dictate the ecology and evolution of the Archaea. *Nature Reviews Microbiology*, **5**, 316.

Valentine, D. L. et al. (2010). Propane respiration jump-starts microbial response to a deep oil spill. *Science*, **330**, 208–11.

Valiela, I. (1995) *Marine Ecological Processes*, Springer, New York, NY.

Van Hamme, J. D., Singh, A. and Ward, O. P. (2003). Recent advances in petroleum microbiology. *Microbiology and Molecular Biology Reviews*, **67**, 503–49.

Van Loosdrecht, M. C. M., Lyklema, J., Norde, W., Schraa, G. and Zehnder, A. J. B. (1987). The role of bacterial cell wall hydrophobicity in adhesion. *Applied and Environmental Microbiology*, **53**, 1893–7.

Van Loosdrecht, M. C. M., Lyklema, J., Norde, W. and Zehnder, A. J. B. (1989). Bacterial adhesion- A physicochemical approach. *Microbial Ecology*, **17**, 1–15.

Van Mooy, B. A. S., Rocap, G., Fredricks, H. F., Evans, C. T. and Devol, A. H. (2006). Sulfolipids dramatically decrease phosphorus demand by picocyanobacteria in oligotrophic marine environments. *Proceedings of the National Academy of Sciences of the United States of America*, **103**, 8607–12.

Van Nieuwerburgh, L., Wanstrand, I. and Snoeijs, P. (2004). Growth and C: N: P ratios in copepods grazing on N- or Si-limited phytoplankton blooms. *Hydrobiologia*, **514**, 57–72.

Vandenkoornhuyse, P., Ridgway, K. P., Watson, I. J., Fitter, A. H. and Young, J. P. W. (2003). Co-existing grass species have distinctive arbuscular mycorrhizal communities. *Molecular Ecology*, **12**, 3085–95.

Vandevivere, P., Welch, S. A., Ullman, W. J. and Kirchman, D. L. (1994). Enhanced dissolution of silicate minerals by bacteria at near- neutral pH. *Microbial Ecology*, **27**, 241–51.

Vardi, A., Van Mooy, B. A. S., Fredricks, H. F., Popendorf, K. J., Ossolinski, J. E., Haramaty, L. and Bidle, K. D. (2009). Viral glycosphingolipids induce lytic infection and cell death in marine phytoplankton. *Science*, **326**, 861–5.

Venter, J. C. et al. (2004). Environmental genome shotgun sequencing of the Sargasso Sea. *Science*, **304**, 66–74.

Verberkmoes, N. C., Denef, V. J., Hettich, R. L. and Banfield, J. F. (2009). Systems biology: Functional analysis of natural microbial consortia using community proteomics. *Nature Reviews Microbiology*, **7**, 196–205.

Vieira-Silva, S. and Rocha, E. P. C. (2010). The systemic imprint of growth and its uses in ecological (meta)genomics. *PLoS Genet*, **6**, doi:10.1371/journal.pgen.1000808.

Visscher, P. T. and Stolz, J. F. (2005). Microbial mats as bioreactors: populations, processes, and products. *Palaeogeography, Palaeoclimatology, Palaeoecology*, **219**, 87–100.

Von Dassow, P., Petersen, T. W., Chepurnov, V. A. and Armbrust, E. V. (2008). Inter-and intraspecific relationships between nuclear DNA content and cell size in selected members of the centric diatom genus *Thalassiosira* (Bacillariophyceae). *Journal of Phycology*, **44**, 335–49.

Von Elert, E., Martin-Creuzburg, D. and Le Coz, J. R. (2003). Absence of sterols constrains carbon transfer between cyanobacteria and a freshwater herbivore (*Daphnia galeata*). *Proceedings of the Royal Society: Biological Sciences*, **270**, 1209–14.

Voroney, R. P. (2007). 'The soil habitat', in Paul, E. A., ed. *Soil Microbiology, Ecology, and Biochemistry*, 3rd edn, pp. 25–49. Elsevier, Amsterdam.

Voytek, M. A. and Ward, B. B. (1995). Detection of ammonium-oxidizing bacteria of the beta-subclass of the class Proteobacteria in aquatic samples with the PCR. *Applied and Environmental Microbiology*, **61**, 1444–50.

Vraspir, J. M. and Butler, A. (2009). Chemistry of marine ligands and siderophores. *Annual Review of Marine Science*, **1**, 43–63.

Vreeland, R. H., Rosenzweig, W. D. and Powers, D. W. (2000). Isolation of a 250 million-year-old halotolerant bacterium from a primary salt crystal. *Nature*, **407**, 897–900.

Wagner, E. K., Hewlett, M., J., Bloom, D. C. and Camerini, D. (2008). *Basic Virology*, Blackwell Publishing, Malden, MA.

Wagner, M., Loy, A., Klein, M., Lee, N., Ramsing, N. B., Stahl, D. A. and Friedrich, M. W. (2005). 'Functional marker genes for identification of sulfate-reducing prokaryotes', in Leadbetter, J.R., ed, *Methods in Enzymology* vol 397, pp. 469–89. Elsevier, San Diego, CA.

Waidner, L. A. and Kirchman, D. L. (2007). Aerobic anoxygenic phototrophic bacteria attached to particles in turbid waters of the Delaware and Chesapeake estuaries. *Applied and Environmental Microbiology*, **73**, 3936–44.

Waldrop, M. P., Zak, D. R., Blackwood, C. B., Curtis, C. D. and Tilman, D. (2006). Resource availability controls fungal diversity across a plant diversity gradient. *Ecology Letters*, **9**, 1127–35.

Walker, C. B. et al. (2010). *Nitrosopumilus maritimus* genome reveals unique mechanisms for nitrification and autotrophy in globally distributed marine crenarchaea. *Proceedings of the National Academy of Sciences of the United States of America*, **107**, 8818–23.

Walker, J. J. and Pace, N. R. (2007). Endolithic microbial ecosystems. *Annual Review of Microbiology*, **61**, 331–47.

Wandersman, C. and Delepelaire, P. (2004). Bacterial iron sources: From siderophores to hemophores. *Annual Review of Microbiology*, **58**, 611–47.

Wang, K. H., McSorley, R., Bohlen, P. and Gathumbi, S. M. (2006). Cattle grazing increases microbial biomass and alters soil nematode communities in subtropical pastures. *Soil Biology and Biochemistry*, **38**, 1956–65.

Wang, X., Le Borgne, R., Murtugudde, R., Busalacchi, A. J. and Behrenfeld, M. (2009). Spatial and temporal variability of the phytoplankton carbon to chlorophyll ratio in the equatorial Pacific: A basin-scale modeling study. *Journal of Geophysical Research*, **114**, doi: 10.1029/2008jc004942.

Ward, B. B. et al. (2009). Denitrification as the dominant nitrogen loss process in the Arabian Sea. *Nature*, **461**, 78–81.

Wardle, D. A. (2006). The influence of biotic interactions on soil biodiversity. *Ecology Letters*, **9**, 870–86.

Wardle, D. A., Williamson, W. M., Yeates, G. W. and Bonner, K. I. (2005). Trickle-down effects of aboveground trophic cascades on the soil food web. *Oikos*, **111**, 348–58.

Warren, L. A. and Ferris, F. G. (1998). Continuum between sorption and precipitation of Fe(III) on microbial surfaces. *Environmental Science & Technology*, **32**, 2331–37.

Watson, S. W., Novitsky, T. J., Quinby, H. L. and Valois, F. W. (1977). Determination of bacterial number and biomass in the marine environment. *Applied and Environmental Microbiology*, **33**, 940–6.

Wawrik, B., Paul, J. H. and Tabita, F. R. (2002). Real-time PCR quantification of *rbcL* (ribulose-1,5-bisphosphate carboxylase/oxygenase) mRNA in diatoms and pelagophytes. *Applied and Environmental Microbiology*, **68**, 3771–9.

Weast, R. C. (ed.) 1987. *CRC Handbook of Chemistry and Physics*, CRC Press, Boca Raton, FLA.

Weber, K. A., Achenbach, L. A. and Coates, J. D. (2006). Microorganisms pumping iron: anaerobic microbial iron oxidation and reduction. *Nature Reviews Microbiology*, **4**, 752–64.

Weinbauer, M. G. (2004). Ecology of prokaryotic viruses. *FEMS Microbiology Ecology*, **28**, 127–81.

Weinbauer, M. G., Fuks, D., Puskaric, S. and Peduzzi, P. (1995). Diel, seasonal, and depth-related variability of viruses and dissolved DNA in the Northern Adriatic Sea. *Microbial Ecology*, **30**, 25–41.

Weinbauer, M. G. and Höfle, M. G. (1998). Significance of viral lysis and flagellate grazing as factors controlling bacterioplankton production in a eutrophic lake. *Applied and Environmental Microbiology*, **64**, 431–8.

Weinbauer, M. G. and Peduzzi, P. (1994). Frequency, size and distribution of bacteriophages in different marine bacterial morphotypes. *Marine Ecology-Progress Series*, **108**, 11–20.

Welch, R. A. et al. (2002). Extensive mosaic structure revealed by the complete genome sequence of uropathogenic *Escherichia coli*. *Proceedings of the National Academy of Sciences of the United States of America*, **99**, 17020–4.

Welch, S. A., Barker, W. W. and Banfield, J. F. (1999). Microbial extracellular polysaccharides and plagioclase dissolution. *Geochimica et Cosmochimica Acta*, **63**, 1405–19.

Welch, S. A. and Ullman, W. J. (1993). The effect of organic acids on plagioclase dissolution rates and stoichiometry. *Geochimica et Cosmochimica Acta*, **57**, 2725–36.

White, J., Prell, J., James, E. K. and Poole, P. (2007). Nutrient sharing between symbionts. *Plant Physiology*, **144**, 604–14.

Whitman, W. B. (2009). The modern concept of the procaryote. *Journal of Bacteriology*, **191**, 2000–5.

Whitman, W. B., Coleman, D. C. and Wiebe, W. J. (1998). Prokaryotes: The unseen majority. *Proceedings of the National Academy of Sciences of the United States of America*, **95**, 6578–83.

Widdel, F. and Hansen, T. A. (1992). 'The dissimilatory sulfate and sulfur-reducing bacteria' in Balows, A. et al. (eds). *The Prokaryotes*, 2nd ed, pp. 583–624. Springer-Verlag, New York, NY.

Wier, A. M., Sacchi, L., Dolan, M. F., Bandi, C., Macallister, J. and Margulis, L. (2010). Spirochete attachment ultrastructure: Implications for the origin and evolution of cilia. *Biological Bulletin*, **218**, 25–35.

Wierzchos, J. et al. (2010). Microbial colonization of Ca-sulfate crusts in the hyperarid core of the Atacama Desert: implications for the search for life on Mars. *Geobiology*, **9**, 44–60.

Wilhelm, S. W., Brigden, S. M. and Suttle, C. A. (2002). A dilution technique for the direct measurement of viral production: A comparison in stratified and tidally mixed coastal waters. *Microbial Ecology*, **43**, 168–73.

Williams, P. J. L. (2000). 'Heterotrophic bacteria and the dynamics of dissolved organic material' in Kirchman, D. L., ed. *Microbial Ecology of the Oceans*, pp. 153–200. Wiley-Liss, New York, NY.

Williams, P. J. L. and del Giorgio, P. A. (2005). 'Respiration in aquatic ecosystems: history and background', in del Giorgio, P. A. and Williams, P. J. L., eds. *Respiration in Aquatic Ecosystems*, pp. 1–17. Oxford University Press, New York, NY.

Williamson, K. E., Radosevich, M. and Wommack, K. E. (2005). Abundance and diversity of viruses in six Delaware soils. *Applied and Environmental Microbiology*, **71**, 3119–25.

Williamson, S. J., Houchin, L. A., McDaniel, L. and Paul, J. H. (2002). Seasonal variation in lysogeny as depicted by prophage induction in Tampa Bay, Florida. *Applied and Environmental Microbiology*, **68**, 4307–14.

Winget, D. M. and Wommack, K. E. (2009). Diel and daily fluctuations in virioplankton production in coastal ecosystems. *Environmental Microbiology*, **11**, 2904–14.

Winkelmann, M., Hunger, N., Hüttl, R. and Wolf, G. (2009). Calorimetric investigations on the degradation of water insoluble hydrocarbons by the bacterium *Rhodococcus opacus* 1CP. *Thermochimica Acta*, **482**, 12–16.

Winter, C., Bouvier, T., Weinbauer, M. G. and Thingstad, T. F. (2010). Trade-offs between competition and defense specialists among unicellular planktonic organisms: the "Killing the Winner" hypothesis revisited. *Microbiology and Molecular Biology Reviews*, **74**, 42–57.

Woese, C. R. and Fox, G. E. (1977). Phylogenetic structure of prokaryotic domain - Primary kingdoms. *Proceedings of the National Academy of Sciences of the United States of America*, **74**, 5088–90.

Wommack, K. E. and Colwell, R. R. (2000). Virioplankton: Viruses in aquatic ecosystems. *Microbiology and Molecular Biology Reviews*, **64**, 69–114.

Woods, R. J., Barrick, J. E., Cooper, T. F., Shrestha, U., Kauth, M. R. and Lenski, R. E. (2011). Second-order selection for evolvability in a large *Escherichia coli* population. *Science*, **331**, 1433–6.

Wootton, E. C., Zubkov, M. V., Jones, D. H., Jones, R. H., Martel, C. M., Thornton, C. A. and Roberts, E. C. (2007). Biochemical prey recognition by planktonic protozoa. *Environmental Microbiology*, **9**, 216–22.

Worden, A. Z. and Not, F. (2008). 'Ecology and diversity of picoeukaryotes', in Kirchman, D. L., ed. *Microbial Ecology of the Oceans*, 2nd ed, pp. 159–205. John Wiley & Sons, Hoboken, NJ.

Wu, D. et al. (2009). A phylogeny-driven genomic encyclopaedia of Bacteria and Archaea. *Nature*, **462**, 1056–60.

Wuchter, C., Schouten, S., Coolen, M. J. L. and Damste, J. S. S. (2004). Temperature-dependent variation in the distribution of tetraether membrane lipids of marine Crenarchaeota: Implications for TEX86 paleothermometry. *Paleoceanography*, **19**, PA4028, doi: 10.1029/2004pa001041.

Wylie, J. L. and Currie, D. J. (1991). The relative importance of bacteria and algae as food sources for crustacean zooplankton. *Limnology and Oceanography*, **36**, 708–28.

Yang, C. H., Crowley, D. E., Borneman, J. and Keen, N. T. (2001). Microbial phyllosphere populations are more complex than previously realized. *Proceedings of the National Academy of Sciences of the United States of America*, **98**, 3889–94.

Yavitt, J. B. and Lang, G. E. (1990). Methane production in contrasting wetland sites-Response to organic-chemical components of peat and to sulfate reduction. *Geomicrobiology Journal*, **8**, 27–46.

Ye, R. W., Averill, B. A. and Tiedje, J. M. (1994). Denitrification-production and consumption of nitric oxide. *Applied and Environmental Microbiology*, **60**, 1053–8.

Yooseph, S. et al. (2010). Genomic and functional adaptation in surface ocean planktonic prokaryotes. *Nature*, **468**, 60–6.

Yu, R., Kampschreur, M. J., Loosdrecht, M. C. M. V. and Chandran, K. (2010a). Mechanisms and specific directionality of autotrophic nitrous oxide and nitric oxide generation during transient anoxia. *Environmental Science & Technology*, **44**, 1313–19.

Yu, X. et al. (2010b). A geminivirus-related DNA mycovirus that confers hypovirulence to a plant pathogenic fungus. *Proceedings of the National Academy of Sciences of the United States of America*, **107**, 8387–92.

Yurkov, V. V. and Beatty, J. T. (1998). Aerobic anoxygenic phototrophic bacteria. *Microbiology and Molecular Biology Reviews*, **62**, 695–724.

Zehnder, A. J. B. and Brock, T. D. (1979). Methane formation and methane oxidation by methanogenic bacteria. *Journal of Bacteriology*, **137**, 420–32.

Zehr, J. P. et al. (2008). Globally distributed uncultivated oceanic N_2-fixing cyanobacteria lack oxygenic Photosystem II. *Science*, **322**, 1110–12.

Zehr, J. P. and Paerl, H. W. (2008). 'Molecular ecological aspects of nitrogen fixation in the marine environment', in Kirchman, D. L., ed. *Microbial Ecology of the Ocean*, 2nd ed, pp. 481–525. Wiley, New York, NY.

Zhou, J., Xia, B., Huang, H., Palumbo, A. V. and Tiedje, J. M. (2004). Microbial diversity and heterogeneity in sandy subsurface soils. *Applied and Environmental Microbiology*, **70**, 1723–34.

Zientz, E., Feldhaar, H., Stoll, S. and Gross, R. (2005). Insights into the microbial world associated with ants. *Archives of Microbiology*, **184**, 199–206.

Zöllner, E., Hoppe, H.-G., Sommer, U. and Jürgens, K. (2009). Effect of zooplankton-mediated trophic cascades on marine microbial food web components (bacteria, nanoflagellates, ciliates). *Limnology and Oceanography*, **54**, 262–75.

Zubkov, M. V. (2009). Photoheterotrophy in marine prokaryotes. *Journal of Plankton Research*, **31**, 933–8.

Zubkov, M. V. and Tarran, G. A. (2008). High bacterivory by the smallest phytoplankton in the North Atlantic Ocean. *Nature*, **455**, 224–6.

Zumft, W. G. (1997). Cell biology and molecular basis of denitrification. *Microbiology and Molecular Biology Reviews*, **61**, 533–616.

Zwart, G., Crump, B. C., Agterveld, M., Hagen, F. and Han, S. K. (2002). Typical freshwater bacteria: an analysis of available 16S rRNA gene sequences from plankton of lakes and rivers. *Aquatic Microbial Ecology*, **28**, 141–55.

Index

Page numbers in *italics* refer to figures or tables

Printed and bound by CPI Group (UK) Ltd, Croydon, CR0 4YY